Water science reviews 3

Water dynamics

Water Science Reviews 3

Water dynamics

EDITED BY

FELIX FRANKS

The right of the University of Cambridge to print and sell all manner of books was granted by Henry VIII in 1534. The University has printed and published continuously since 1584.

CAMBRIDGE UNIVERSITY PRESS

CAMBRIDGE

NEW YORK NEW ROCHELLE MELBOURNE SYDNEY

CAMBRIDGE UNIVERSITY PRESS
Cambridge, New York, Melbourne, Madrid, Cape Town, Singapore,
São Paulo, Delhi, Dubai, Tokyo, Mexico City

Cambridge University Press
The Edinburgh Building, Cambridge CB2 8RU, UK

Published in the United States of America by Cambridge University Press, New York

www.cambridge.org
Information on this title: www.cambridge.org/9780521350143

First published 1988

A catalogue record for this publication is available from the British Library

Library of Congress Cataloguing in Publication data
Water Science Reviews.—1—Cambridge; New York:
Cambridge University Press, 1985-
v.:ill.; 24 cm.
Annual.
Editor: 1985- F. Franks
1. Water chemistry—Periodicals. 2. Water—Periodicals.
I. Franks, Felix.
GB855.W38 546′.22′05—dc19 86-643278
AACR 2 MARC-S

ISBN 978-0-521-35014-3 Hardback
ISBN 978-0-521-09111-4 Paperback

Contents

1

The nature of the hydrated proton

Part Two: Theoretical studies; the liquid state

C. I. RATCLIFFE* AND D. E. IRISH†

**Division of Chemistry, National Research Council Canada, Ottawa, Ontario, Canada K1A 0R6*
†Department of Chemistry, University of Waterloo, Waterloo, Ontario, Canada N2L 3G1

In Part One [1] the *raison d'être* for this review was given as follows: 'To what extent can one "see" that most important of all the cations – the hydrated proton?' In Part One our search for an answer encompassed a review of studies of the proton, in association with H_2O, in the solid state and the gaseous state by a variety of techniques: X-ray and neutron scattering, vibrational spectroscopy, NMR studies, proton conduction and mass spectrometry. This search is continued here. We shall deal with a review of theoretical studies and with the more difficult liquid state. The literature relating to Part One has also been brought up to date.

1.1 Theoretical studies

The major part of this section will be concerned with molecular orbital (MO) calculations, though other kinds of model calculations will be described where appropriate. There is now a large body of literature concerned with semi-empirical and *ab initio* MO calculations applied to numerous and diverse aspects of the hydrated proton:

(1) Molecular energies of the ground and excited states.
(2) Molecular geometry.
(3) Charge distribution.
(4) Inversion barriers.
(5) Proton affinities.
(6) Force constants, vibrational frequencies and IR intensities.
(7) NMR chemical shift tensors and isotropic shifts, and quadrupole coupling constants of ^{17}O and ^{2}H.
(8) Heats of reaction with other small molecules and equilibrium constants.
(9) Proton transfer and the shape of the potential of the central strong hydrogen bond of $H_5O_2^+$.
(10) Relative stabilities of the higher hydrates in different configurations.

We will not go into a lengthy discussion of the accuracy and merits of the different methods of calculation, but rather attempt to relate the contribution of these results to our understanding of the hydrated proton. The majority of the calculations apply most directly to the gas phase, since they concern isolated H_3O^+, $H_5O_2^+$ or higher hydrated clusters, though the results can to a certain extent be extrapolated to the liquid phase. In a few cases, which were mentioned in Part One [2–4] and will not be discussed further here, calculations have been applied to specific crystal environments.

1.1.1 H_3O^+

The list of calculations on the oxonium ion is lengthy [5–93, 166–9], and, no doubt, not exhaustive. The interested reader may find useful the bibliographies of *ab initio* calculations by Richards, Scott, Sackwild & Robins [91] and Ohno & Morokuma [92, 93]. Hund, in 1925, was really the first to attempt theoretical calculations on H_3O^+. [5, 6] Using a semi-empirical method, in which the system was treated as interacting O^{2-} and $3H^+$, he calculated the proton affinity of H_2O to be -753 ± 167 kJ mol^{-1}. *Ab initio* MO studies did not begin until 1961 with the work of Grahn [8], but there were also a few other semi-empirical quantum mechanical calculations prior to and around this time. [7, 9, 11] In those early years the calculations centred mainly on obtaining the equilibrium configuration of atoms which gave an energy minimum, and the proton affinity of H_2O. Interest in these parameters has continued up to the present, through numerous approaches and refinements, as indicated in table 1.1. The early calculations nearly always arrived at a planar minimum energy configuration for H_3O^+, which confounded experimental evidence from solids for a pyramidal form. It was found in 1973 that the inclusion of polarisation functions reduced the minimum energy and gave pyramidal equilibrium configurations [21, 22]; i.e. studies which included only s and p functions on oxygen and s on hydrogen all gave planar H_3O^+, whereas introduction of p polarisation functions on hydrogen gave slightly non-planar structures, and introduction of d functions on oxygen proved to be even more important in stabilising the C_{3v} form relative to the planar D_{3h} form. Further improvements in energy minimisation were then realised with the inclusion of configuration interaction (CI) to take account of electron correlation effects.

As can be seen from table 1.1 the geometrical parameters determined in more recent years (since 1980) nearly all fall within small ranges of values: O–H = 96.2–98.0 pm, $\angle$HOH = 111.6–114.4° (see figure 1.1). The lowest minimum energies were obtained by Lischka & Dyczmons [22], Rodwell & Radom [44], and followed most recently by Botschwina [90], whose optimised equilibrium geometry was O–H = 97.45 pm and $\angle$HOH = 111.9°. The bond length is significantly shorter than is found for H_3O^+ in solids (average values from neutron diffraction studies in Table 3 of Part One are

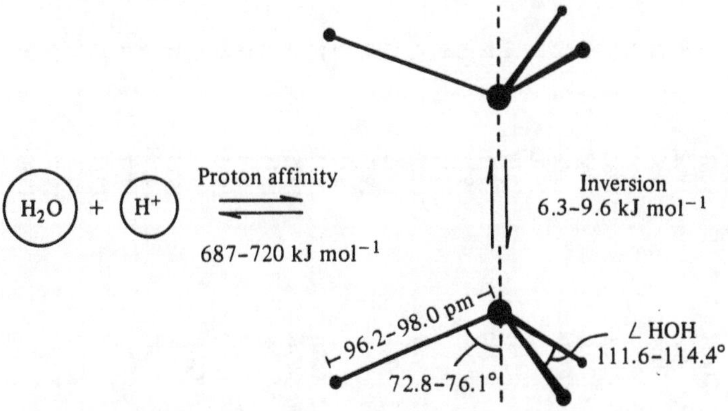

Fig. 1.1. The structure of H_3O^+, inversion barrier and proton affinity of water from *ab initio* calculations. The ranges of values given are from the more accurate calculations. (The range of inversion barriers are from the five lowest total energy calculations.)

O–H = 101.7 pm and ∠HOH = 111.8°) as might be expected in the absence of hydrogen bonding. Unfortunately there are only a few values of the geometrical parameters available from gas phase experiments, and these are derived from fits of the vibration–rotation spectra and are thus dependent on the details of the model inversion potential. These experimental values fall in the range O–H = 97.6–98.6 pm, ∠HOH = 110.7–115.3°. [94–7] The *ab initio* values are, therefore, in quite good agreement.

Once a pyramidal equilibrium geometry had been established, it was possible to calculate the inversion barrier from the difference in energy between the C_{3v} pyramid and the D_{3h} planar transitional configuration. Since this is a very small difference between two very large energies, one might expect appreciable differences in the values obtained using different basis sets. In fact an inversion barrier of 10 kJ mol^{-1} is of the order of 0.005% of the total electronic energy of H_3O^+. Over the years the calculated values have fallen in the range 2.6–14.6 kJ mol^{-1} (table 1.1). Rodwell & Radom's [44] value was 9.6 kJ mol^{-1}, and Botschwina's [90] value was 8.13 kJ mol^{-1}. Again, experimental results are few; two values have been obtained from gas phase IR, namely 10.82 and 8.04 kJ mol^{-1}. [94, 95]

The proton affinity (PA) or ΔH for the reaction $H_2O + H^+ \rightarrow H_3O^+$ was, in the early estimates, taken as the difference (ΔE) in the total energies of H_2O and H_3O^+. The minimised energies are for the rigid ground states and hence really apply to zero degrees Kelvin. Before any meaningful comparison can be made with experimental values the calculated values should be corrected for zero point energy (ZPE) and thermal differences [e.g. 50, 57]:

$$PA = \Delta E - \Delta ZPE + \tfrac{5}{2}RT$$

Values which have had a correction applied are indicated in table 1.1 by the letter C preceding the value. The most recent experimental values of the PA of

Table 1.1. *Selected results of molecular orbital calculations on H_3O^+ by year*

Method	Total energy (au)[a]	O–H[b] (pm)	∠HOH[b] (deg)	Inversion barrier (kJ mol^{-1})	Proton affinity[c] (−kJ ,ol^{-1})	Year	Ref.
STO (SE)		94.7	(104.5)			1959	7
SCF LCAO	−76.182	95.81	120		1104	1961	8
United atom (SE)	−75.412	90.5	120		782	1962	9
Valence bond	−76.174	95.8	120		1054	1962	11
SCF LCAO OCE CI	−76.0184	95.0	114.4			1965	13
SCF LCAO Gaussian	−76.3213	95.3	120		753	1965	12
SCF OCE	−76.2330	96.3	110.9			1967	14
SCF LCAO large Gaussian set	−76.3066				C728	1968	15
CNDO/2 (SE)		105.0	111			1970	17
SCF GTO DZ		95.8	120		C745	1970	16
SCF 4-31G	−76.2006	96.4	120		C740	1971	18
SCF STO-3G	−75.33044	99.0	113.9	4.2		1971	19
SCF 4-31G	−76.20060	96.4				1971	19
Near HF SCF	−76.3386	97.9	113.5	3.93	739	1973	23
+IEPA-PNO	−76.6015				728	1973	23
SCF (several bases)	−76.34042	96.5	113.5	2.6–6.3		1973	22
+IEPA-PNO	−76.60166	98.3	111.6	14.2–14.6		1973	22
SCF LCAO	−76.3326	96.3	112.5	7.9	C701	1973	21
SCF STO-DZ	−76.30141	97.0	120		776	1973	24
SCF	−76.32436	(96.3)	114.3			1974	25
6-31G*	−76.28656	96.9	113.2			1974	28
SCF LCAO	−76.329776	95.9	113.5	5.4	C707	1975	30
+CI	−76.541803	97.2	111.6	8.6	C701	1975	30
SCF PNO-CI	−76.54453	(97.9)	(111.4)	3.3		1975	31
CEPA-PNO	−76.55526			6.3		1975	31
SCF LCAO large Gaussian set	−76.3233	96.3	114.2	5.0		1977	33
SCF CI	−76.33822	96.4	113			1977	34
MBRSPT + CP					C712	1978	37
SCF DZ + P	−76.329995				C687	1978	38
CI	−76.579539	97.8	111.8	6.3		1980	40

SCF large basis	−76.58579	97.3	111.6	9.6		1981	44
Near HF/MP3							
SCF various basis sets					736–57	1981	42
6-311G**/MP4	−76.58079					1981	42
STO-6G, 4-31G*		98.9	113.9		C733	1982	50
6-31G**, MP4					C720	1982	49
STO-3G, DZ + D	−76.31231	96.7	113.2		C695	1983	57
HF MP4/SDTQ	−76.57585				C689	1983	56
SCF CI		97.8	111.6	8.5		1983	53
SCF DZ + P	−76.32970	96.3	114.4	4.5		1983	65
SCF extended	−76.33808	96.2	114.1			1983	65
6-31G*	−76.28934	96.9	113.1		C711	1983	59
SCEP-CEPA		98.0	111.6	9.4		1983	55
SCF CI	−76.563579	97.5	111.8	9.2		1984	68
SCF-CEPA-1	−76.621730	97.45	111.9	8.13	C687	1986	90

[a] 1 au = 2625.15 kJ mol^{-1}.
[b] Values in brackets were held fixed.
[c] Proton affinity for H_2O (i.e. $H_2O + H^+ \rightarrow H_3O^+$), values corrected for ZPE and thermal effects are preceded by C.

Key to table 1.1

CEPA	Coupled electron pair approximation
CI	Configuration interaction
CNDO	Complete neglect of differential overlap
CP	Counter-poise correction
DZ	Double-zeta
GTO	Gaussian type orbitals
HF	Hartree–Fock
IEPA	Independent electron pair approximation
LCAO	Linear combination of atomic orbitals
MBRSPT	Many-body Rayleigh–Schrödinger perturbation theory
MP3	Möller–Plesset perturbation theory to third order
OCE	One-centre expansion
P	Polarisation
PNO	Pair-natural orbitals
SCEP	Self-consistent electron pairs
SCF	Self-consistent field
SDTQ	Single, double, triple and quadruple excitation
SE	Semi-empirical
STO	Slater type orbitals

For explanations of the notations for such basis sets as 6-31G** the reader is referred either to the reference given or to two modern texts on *ab initio* calculations. [366, 367]

H_2O (since 1980) all fall in the small range -691 to -697 kJ mol^{-1} [98–101], though going back to 1977 one can find values ranging up to -713 kJ mol^{-1}. [102] The values obtained from reasonably accurate *ab initio* calculations agree quite well, falling in the range -687 to -720 kJ mol^{-1}. Values are portrayed on figure 1.1.

It is appropriate at this point to mention the polarisation model developed by Stillinger & David [36], which treats O^{2-} and H^+ particles as the structural and dynamic elements. The results of calculations for H_3O^+ were surprisingly good (cf. *ab initio* results in table 1.1). They found a C_{3v} pyramidal structure with O–H = 104.1 pm and $\angle$HOH = 107.9°, a PA of H_2O of -692 kJ mol^{-1}, and an inversion barrier of 16.6 kJ mol^{-1}.

The MO calculations can also be used to give an idea of the charge distribution in H_3O^+. The most recent evaluation of this [58] found $q_H = +0.518$ au and $q_O = -0.555$ au for a calculation using a 6-31G** basis.

There have been numerous reports of calculated vibrational force constants or frequencies. [12, 13, 25, 30, 32, 38, 47, 53, 55–7, 65, 68, 86] The most recent of these have been summarised in [86]. In general, because of inadequacies in the theory and neglect of anharmonicity, the calculated harmonic frequencies are too high. However, scaling factors have been proposed for particular basis sets, based on comparison of experimental versus theoretical results for many molecular species. DeFrees & McLean [86] give harmonic frequencies (MP2/6-31G* calculations) $\nu_1 = 3519$, $\nu_2 = 963$, $\nu_3 = 3632$ and $\nu_4 = 1731$ cm^{-1} respectively, which when uniformly scaled (by 0.96) give $\nu_1 = 3378$, $\nu_2 = 924$, $\nu_3 = 3487$ and $\nu_4 = 1662$ cm^{-1} as the predicted frequencies. Some authors have also attempted to predict the doubled mode frequencies caused by inversion; their results for the inversion mode ν_2 are in reasonable agreement with experiment for the $\nu_2(1- \leftarrow 0+)$ transition (observed = 954.4 cm^{-1}, calculated = 992, 961 and 985 cm^{-1} [53, 55, 68]). Since there is not, as yet, a complete set of experimental frequencies for H_3O^+ in the gas phase, a full comparison cannot be made, though the calculated results are obviously quite reasonable (see gas phase section, Part One). They also compare reasonably well with the solid phase frequencies for H_3OSbCl_6 (see solids section, Part One), with the exception of the ν_2 mode, which might be expected to be most affected by external interactions. Colvin, Raine, Schaefer & Dupuis [65] also calculated the IR intensities for the vibrational modes of H_3O^+, H_2DO^+, HD_2O^+ and D_3O^+. They predicted intensities for H_3O^+ roughly in the ratios 1:13.9:13.5:3.2 for ν_1, ν_2, ν_3, and ν_4 respectively (i.e. $\nu_3 > \nu_1$ and $\nu_2 > \nu_4$). In connection with Raman vibrational mode intensities Lopez Bote & Montero [79] have calculated the Raman tensor of H_3O^+ based on bond polarisability parameters.

A few papers have dealt with calculations of NMR or NQR properties of H_3O^+. [10, 27, 52, 63, 84, 85, 87] The majority [10, 52, 63, 84, 87] concerned the ^{1}H chemical shielding tensor and isotropic chemical shift; σ_{iso} values range from 19.8 to 26.0 ppm. In general these reproduce the ^{1}H shifts relative to

H_2O moderately well; the observed values fall 2–3 ppm further downfield, but this may be due in part to medium effects (i.e. the experimental shifts are for solids and liquids whereas the calculations are for isolated ions). The calculated ^{17}O shifts [52, 63, 87] fall in the range $\sigma = 294$–330 ppm. The observed line falls 21.6 ppm downfield from molecular H_2O (and 9 ppm downfield from bulk water). [103] This is contrary to the predictions of Galasso [63] (57 ppm upfield from H_2O) and Fukui, Miura & Tada [52] (34.6 ppm upfield). Fukui, Miura, Yamazaki & Nosaka [87] later predicted a value 27.6 ppm downfield from H_2O, in much better agreement. Galasso [63] has also calculated spin–spin coupling constants; $1J(O–H) = -161$ Hz (observed = 106 Hz) and $2J(H–H) = -9.7$ Hz and more recently Fronzoni & Galasso [85] studied the effects of electron correlation on J couplings.

Dixon, Overill & Claxton [27] have studied the effects of varying bond length and angle on the calculated ^2H and ^{17}O quadrupole coupling constants (QCC) of D_3O^+. The ^2H QCC is quite sensitive to bond length (QCC decreases, and η, the asymmetry parameter, increases as O–H distance increases) but is almost independent of bond angle; the QCC, which is 288 kHz for the optimum geometry ($\eta = 0.128$), is reduced by only 2 kHz from $\angle$ HOH = 100° to 120° (η increases significantly). For ^{17}O the QCC is relatively insensitive to bond length and more sensitive to angle ($\eta = 0$ because of the C_{3v} symmetry). At the optimum geometry QCC = −10 MHz, whereas experiment gives QCC = +7.513 MHz, which is opposite in sign to that calculated, with $\eta = 0.104$ (values for $H_3O^+HSO_4^-$ where H_3O^+ is not C_{3v} [45]).

Apart from the calculations for the protonation of H_2O a number of *ab initio* calculations have dealt with protonation reactions of H_3O^+ with other simple molecules or ions including NH_3, NH_2, OH^-, HF, HCl, PH_3, H_2S, C_2H_2, C_2H_4, N_2, O_2, CO, HCN, N_2H_2, HCOOH, CH_4^{2+}, H^+ [29, 35, 38, 39, 41, 46, 51, 59, 60, 64, 66, 67, 69, 71, 73, 80–3, 89, 167–9], with the general aim of determining energies or heats of reaction, and in some cases the equilibrium constant. (Again, as with PA, one must take account of ZPE and thermal parameters before reasonable comparison with experiment can be made.) The equilibria usually favour the stability of the reaction products over H_3O^+ and reactant: e.g.

$$NH_3 + H_3O^+ \rightleftarrows H_2O + NH_4^+ \qquad \Delta E = -171.3\ \text{kJ mol}^{-1} \qquad [38]$$

$$H_3O^+ + OH^- \rightleftarrows 2H_2O \qquad \Delta E = -1023.7\ \text{kJ mol}^{-1} \qquad [35]$$

$$(H_2O)_2^+ \rightleftarrows H_3O^+ + OH \qquad \Delta E = +93.6\ \text{kJ mol}^{-1} \qquad [64]$$

However, one should note that these reactions apply to isolated reactant and product species. The calculations on $(H_2O)_2^+$ also show that although the dimer cation is more stable than dissociated H_3O^+/OH its minimum energy structure is actually $H_3O^+ \cdot OH$. [64, 67] On the other hand $H_2O + HX$ (X = F, Cl) is slightly less stable than the associated hydrogen-bonded species

$H_2O \cdot HX$, which is much more stable than the separated ions H_3O^+/X^-. [39] *Ab initio* calculations on the interesting H_4O^{2+} ion (i.e. protonated H_3O^+ [89]) show that the tetrahedral form of this species is at an energy minimum, although it is 248 kJ mol^{-1} less stable than separate H_3O^+ and H^+, but the barrier to dissociation is considerable; 184 kJ mol^{-1}. The reaction of $H_3O^+ + H_2O$ will be considered in the next section.

Attention has also been paid to the processes involving H_3O^+ occurring at electrodes [61, 62 and references therein, 77, 78], though we will only briefly mention this here. Knowles [61] used a quantum mechanical model to study the charge transfer reaction

$$M^- + H_3O^+(aq) \rightarrow MH + H_2O$$

which occurs in the electrochemical generation of H_2 from acid solution, and compared the activation energy for this process with that for

$$MH^- + H_3O^+(aq) \rightarrow MH_2 + H_2O$$

as a function of the M–H coupling. (MH represents H chemisorbed on the metal electrode.) The model accounted for M–H and H–H interactions and solvation effects. Pataki *et al.* [77] used *ab initio* calculations to study 12 possible reaction mechanisms which might explain the simplified process $H_2 \rightarrow 2H^+ + 2e^-$ at the anode of the hydrogen/oxygen acidic fuel cell. They concluded that the following reaction steps provided the most likely mechanism, though several other schemes were not excluded:

$$H_2 + 2H_2O + 2H_3O^+ \rightarrow 2H + 2H_2O + 2H_3O^+ \rightarrow H + H_2O + H_3O + 2H_3O^+$$
$$\rightarrow H + H_2O + 3H_3O^+ + e^- \rightarrow H_3O + 3H_3O^+ + e^-$$
$$\rightarrow 4H_3O^+ + 2e^-$$

They also investigated some reaction schemes involving H_3O^+ in the hydrogen/oxygen alkaline fuel cell. [78]

1.1.2 $H_5O_2^+$

There have been a number of semi-empirical MO calculations on $H_5O_2^+$, [17, 26, 104–6], but since this type of calculation is notoriously poor in modelling hydrogen bonds we will concentrate on the more rigorous *ab initio* MO calculations. [16, 18, 24, 26, 41, 59, 80, 83, 84, 106–27] *Ab initio* studies have been particularly useful in looking at $H_5O_2^+$, beginning in 1970 with the work of Kollman & Allen [16] and Kraemer & Diercksen [107], since they help in understanding properties which arise because of the presence of the strong central hydrogen bond.

Relative to H_3O^+ it is much more costly and time consuming to do a complete geometry search for an energy minimum of $H_5O_2^+$, because of the increased number of atoms and internal degrees of freedom. Many of the earlier calculations optimised the geometries within certain symmetry constraints. Symmetries were favoured where each oxygen atom had a planar configuration of three hydrogen atoms [e.g. 16, 24, 107, 109, 112, 115, 120]. However, more recent calculations have shown that for some properties, such as the hydration energy and the central hydrogen-bonded proton potential, this is not a serious drawback. Newton & Ehrenson [18] also investigated several unconventional structures for $H_5O_2^+$ using the 4-31G basis set: (*a*) a bifurcated ring structure in which an H_2O accepts two hydrogen bonds from an H_3O^+; (*b*) a structure in which an H_2O donates a proton to hydrogen bond with the lone pair of H_3O^+; (*c*) a charge dipole structure in which the oxygen of H_2O sits below the base of the H_3O^+ pyramid. However, these were all found to be substantially higher in energy than the normal central hydrogen-bonded structure, by 65, 222 and 106 kJ mol^{-1} respectively. (Note that the H_2O proton donor (case (*b*)) is even less stable than separate $H_3O^+ + H_2O$.)

Only in recent years have structures been calculated which have optimised to non-planar coordination of the oxygens, i.e. more in keeping with most of the unambiguous solid structures (see Part One). Note here particularly [83, 118, 119, 122, 126] which are the most accurate to date. Yamabe, Minato & Hirao [118] obtained an asymmetric system in which the central hydrogen bond has $O \cdots O = 240.92$ pm, $O\text{–}H' = 106.7$ pm and $\angle OH'O = 177.0°$ where H′ is the proton involved in the central strong hydrogen bond of $H_5O_2^+$. (Note that on the basis of our earlier criteria (Part One) these distances would make this an $H_3O^+H_2O$ system rather than $H_5O_2^+$. We would remark, however, that it seems odd that this is the only calculation which has minimised to such an asymmetric but short hydrogen bond. Nor does this geometry fit the well known O–H vs. $O \cdots O$ correlation [128–30] from which an O–H distance of 106.7 pm would correspond to $O \cdots O = 250$ pm.) Potier, Leclercq & Allavena [119] did a very careful study of the effects of variations in the angular conformations. Their lowest energy structure (see figure 1.2) has $O \cdots O = 238.7$ pm with the protons placed slightly off the centre of the $O \cdots O$ line ($\angle OH'O = 178.4°$). The terminal H_2O groups have $\angle HOH = 110.5°$ and the bisector of this angle is at 30° to the $O \cdots O$ direction. However, the energy needed to make the coordinations of the oxygens planar is only 5 kJ mol^{-1}. Rotation of the two end H_2O groups (dihedral angle ϕ) with respect to each other shows that the least favourable conformation is *cis* ($\phi = 0°$) which is about 5.3 kJ mol^{-1} above the lowest energy conformation which is *gauche* ($\phi = 105°$). However, the *gauche* form is only about 2.5 kJ mol^{-1} more stable than the *trans* form ($\phi = 180°$). A recent study [122] reported on a geometry search carried out using a gradient optimisation technique with the 6-31G(d) and 6-31+G(d) basis sets. The optimised structures were then used in calculations which included fully polarised basis sets, 6-31G(d, p) and 6-31+G(d, p), and electron correlation. All the minimum energy structures calculated (with different symmetry constraints) gave non-planar coordination of the oxygen atoms. The lowest energy was obtained using 6-31+G(d, p)/MP4SDQ for a C_2 structure with a

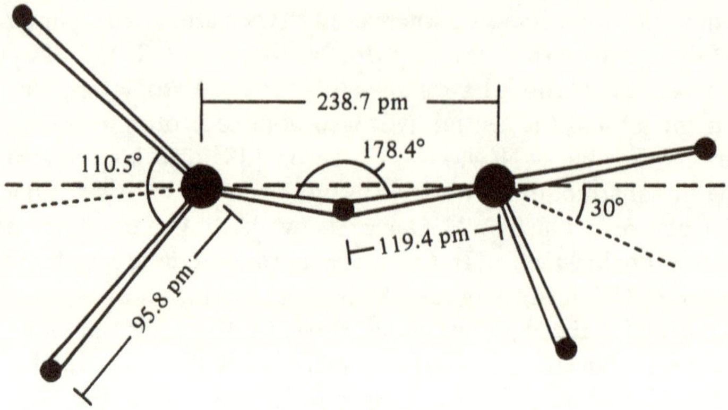

Fig. 1.2. The structure of $H_5O_2^+$ from the lowest energy conformation calculated by Potier *et al.* [119] The structure is C_2, and *gauche* (viewed down the O···O axis the bisectors of the two terminal H_2O units are rotated 105° with respect to each other). The pyramidal geometry around the oxygen atoms is indicated by the 30° angle between the O···O axis and the bisector of the H_2O group.

symmetrical O··H'··O = 238.2 pm and ∠OH'O = 177.5°. What is perhaps of more interest, however, is the remarkably small energy difference of 4.2 kJ mol^{-1} between this and a C_s structure with O–H'··O = 248.2 pm and O–H' = 103.9 pm calculated at the same level of theory. (At lower levels of theory without correlation this C_s structure is actually more stable by 0.9 kJ mol^{-1}.)

In the most recent study [126] the calculations have been taken to yet another level of theory to obtain an even lower energy. Unfortunately few geometrical parameters were given in this last paper, but it is included in table 1.2 to indicate that the search for the minimum is not yet over. Once again we may consider for comparison the results of Stillinger & David's non-*ab initio* polarisation model. [36] As might be expected, this does not appear to model the strong hydrogen bond very well (symmetrical O··H'··O = 256.5 pm), though it does produce pyramidal coordination for the oxygens and a hydration energy for H_3O^+ of −151 kJ mol^{-1}. Their minimised structure has C_{2h} symmetry.

What clearly emerges from all these calculations is that the potential energy surface around the minimum energy configuration must be rather flat; many different configurations appear to have only very small energy differences. This helps to explain the great variety of conformations observed for $H_5O_2^+$ ions in different solids; the configuration must readily adjust to the requirements of the lattice, since the external barriers formed by hydrogen bonding to anions in a lattice are likely to be much larger than the small internal conformational barriers. Most of the calculations obtain lowest energy O···O distances of about 239 ppm, which is a little less than the range of 241–244 pm observed in crystals (from neutron diffraction, see Part One).

Table 1.2. *Selected results of molecular orbital calculations on $H_5O_2^+$ by year*

Method	Symmetry	Total energy (au)	Heat of hydration[a] (kJ mol^{-1})	Ref.	Year
LCGO + P	D_2	−152.42846	−134.9	107	1970
DZ + P	D_{2d}	−152.36230	−154.4	16	1970
4-31G	D_{2d}	−152.1791	C−184.5	18	1971
STO DZ	D_{2d}	−152.37793	−185.7	24	1973
6-31G*	C_{2h}	−152.35070	−133.0	59	1983
DZ + P	D_{2d}	Not given	−132.2	120	1984
LCGO + P	C_2	−152.427208	−134.1	119	1984
DZ + P/SAC	C_1	−152.84381	C−148.5	118	1984
6-31 + G(dp)/MP4SDQ	C_2	−152.81400	C−138.1	122	1985
MP2/6-311 + + G(2d, 2p)	C_2	−152.961025	C−142.7	126	1986
6-311 + + G(3df, 3pd)	C_2	−153.005412	C−146.4	126	1986

[a] $H_3O^+ + H_2O \rightleftharpoons H_5O_2^+$, values corrected to give ΔH rather than ΔE are preceded by C.

This small difference is perhaps not surprising since in the crystalline materials the peripheral hydrogen atoms of the $H_5O_2^+$ are involved in hydrogen bonds.

From the compilation in table 1.2 it can be seen that the most recent and more accurate calculations give values for the heat of hydration of H_3O^+ in the reaction

$$H_3O^+ + H_2O \rightleftharpoons H_5O_2^+$$

in the neighbourhood of −142 kJ mol^{-1}. The comparison with the experimental values available, $\Delta H = -132$ [131] and $\Delta H = -138$ [132] kJ mol^{-1} is good. Furthermore, this is additional evidence that the formation of $H_5O_2^+$ is energetically quite favourable.

Many of the *ab initio* studies had the principal objective of studying the O··H′··O hydrogen-bond potential and proton transfer [16, 83, 107–14, 116, 117, 119–21, 123–5]; indeed, $H_5O_2^+$ is often chosen as a model for double-well potential systems. In 1970 Kollman & Allen [16] and Kraemer & Diercksen [107] were the first to study the potential as a function of the displacement of the hydrogen atom between the two oxygen atoms. Kollman & Allen's results indicated a single narrow minimum at 230.2 pm which broadened to a rather flat single minimum at 238 pm and which had already developed into a double minimum by 248.7 pm. Kraemer & Diercksen's potential at 239 pm was broad with just a hint of a central barrier. Meyer, Jakubetz & Schuster [109] later showed that the barrier in the double-minimum potential (their calculations were for O···O = 262–274 pm) is very sensitive to the inclusion of electron correlation.

De la Vega *et al.* [105, 133, 134] and Cao, Allavena, Tapia & Evleth [83] have used non-*ab initio* methods of describing proton tunnelling in double-well potentials including the $H_5O_2^+$ system. The potentials used in some cases [83, 133, 134] were, however, generated using *ab initio* calculations.

More recently (since 1980) Scheiner and coworkers [112–4, 116, 117, 121, 123–5] have undertaken a large amount of work to study the potential and proton tunnelling in the central O··H′··O bond. They used principally a 4-31G basis set, though the effects of using larger basis sets and electron correlation were also investigated. Their results may be summarised as follows: geometry optimisation, in particular the conformation of the two end H_2O units, or the inclusion of two additional H_2O molecules attached to each end of the $H_5O_2^+$, have little affect on the barrier to proton transfer for a fixed O···O distance. There is a strong dependence on the O···O distance; probably the most significant finding is that below about 240 pm the potential has a single minimum. Above 240 pm the potential has a double minimum and the barrier increases rapidly with increasing separation of O···O, while at the same time the shorter O–H′ bond distance decreases (figure 1.3). Angular bending of the hydrogen bond also causes an increase in the barrier though for

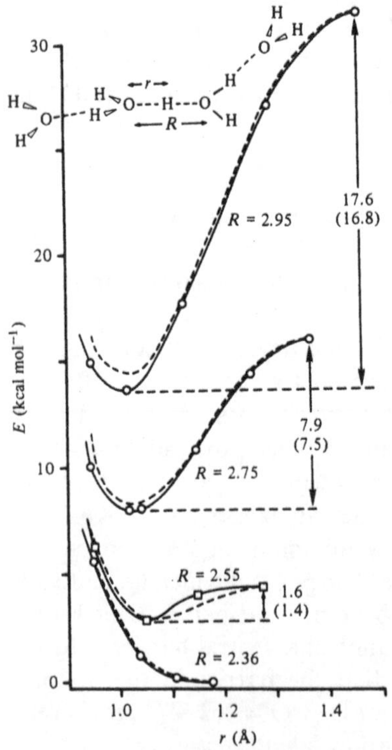

Fig. 1.3. Proton transfer potentials for $H_5O_2^+$ (dashed curves and energy barriers in parentheses) and $H_5O_2^+(H_2O)_2$ (solid curves) calculated by Scheiner. [114] All hydrogen bonds were kept linear. The curves for $H_5O_2^+$ are superimposed on the curves for $H_5O_2^+(H_2O)_2$ such that the mid-point energies coincide. NB. In this original figure energy values were given in kcal mol^{-1} (conversion: 1 cal = 4.1835 J).

small deformations (<20°) this increase is small. This latter effect is more pronounced for larger O···O separations. There is an overall loss of electron density on the H_2O unit accepting the proton during transfer [113] and a gain on the H_2O unit of the donor. The largest electron density changes occur near the centre of the hydrogen bond as the transferring proton pulls substantial density with it. Larger basis sets lead to higher barriers, while substantial reductions result from inclusion of correlation effects. Also the overall positive charge of the system is probably of no great importance since $H_3O_2^-$ has a similar barrier for the same O···O separation.

Since the above results show that the principal influence on the barrier is the O···O separation, Scheiner *et al.* suggest that it is simplest to use a planar configuration of $H_5O_2^+$ when studying the properties of the O··H'··O bond. They used this form when studying the effects of external ions on the potential, [123, 125], and a fixed O···O distance of 274 pm. With an ion placed in the plane perpendicular to the O–H'–O axis at the mid-point of the hydrogen bond the potential remained symmetrical and the barrier height and proton tunnelling rates were only slightly affected. However, with the ion placed in the plane perpendicular to the O··H'··O axis at one of the oxygen atoms the double-well potential became considerably asymmetric and the proton transfer rates from one well to the other were reduced by several orders of magnitude. (For the isolated $H_5O_2^+$ ion this rate is of the order of $1.95 \times 10^{12}\,s^{-1}$ for the case where O···O = 274 pm.) These effects were even greater when the ion was placed at a point on the extension of the O··H'··O axis. For example, for Na^+ at 500 pm one well is higher than the other by about 42 kJ mol^{-1}. Ions or point charges were placed either 500 pm away from the oxygen atom or bond mid-point (for Na^+ or positive charge) or 600 pm away (for Cl^- or negative charge). At shorter ion distances (about 350 pm) along the hydrogen-bond axis the double-well character is eventually lost. The results also show that the hydrogen bond is very polarisable and can almost double the electric field produced by the external ion. This is in agreement with the results of Zundel [156, 157] which will be discussed later.

Force constants, and in some cases vibrational frequencies of $H_5O_2^+$, have only been calculated using MO calculations in a few instances. [16, 18, 126, 127, 135] Yukhnevich, Kokhanova, Pavyluchko & Volkov [135] used CNDO calculations to obtain the potential surface and hence the vibrational force constants for the O··H'··O bridge in $H_5O_2^+$, and then proceeded to calculate the vibrational frequencies of the ion assuming force constants for the peripheral O–H bonds and angles. Williams [127] obtained vibrational frequencies (which were not scaled) using a 4-31G basis. Perhaps the most useful information from this work comes from a comparison of frequencies with those of the water dimer calculated using the same basis set. [136] The peripheral water O–H stretching frequencies of $H_5O_2^+$ were found to be lower and the H_2O bending modes of $H_5O_2^+$ were higher than the similar modes in $(H_2O)_2$. The most recent, and probably most accurate, work is that of Frisch,

Del Bene, Binkley & Schaefer [126], though the frequencies given were not scaled in any way: they obtained stretching frequencies for $O\cdot\cdot H'\cdot\cdot O$ of 805 and 621 cm^{-1}, and peripheral O–H stretching frequencies in the range 3776–3877 cm^{-1}. There is a lack of unambiguous experimental assignments for comparison but these calculations do show that the stretching frequencies for the central hydrogen bond are very low.

Ab initio calculations have also been used to calculate ^1H NMR chemical shifts for $H_5O_2^+$. [84, 115] The results are dependent on the particular geometry used but qualitatively they show the same trends; the central proton is strongly deshielded relative to H_2O and has a chemical shift anisotropy almost twice that of the peripheral protons. The peripheral 'H_2O' protons themselves have isotropic chemical shifts closer to that of H_3O^+ than that of H_2O. (For one set of calculations using a 6-31G** basis set and for $O\cdot\cdot H'\cdot\cdot O = 256.5$ pm and $O–H' = 104$ pm for the $H_5O_2^+$ the following absolute (σ) isotropic shifts (in ppm) were obtained: H_2O, 31.17; H_3O^+, 24.29; $H_5O_2^+$, H', 15.55, H peripheral, 26.13 and 23.46.)

1.1.3 *Higher hydrates*

Theoretical studies of the proton hydrated beyond $H_5O_2^+$ are fewer in number. [17, 18, 32, 36, 114, 127, 137–46, 365] As early as 1961 Grahn [137] attempted to model the hydration of H_3O^+ using a semi-empirical SCF–MO–LCAO method. He paid particular attention to $H_9O_4^+$ and, assuming a planar configuration with $O–H\cdot\cdot O = 255$ pm, he calculated a value of 188 kJ mol^{-1} for the energy of one of the hydrogen bonds, of which 167 kJ mol^{-1} was electrostatic energy. The changes in charge distribution also suggested that the ability of the $H_9O_4^+$ to bind further water molecules should be greater than a normal water molecule. Salaj [138] used a simple point charge model and calculated the equilibrium $O\cdot\cdot O$ distances in $H_9O_4^+$ as 245 pm and the binding energy of the hydrogen bond as 194 kJ mol^{-1}. De Paz, Ehrenson & Friedman [17] used a semi-empirical CNDO/2 method to study $H_3O^+(H_2O)_n$ ($n = 0$–3). They showed that chain structures were preferred over cyclic structures. Proton centred structures were found to be even less stable, e.g. a proton coordinated by three water molecules to give $H_7O_3^+$ was 68 kJ mol^{-1} less stable than the chain structure.

By far the most frequently quoted work is that of Newton & Ehrenson [18] who studied $H_3O^+(H_2O)_n$ for $n = 0$–4 using the 4-31G basis set in *ab initio* calculations. They considered several configurations and combinations of units for each case though here we will mention principally their lowest energy structures. Once again symmetry constraints were used to simplify the calculations. $H_7O_3^+$ was constrained to a chain form with C_{2v} symmetry optimised to $O\cdot\cdot O$ distances of 246 pm; this was essentially H_3O^+ with two symmetrically attached waters. (This was 8.7 kJ mol^{-1} more stable than the structure consisting of an $H_5O_2^+$ with one attached water, and most of the other structures considered were considerably less stable.) For $H_9O_4^+$ the most stable form considered was a D_{3h} system consisting of a central H_3O^+ with three attached waters and O··O = 254 pm. This can be compared with the value 252 pm obtained from the peak of the O··O radial distribution determined from diffraction experiments on acid solutions. [147] A pyramidal form was only 5.8 kJ mol^{-1} less stable. It should be remembered, however, that the 4-31G basis set does give the wrong (planar) form for H_3O^+ in the first place. A straight chain form of $H_9O_4^+$ was 19.7 kJ mol^{-1} less stable. For $H_{11}O_5^+$ the most stable form considered consisted of a water attached via its oxygen atom to a peripheral hydrogen atom of the stable D_{3h} $H_9O_4^+$ unit.

The more general observations and conclusions from these calculations are probably of greater interest. The calculations show a significant lengthening of the O–H–O bonds as the number of H_2O units is increased (236 pm for $H_5O_2^+$ to 254 pm for $H_9O_4^+$). In general the H_3O^+ unit, alone or in hydrated species, does not accept a hydrogen bond from another H_2O. Both these results are in agreement with observations made earlier concerning solids (Part One). They found that it requires about 54 kJ mol^{-1} to move the excess proton in $H_7O_3^+$ and $H_9O_4^+$ from the central oxygen to a peripheral oxygen, and that the hydrogen bonds involved have asymmetric single-well potentials. They thus concluded that proton hopping is not an intrinsic property of the *isolated* $H_7O_3^+$ and $H_9O_4^+$ species. Their heats of hydration for successive steps are given in table 1.3, and are consistently larger than the experimental gas phase values. There is no indication of a spreading of the positive charge over the whole complex; in $H_7O_3^+$, $H_9O_4^+$ and $H_{11}O_5^+$ the positive charge is found to be at least 80% localised on the central ionic H_3O^+ or $H_5O_2^+$ unit. It was also suggested that the major component of the binding energy can be viewed as due to charge–dipole interaction, either between H_3O^+ and H_2O or between H^+ and two H_2Os.

Williams [127] has recently calculated a geometry-optimised structure of $H_7O_3^+$ using the 4-31G basis set, though his principal objective was to obtain vibrational frequencies. Once again the 4-31G calculation optimised to a planar configuration about the central H_3O^+ unit, with O··O distances in

Table 1.3. *Heats of hydration in kJ mol^{-1} for successive stages of the reaction* $H_3O^+(H_2O)_{n-1} + H_2O \rightarrow H_3O^+(H_2O)_n$

	4-31G *Ab initio*		Polarisation model	Experimental
(n − 1), n	[18]	[32]	[36]	[165]
−1, 0	−740		−692	−697
0, 1	−156	−155	−154	−132
1, 2	−111	−109	−88	−81.6
2, 3	−94	−92	−86	−74.9
3, 4	−64	−67	−66	−53.1
4, 5		−63		−48.5
5, 6				−44.8

close agreement with Newton & Ehrenson. [18] The structure and vibrational frequencies support the formulation as $H_3O^+(H_2O)_2$.

In a later paper Newton [32] extended the 4-31G calculations to obtain O–H stretching frequencies for the species $H_9O_4^+$, $H_{11}O_5^+$ and $H_{13}O_6^+$. (This work was stimulated by the gas phase IR studies of Schwarz. [148]) He found that for all three species the lowest energy structure of the conformations considered had H_3O^+ as the central unit. However, in the case of $H_{13}O_6^+$ with $H_5O_2^+$ as the central unit and four waters attached symmetrically this was only 9.2 kJ mol^{-1} higher in energy than the H_3O^+ centred form. He then suggested that this $H_5O_2^+$ centred structure of $H_{13}O_6^+$ might be an intermediate in the proton transfer process, since at the equilibrium geometry the central O··H′··O bond is only 237 pm long and the proton transfer potential is a single symmetric minimum; the proton could move to either oxygen atom as the activated complex transforms back to the more stable structure.

Newton also investigated a possible addition of a fourth water molecule to $H_9O_4^+$ at the oxygen atom of the H_3O^+ unit, i.e. as another inner-shell water. (It is common practice when discussing these higher hydrated forms to refer to water molecules as being in the first, second or higher hydration shells (or spheres) surrounding a central H_3O^+ unit.) This was found to be quite unstable in a hydrogen-bonding configuration but energetically neutral in a charge-dipole configuration (see figure 1.4). Thus the charge-dipole case suggests it would be feasible to have a fourth inner-shell water. Such a model was found to be compatible with X-ray diffraction results on acid solutions [147] (discussed later).

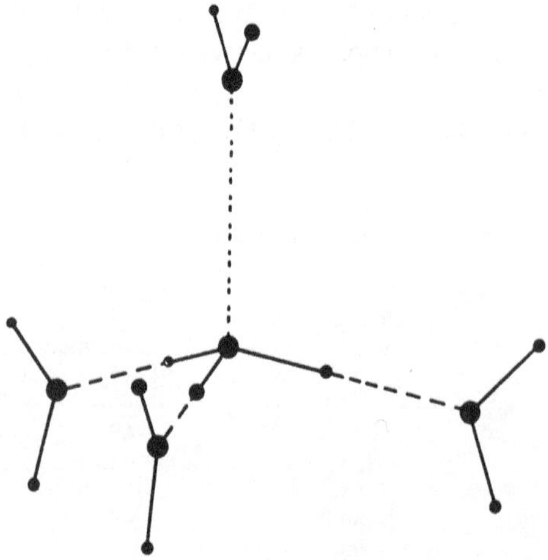

Fig. 1.4. The structure of $H_3O^+(H_2O)_4$ showing the $H_9O_4^+$ configuration calculated by Newton [32] with a fourth water in the energetically neutral charge–dipole configuration.

In the calculations of the stretching frequencies Newton first obtained the force constants. (Values for the O–H stretching force constants of the lower hydrated species were obtained in the earlier calculations. [18]) He neglected coupling of the O–H stretching modes of the central H_3O^+ unit to other modes and also neglected the coupling between the stretching modes of hydrogen-bonded and free O–H bonds on the second hydration shell waters. He then found a linear-least-squares correlation between the calculated values of $(F_sG_s)^{\frac{1}{2}}$ (where F_s = force constant and G_s = reciprocal effective masses) and the experimental gas phase frequencies of Schwarz. [148] We gave a comparison of the frequencies he was then able to obtain from this correlation with the observed values in Table 9 of Part One. (NB. These are not directly calculated frequencies. In effect a scaling factor has been applied similar to the one mentioned earlier in the H_3O^+ section.) Another feature of interest was that the calculations gave $\nu_s < \nu_a$ (i.e. $\nu_3 > \nu_1$) in isolated H_3O^+ but $\nu_a < \nu_s$ for the H_3O^+ unit in $H_9O_4^+$ (s = symmetric, a = antisymmetric).

Newton also remarked upon the cooperative interactions between hydrogen bonds; e.g. if a water molecule X in the first hydration shell of H_3O^+ acts as a hydrogen donor to another water in the second shell, then X's hydrogen bond acceptance from the H_3O^+ is enhanced. This is illustrated in figure 1.5.

While Newton's results for the higher hydrates represent a remarkable step forward, it should still be borne in mind that these are not definitive and one must wonder how much some of the conclusions are affected by the

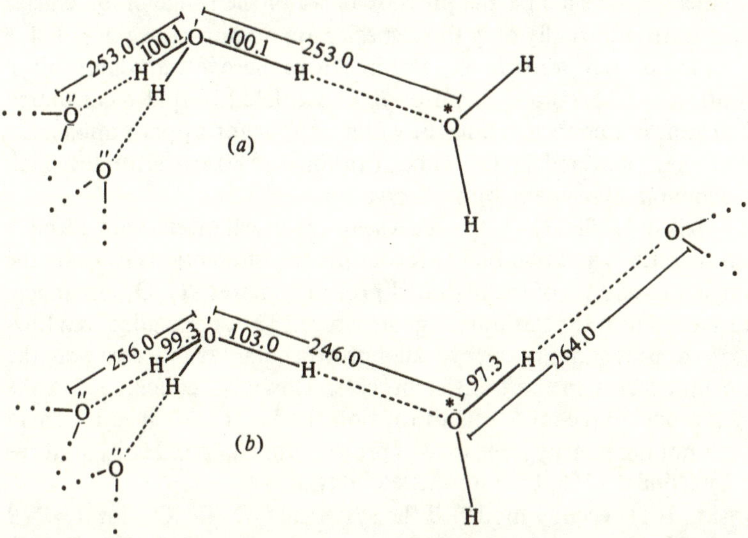

Fig. 1.5. Cooperative interactions between hydrogen bonds in $H_{11}O_5^+$ according to Newton [32]. When O* hydrogen bonds to a second-shell H_2O, as in (*b*), O′ ··· O* decreases, O′–H in the O′–H ··· *O bond increases and O′ ··· O″ increases (distances in pm).

assumptions and constraints which have been applied. In the first place the structures are not fully geometry-optimised and one should anticipate numerous possible configurations which are very close in energy (recall the difficulties experienced in trying to fully optimise even $H_5O_2^+$ in the most rigorous calculations described earlier). The results also apply strictly to isolated clusters and if one extrapolates from these results to suggest behaviour in the liquid phase the conclusions must be described as conjectural.

Stillinger & David's non-*ab initio* polarisation model [36], mentioned earlier, gave non-symmetrical, lowest energy structures for the higher proton hydrates (note that they attempted complete geometry searches). $H_7O_3^+$ had O–H–O bond lengths of 257.2 pm and 276.1 pm and was thus rather more like $H_5O_2^+ \cdot H_2O$ than $H_3O^+ \cdot (H_2O)_2$ as found by Newton & Ehrenson. [18] Their lowest energy $H_9O_4^+$ structure consisted of an $H_5O_2^+$ unit with two waters, one attached at each end. This was 19.7 kJ mol^{-1} more stable than a structure with H_3O^+ and three attached waters (i.e. the reverse of Newton & Ehrenson's [18] result). Similarly their structure for $H_{11}O_5^+$ consisted of $H_5O_2^+$ with three waters attached. Their heats of hydration are included in table 1.3 for comparison.

Suck *et al.* [139] used a statistical mechanical treatment to calculate the distribution of different $(H_2O)_nH^+$ ions and their mobility as a function of humidity, in connection with studies of ion clusters which are relevant in the atmosphere. They emphasized that the results obtained are not quantitative because of uncertainties in the input parameters, but they do give a good qualitative picture. At each partial pressure of water the range of ion cluster sizes is quite narrow; usually only three species are dominant, e.g. $n = 3, 4, 5$ or $5, 6, 7$. As the partial pressure of water increases there is a gradual shift of the distribution to larger sizes. Ferguson & Fehsenfeld [140] also calculated hydrated proton ion concentrations in water vapour for upper atmosphere conditions. They obtained $H_5O_2^+$ concentrations at 80 km altitude which were in reasonable agreement with observation.

Fang, Godzik & Hofacker [146] used a semi-empirical model with effective one proton potentials and two-body proton–proton interactions to study the proton transfer dynamics of the hydrated proton complex $H_{21}O_{10}^+$ in an ice-like conformation (not the clathrate cage structure). They concluded that low-lying proton transfer states in the hydrated proton complex in water were the most important factor in the transfer process. They also concluded on the basis of their model that continuous absorption bands in the IR spectra of acid solutions are not necessarily indicative of proton tunnelling in $H_5O_2^+$ units as proposed by Zundel [156, 157] (discussed later).

Kochanski [141] recently modelled the systems $H_3O^+(H_2O)_n$ for $n = 1$–9 with Monte Carlo calculations using approximate H_3O^+/H_2O and H_2O/H_2O interaction potentials. The cluster formation energies for successive hydration steps compared reasonably well with experimental values at $n > 4$ but the differences were greater for smaller clusters. Perhaps the most interesting finding was the tendency at larger n values for a fourth water molecule to enter the first hydration shell around H_3O^+ and for a third shell to begin before the second was complete.

The most recent Monte Carlo calculations of Fornili, Migliore & Palazzo [142] are of even more interest and have some significance for understanding the hydrated proton in solution. Atom–atom pair potentials for the H_3O^+/H_2O interaction were first evaluated from *ab initio* calculations using an extended basis set. These potentials were then used in the Monte Carlo simulations, which involved one H_3O^+ and 215 water molecules contained in a cube of side 1860 pm (thus corresponding to the density of an acid solution). The temperature of the system was held at 300 K. The MCY potential [149] was used for H_2O/H_2O interactions. After equilibration the $H_3O^+ \cdot\cdot H_2O$ interaction energy averaged over the first hydration shell was -105 ± 0.8 kJ mol^{-1}. As would be expected the oxygen atoms of the waters were closer to the H_3O^+ than the hydrogen atoms due to the positive charge. The radial distribution functions obtained clearly showed at least two hydration shells around the H_3O^+ with an integral value close to four water molecules in the first shell. The novel feature of this inner hydration shell is that all four water molecules are predominantly located to one side of the plane which passes through the H_3O^+ oxygen atom parallel to the plane of the three hydrogen atoms (see figure 1.6). The fourth water molecule thus appears to compete for hydrogen bonding to the hydrogen atoms of the H_3O^+, instead of sitting above the oxygen atom of H_3O^+ (i.e. on the opposite side of the

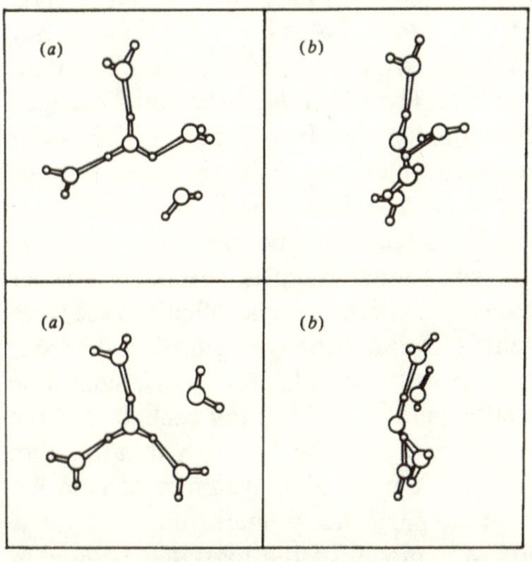

Fig. 1.6. Two representative results of the Monte Carlo calculations of Fornili *et al.* [142] showing the closest water molecules surrounding H_3O^+: (*a*) top view; (*b*) side view. Note in particular the position of the fourth water molecule.

plane) in the energetically-neutral charge-dipole configuration found in Newton's calculations on $H_3O^+(H_2O)_4$. [32] This leads to distortion and mobility of the hydrogen bonds. The radial distributions are consistent with the experimental X-ray and neutron diffraction results. [147] One major assumption (and possible drawback) of these calculations is that the water molecules can attain equilibrium configurations before the excess proton is able to transfer from H_3O^+ to another water molecule (the lifetime of H_3O^+ in aqueous solution is estimated to be of the order of 10^{-12}–10^{-11} s (see later)).

Theoretical studies have also been applied to try to explain the enhanced stabilities of the so-called 'magic number' hydrated proton clusters $(H_2O)_{21}H^+$ and $(H_2O)_{28}H^+$. [143–5] The experimental observation of these clusters in the gas phase was discussed previously in Part One. Hermann, Kay & Castleman [150], in 1982, suggested that the stability of $(H_2O)_{21}H^+$ might be explained by a clathrate structure, consisting of a regular pentagonal dodecahedron of 20 water molecules and the excess proton linked together by hydrogen bonds, with the 21st water molecule inside the cage. Such cages are commonly found in solid clathrate hydrates. [151] Holland & Castleman [143] investigated this model theoretically and concluded that self-stabilisation of the regular cage could occur through rapid exchange of the excess proton among all oxygen sites involving proton tunnelling and ion induced reorientation. Buffey & Brown [145] obtained minimum energy structures for $(H_2O)_nH^+, n=20, 21, 22$, using two-body potentials and found that the $n=21$ case was slightly more stable than $n=20, 22$. This work was extended in the Monte Carlo calculations of Nagashima, Shinohara, Nishi & Tanaka [144] who used the MCY potential [149] for the H_2O/H_2O interaction and the potential of Buffey & Brown [145] for H_3O^+/H_2O. They interpreted their results as follows: the $(H_2O)_{21}H^+$ species is more stable than its neighbouring cluster numbers and the H_3O^+ ion is inside and hydrogen bonded to a highly distorted pentagonal dodecahedral cage of 20 water molecules. However, the greater stability which they claim their calculations show for this species is, in fact, very marginal (generally only about 1.4 kJ mol^{-1} and in the best case 2.51 kJ mol^{-1}) and since the numbers apparently do not take account of vibrational energies one must wonder whether they have any real significance. Furthermore it is difficult to see from the example results shown (in their fig. 5) how the configuration of water molecules can be construed as a pentagonal dodecahedron. Once again one must also question whether it is really valid to exclude any exchange of the excess proton. Clearly more theoretical work could be done to help rationalise the stability of magic number clusters. A molecular dynamics calculation including proton migration might prove particularly interesting.

Zundel is the greatest exponent of a model of the hydrated proton in solution in which an '$H_5O_2^+$ grouping' (and also to some extent $H_9O_4^+$) is the most important central unit for explaining a number of observed properties.

Over the years Zundel and coworkers have developed the idea and attempted to back it up with numerous experiments (to be discussed later) and theoretical studies. Although many of the calculations concern $H_5O_2^+$ directly we will discuss them here separately since Zundel has generally attempted to relate the results to the liquid phase (i.e. $H_5O_2^+$ in the presence of other water molecules).

Weidemann & Zundel [152, 153] calculated the effects of an electrical field on the tunnelling proton in a symmetrical double-well potential. In particular they showed that the polarisation in such a system is very large and about 100 times greater than that of H_3O^+ alone. They concluded that the polarisation of the hydrogen bond in the $H_5O_2^+$ grouping is the field-dependent mechanism best able to explain the anomalous (high) proton conductivity in acid solutions. Even though the shift in the weights of the proton boundary structures (i.e. the structures with the proton on one or the other of the two terminal H_2O groups) would only be very small for the field strengths normally applied in conductivity measurements, it is still sufficient for the structural diffusion of protons to have a preferred direction. At its simplest level this highly polarisable proton model can be described (or visualised) for the symmetrical double-well case as follows: the energy levels in a double-minimum potential are always split into two levels because of tunnelling. The two lowest levels are symmetric and antisymmetric states and since the tunnelling splitting is very small only a very small electrical field is required to mix the excited state with the ground state. This mixing of the two states then results in an asymmetric charge distribution, i.e. polarisation. Figure 1.7 shows the levels and wavefunctions schematically.

Janoschek and Zundel *et al.* [108, 111] used MO calculations on $H_5O_2^+$ to study various properties of the strong symmetrical hydrogen bond with and without applied electrical fields, such as potential curves, dipole moment, polarisability, hyperpolarisability and O–H transition moments. Again it was found that the polarisability is about 100 times larger than usual polarisabilities, and further the polarisability of broad, flat, single minima and unsymmetrical hydrogen bonds may also be considerably larger than usual. This large polarisability causes proton dispersion interaction between symmetrical hydrogen bonds, induced dipole interaction between the hydrogen bond and anions and dipolar solvent molecules, and interaction between vibrational transitions in the hydrogen bond and other vibrations. The transition moments were also found to be relatively large compared to those of O–H stretching vibrations in asymmetric hydrogen bonds. All of these factors apparently contribute to the strong continuous absorption band observed in the IR spectra of acid solutions. The smaller polarisability of unsymmetrical hydrogen bonds causes spectral band broadening. These calculations were later extended [110] to take into account the coupling of the proton motion with the O··H′··O stretching vibration. The high polarisability was largely unaffected by the coupling. It was found that a

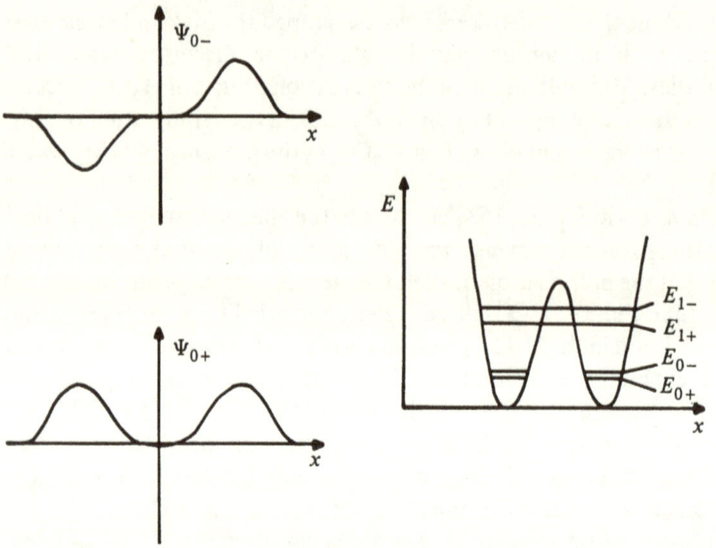

Fig. 1.7. Schematic representation of a symmetric double-well potential and the wavefunctions of the two lowest levels E_{0+} (even) and E_{0-} (odd). One can see that any mixing of these two wavefunctions in the presence of an electric field would produce an asymmetric charge distribution (i.e. polarisation). (After Zundel [157].)

major factor contributing to the multitude of energy level differences necessary to give the continuous IR absorption was the range of induced dipole interactions of the hydrogen bonds with fields from their environment. Similar calculations were carried out for $D_5O_2^+$. [154] At smaller fields the polarisability of the O··D′··O bond is greater, primarily due to a smaller tunnelling splitting of the lowest levels. Also the calculated intensity of the IR continuum for deuterium is less than for hydrogen in agreement with experiment.

Later, calculations were carried out to simulate the IR absorbance of strong aqueous acid solutions [155] using the same procedure as before but applying both a distribution of the local electric field strength and a distribution of O–O bond lengths such as might be found in solution. It was found that both kinds of distribution are important and the results showed similarities to the experimental spectra in terms of wavenumber dependence, intensity and structure. The intensity decreased as the mean field strength was increased though the structure of the continuum was unchanged over a large temperature range. In order that the continuum be observed over the whole range of frequencies below 3400 cm^{-1} it was found that the O–O bond length distribution had to include the typical double-minimum potential with a moderate barrier. Much of this work has been summarised by Zundel [156, 157] and by Janoschek. [111] The apparent success of all these calculations in producing a feasible model to explain a number of properties of aqueous acid solutions is remarkable, but one should not forget that they are based on the simple model '$H_5O_2^+$ grouping'.

Zundel's highly polarisable hydrogen bond model has not gone completely unchallenged, however. It has been criticised by Librovich *et al.* [158, 159 and comments 135, 160, 161] They proposed [158, 159] an alternative 'phonon theory' for the continuous IR absorption, which depends on a strong interaction of the stretching vibrations of the bridging-protons with a large distribution of low frequency vibrations of the surrounding solution to give many combinational transitions in the form of a phonon sideband; hence the continuum. It is the current reviewers' opinion that on balance the highly polarisable hydrogen-bond model seems more plausible and better able to explain numerous experimental observations; however, this is not to say that the phonon model is untenable and indeed both mechanisms may operate. There are also other proposed models concerning vibrational spectra of O··H··O bonds in general, see for example [162] and references therein.

Laforgue *et al.* [163] also used 4-31G basis-set calculations of proton transfer in the $H_5O_2^+$ unit in solution to support their proposed model of anomalous conduction in acid solutions. They concluded that $H_5O_2^+$ appears to be the essential structure in both the normal and anomalous conduction mechanisms; their calculations suggested that the lifetime of $H_5O_2^+$ is probably long enough to allow several proton exchanges (hence anomalous conduction) and that as one $H_5O_2^+$ unit is destroyed a new one forms (hence the normal conduction component). Furthermore, it was concluded that both tunnelling and concerted thermal jumps were important in anomalous conduction.

Further criticism of Zundel's model has come from Fang *et al.*, [146] mentioned earlier, and Giguère. [164] One cannot help feeling that to some extent this criticism is levelled because of opposition to the notion of there being $H_5O_2^+$ groups in solution. Indeed, although Zundel refers to $H_5O_2^+$ groups, most of his calculations consider O··H′··O lengths in the 250–270 pm range, which on the basis of our criteria for $H_5O_2^+$ in solids (Part One) would clearly be labelled as ($H_3O^+H_2O$). Nevertheless this pair is essential to the polarisability theory. (This really may be no more than a problem of semantics.)

At this point it is also worth reconsidering Newton's [32] suggested model for proton transfer, in which an $H_3O^+H_2O$ pair, in an $H_{13}O_6^+$ cluster, comes close enough together to form an $H_5O_2^+4H_2O$ transition state with a symmetrical, single-minimum potential for the bridging proton. As the O··O distance in the transition state then increases, the proton can move either way, i.e. transfer or not transfer. In connection with this it is worth mentioning that the observed peak in the radial distribution of $H_3O^+\cdots OH_2$ distances in hydrochloric acid solutions is at 252 pm, [147] a distance which would correspond to a double-minimum configuration for $H_5O_2^+$. Note, however, that the radial distribution does extend to lower O··O distances which would include single minima. It is then of further interest to note that Scheiner's calculations [114] show that at 252 pm the total energy at the potential minima is only about 8 kJ mol^{-1} higher than the total energy at his single minimum distance where $O\cdots O = 240$ pm. (Note that Zundel's model does take account of vibrations and also does apply to broad, flat, single minima. Giguère also suggests an $H_5O_2^+$ transition state of sorts. [164])

So we see in conclusion that there is as yet no complete consensus on the mechanism of proton transfer in solution. All models to date have a common problem, due to necessary constraints on the degree of sophistication and the computing expense, in that they do not model all the components of a real solution. A good model must eventually include both anions and cations and take account of the dynamics of the system. One hopes that this might be done in the not too distant future.

1.2 The liquid state

1.2.1 *The hydrated proton in non-aqueous liquids*

Understandably the majority of studies of the hydrated proton in the liquid phase involve aqueous systems, though there has been a substantial number of studies of non-aqueous solutions and melts which we will describe here. As in the gas phase, these systems largely isolate the hydrated proton from interactions with other water molecules, though now effects due to interactions with the solvent molecules and the anions can frequently be observed. In general, however, the problem of proton exchange with the solvent is circumvented. Strictly speaking, small mole ratio mixtures of water in acids constitute non-aqueous solutions, in the sense that all the water molecules are bound in hydrated proton species, but these will be treated later since they frequently form an integral part of studies of aqueous acid solutions over wide ranges of composition.

One of the earliest studies which gave concrete evidence of the existence of H_3O^+ in solution was the work of Bagster & Cooling [170] on the electrolysis of hydrogen bromide in liquid sulphur dioxide. Anhydrous hydrogen bromide solutions in sulphur dioxide were almost non-conducting. When water was added it dissolved in equimolar proportions to the hydrogen bromide present (known from an earlier study [171]), the solution became conducting, hydrogen and water were liberated at the cathode in equivalent amounts, and bromine was liberated at the anode. All this suggested the presence of ionic oxonium bromide ($H_3O^+Br^-$) in the solution.

In Part One we mentioned the unusual case of $BF_3\cdot 2H_2O$, which is definitely non-ionic in the solid but is about 20% dissociated in the melt:

$$BF_3\cdot 2H_2O \rightleftarrows H_3O^+ + BF_3OH^-$$

Similarly the monohydrate $BF_3\cdot H_2O$ has about 10% ionic dissociation in the melt. [172–174]

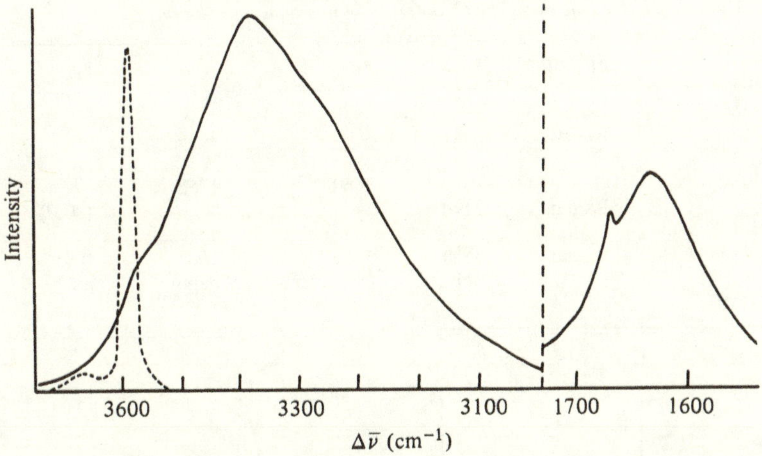

Fig. 1.8. Raman spectrum of an approximately 1 M solution of oxonium chloride in liquid sulphur dioxide at 233 K. (The dashed line shows the spectrum (× 10) of water in liquid sulphur dioxide at 253 K.) (Giguère & Madec [176].)

Vibrational spectroscopic studies of non-aqueous solutions are rare. Giguère *et al.* [175, 176] have studied the IR and Raman spectra of oxonium bromide and chloride (including some deuterated species) in liquid sulphur dioxide solutions at 213 K and 233 K. A Raman spectrum of the chloride is shown in figure 1.8. They identified bands due to the stretching modes of H_3O^+ and D_3O^+ and the ν_4 bending mode of H_3O^+, but the bending modes could not be found for all the isotopic species due to interference from the bands of sulphur dioxide. By means of a comparison with the spectra obtained for water dissolved in liquid sulphur dioxide it was possible to show that H_2O was not present in the acid solutions. [176] In this solvent the bands were found to be broad and intense due to hydrogen bonding both with the solvent and the anions. A splitting of ν_4 was interpreted as an effect of ion pairing in the low dielectric solvent: In the ion pair the symmetry would likely be reduced from C_{3v} to C_s and the degeneracy of the ν_3 and ν_4 modes would be lifted. (Some results are given in table 1.4.)

Huong & Desbat [177, 178] obtained the polarised Raman spectra of $H_3O^+SbCl_6^-$ in methylene chloride solution at room temperature. The polarisation results permitted a fairly conclusive assignment of the four modes, table 1.4 and figure 1.9. The authors did not address the question of ion pairing and the consequent lifting of degeneracies, but the bands are clearly much narrower than for the sulphur dioxide solutions (the halfwidth of the Raman band in the O–H stretching region is only about $\frac{1}{4}$ of that for oxonium chloride in sulphur dioxide), and both this and the high frequencies of the O–H stretching modes suggest that there is no significant hydrogen bonding. Also note that they found $\nu_1 > \nu_3$ in this case. The frequencies for ν_2 and ν_4 corresponded closely with those in crystalline $H_3O^+SbCl_6^-$.

Table 1.4. *Vibrational frequencies of H_3O^+ observed in non-aqueous solution*

System	Spectrum	ν_1	ν_3	ν_4	ν_2
$H_3O^+Cl^-/SO_2$	IR	—	3470	1700	—
233 K	Raman	3385	—	1635	—
[176]					
$D_3O^+Cl^-/SO_2$	IR	—	2660	1255	—
233 K	Raman	2490	—	—	(800?)
[176]					
$H_3O^+SbCl_6^-/CH_2Cl_2$	Raman	3560	3510	1600	1095
298 K		pol.	(shoulder)	depol.	pol.
[177, 178]					

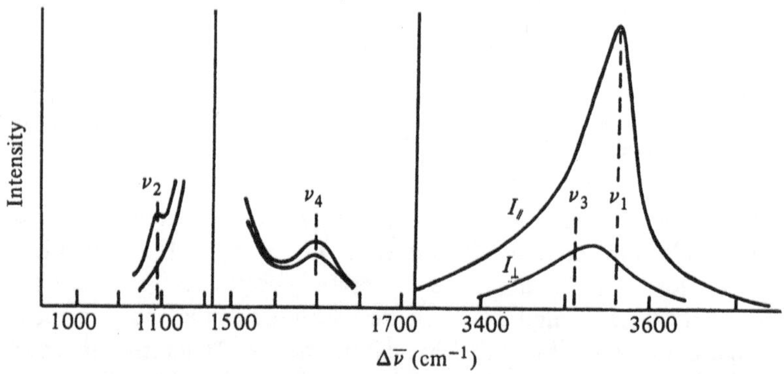

Fig. 1.9. Polarised Raman spectra of $H_3O^+SbCl_6^-$ in methylene chloride at 298 K, showing the bands assigned to H_3O^+ (Desbat & Huong [178]); ν_2 and ν_4 are overlapped by the wings of much more intense lines of the solvent.

Denisov & Golubev [179] have studied the IR of dilute solutions of acids and water in carbon tetrachloride. For CF_3SO_3H:H_2O in the ratio 1:1 the spectrum showed a broad doublet in the range 2100–2700 cm^{-1} and a separate peak at 2900 cm^{-1}. These bands were assigned to stretching modes of a distorted H_3O^+ ion (paired to $CF_3SO_3^-$). When the amount of water was doubled the intense continuum, usually associated with $H_5O_2^+$, was observed in the range below 2500 cm^{-1}. A similar continuum was also observed for HI:2H_2O.

The existence of hydrated proton species in various organic liquids has been shown or inferred from the study of the extraction of strong acids in aqueous solution into basic solvents. [180–2] (Marcus has reviewed work on solvent extraction up to 1962. [183]) Much of the early work showed that in most cases water was extracted along with the strong acid (e.g. $HClO_4$, HCl, HBr, HI) into the organic phase (e.g. dibutyl cellosolve, diisopropyl ketone or tri-n-butyl phosphate) such that the composition in the organic phase was close to 1:4, acid:water. It was inferred from this, and the fact that the anions did not seem to require hydration in the solvent (from work on salt extractions), that $H_9O_4^+$ was a particularly stable species, consisting of H_3O^+ with its primary hydration shell. However, in the case of $HClO_4$ or $HReO_4$ in much less polar solvent mixtures, the extracted composition was 1:1, acid:water consistent with the presence of $H_3O^+ClO_4^-$ (or ReO_4^-) ion pairs. [e.g. 182]

Experiments on extracts of $HFeCl_4$ in diisopropyl ether and dichloroethyl ether have also shown that the water, averaging about 4.5 per extracted acid, is associated with the excess proton and not with the anion.[184–7] Weaker acids such as HNO_3 and CCl_3COOH generally do not extract with about four water molecules and often are present in undissociated form. [181]

Kolthoff & Chantooni [188, 189] studied the protonation of water in acetonitrile by a spectrophotometric method using indicators. Perchloric or picric acids were used as the source of acid protons. They obtained formation constants for the successive hydration steps

$$H_s^+ + H_2O \rightleftarrows H_3O^+ \qquad K_1 = [H_3O^+]/[H_2O][H_s^+]$$

$$H_s^+ + nH_2O \rightleftarrows (H_2O)_nH^+ \qquad K_n = [(H_2O)_nH^+]/[H_2O]^n[H_s^+]$$

where H_s^+ represents H^+ solvated by acetonitrile: $K_1 = 1.6 \times 10^2$, ($pK_1 = 2.2$), $K_2 = 8 \times 10^3$, ($pK_2 = 3.9$), $K_3 = 6 \times 10^4$, ($pK_3 = 4.8$), and $K_4 = 2 \times 10^5$, ($pK_4 = 5.3$). Also for $[H_2O] > 0.6$ M the results indicated once again that the dominant species was $(H_2O)_4H^+$. Skarda, Rais & Kyrs [190] found a hydration number of 5.5 for H^+ (counter-ion $[(1,2\text{-}B_9C_2H_{11})_2Co]^-$) extracted into nitrobenzene. The reason for this higher number is not clear. They also showed by conductivity measurements that the ions were dissociated in this solvent, which has a fairly high dielectric constant.

Talarmin, L'Her, Laouenan & Courtot-Coupez [191, 192] studied water protonation in propylene carbonate using $HClO_4$ and HSO_3CF_3 by means of voltammetry, conductimetry and potentiometry. Surprisingly, although they found evidence for H_3O^+, $H_3O^+(H_2O)$ and $H_3O^+(H_2O)_2$ they could not detect $H_3O^+(H_2O)_3$ in the water concentration range studied (0.01–0.3 M). In the $HClO_4$ solutions the ions were found to be dissociated, opposite to the findings for HSO_3CF_3. They were able to obtain pK values for the hydrated proton species in the $HClO_4$ solutions: 4.8 ± 0.2 (H_3O^+), 6.1 ± 0.2 ($H_5O_2^+$) and 7.5 ± 0.2 ($H_7O_3^+$).

Isotope fractionation of the unhydrated oxonium ion in acetonitrile has been studied by Kurz, Myers & Ratcliff. [193] They estimated a deuterium fractionation factor of 0.79. It was implied from a comparison with the factor 0.69 for aqueous solutions that the L_3O^+ ion (L = H or D) is hydrogen bonded to three acetonitrile molecules in the acetonitrile solution.

Some of the most interesting work on the hydrated proton in non-aqueous solutions has involved NMR, principally of ^1H and ^{17}O nuclei. When water is dissolved in a liquid made up of liquid hydrogen fluoride saturated with boron trifluoride the reaction

$$HF + H_2O + BF_3 \rightleftarrows H_3O^+ + BF_4^-$$

takes place. In such a solution at 198 K MacLean & Mackor [194, 195] observed a ^{1}H NMR line due to H_3O^+, and an ^{19}F NMR quadruplet due to the BF_4^- ion. Above 223 K the ^{1}H signals due to HF and H_3O^+ were merged due to rapid exchange. They also observed a signal due to H_3O^+ in a solution of ethyl alcohol in HF/BF_3 at 203 K, and from the relative values of the chemical shifts given in Hz for the various species in this solution one can estimate a value of the chemical shift for H_3O^+ (relative to tetramethylsilane, TMS) of about 10 ppm.

Commeyras & Olah [196] studied the 'Magic Acid' (SbF_5/HSO_3F) solvent system. When water was dissolved in small amounts (mole ratio H_2O/HSO_3F about 0.01) a ^{1}H NMR signal was observed at 10.25 ppm due to H_3O^+, but as more water was added this peak gradually coalesced with the peak of the acid proton. When acetic acid was dissolved in 'Magic Acid' H_3O^+ was formed via protonation and dissociation:

$$CH_3COOH \xrightarrow{H^+} CH_3COOH_2^+ \xrightarrow{H^+} CH_3CO^+ + H_3O^+$$

The ^{1}H chemical shift for H_3O^+ in such solutions was 10.25–10.35 ppm dependent upon the exact composition of the solution. (It is not clear at what temperature this was measured.) It was suggested that at high concentrations of acetic acid, $H_5O_2^+$ might be present, since there would be fewer acid protons than water molecules in the system, but no distinct NMR signal for this species was observed.

Although it was logical to assign the ^{1}H peak at 10.25 ppm to H_3O^+ in the above experiments, its true identity was not established conclusively until the work of Gold, Grant & Morris. [197] They observed ^{1}H signals at 220 MHz of the hydrated proton species obtained when small amounts of D_2O were dissolved in SbF_5/HSO_3F which was then diluted with liquid sulphur dioxide or sulphuryl chloride. Depending on the exact composition and the temperature (213–253 K) they found the hydrated proton peak in the region 9.8–11.1 ppm (relative to TMS). However, at 220 MHz the peak was resolved into three signals (figure 1.10) which they established to be due to H_3O^+ (a single peak), H_2DO^+ (a 1:1:1 triplet due to coupling with the spin 1 ^{2}H nucleus) and HD_2O^+ (a 1:2:3:2:1 quintet would be expected but only a broad unresolved peak was observed). The relative intensities of the three signals also matched those expected on the basis of a random distribution of isotopes among the chemically equivalent sites. At 229 K in a solution diluted with liquid sulphur dioxide the shift for H_3O^+ was close to 10.28 ppm. H_2DO^+ and HD_2O^+ were about 0.055 and 0.11 ppm less shielded respectively.

Emsley, Gold & Jais [198] observed a ^{1}H signal at 13.0 ppm, which they

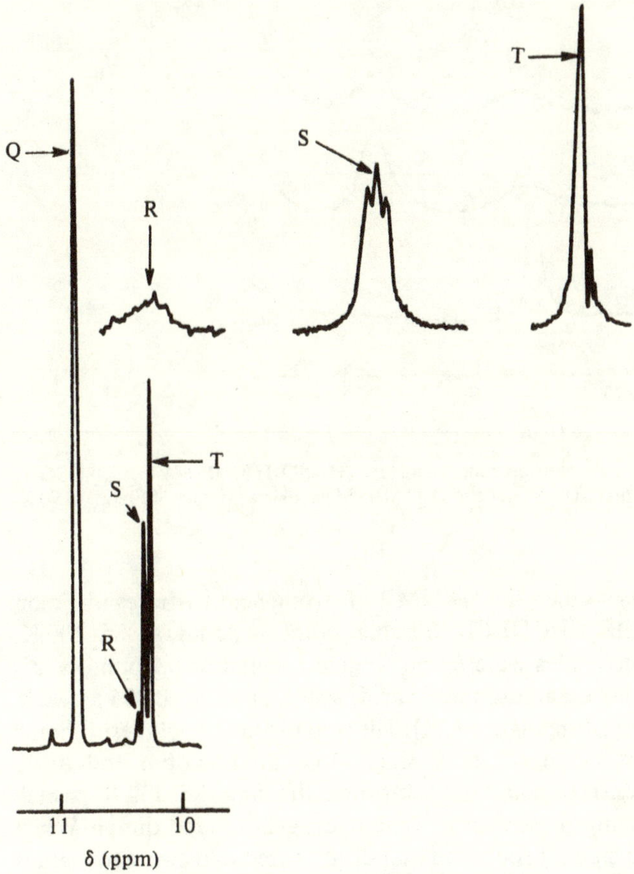

Fig. 1.10. ^1H NMR spectrum (220 MHz) of $HSO_3F/SbF_5/D_2O$ (1 mol:0.7 mol:0.16 mol) in liquid sulphur dioxide (4 vol.) at 229 K: Q, HSO_3F; R, HD_2O^+; S, H_2DO^+; T, H_3O^+. The oxonium ion signals R, S, are also shown on an expanded scale. (Gold *et al.* [197].)

assigned to H_3O^+, in very dilute solutions of water with hydrogen bromide in CF_2Br_2/CD_2Cl_2 at 153 K. One must presume that the downfield shift from the previous results described above is an effect of the quite different medium. (Note that in all these systems the lower temperatures are used to slow down hydrogen exchange and thus resolve signals from the individual species present.)

Rimmelin, Schwartz & Sommer [199] proposed the use of the H_3O^+ ^1H signal as an internal reference for ^1H NMR in super-acid media (HF or HSO_3F in SbF_5), since it appears to be less sensitive to solute–solvent interactions than tetramethylammonium bromide which has frequently been used as a reference. The signal of H_3O^+ in HF/SbF_5 is at 9.8 ± 0.18 ppm and in HSO_3F/SbF_5 is at 9.6 ± 0.07 ppm (relative to TMS).

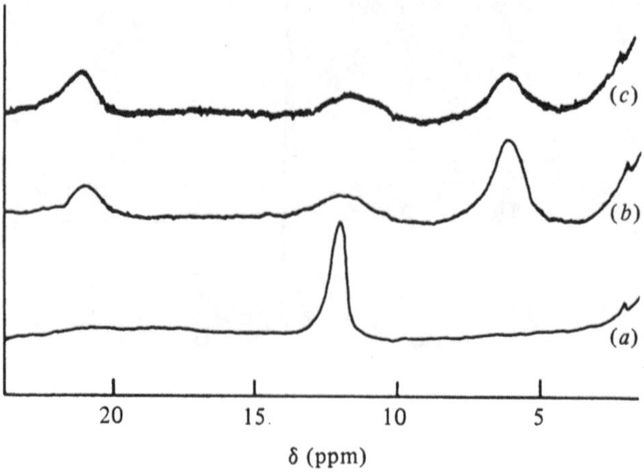

Fig. 1.11. ^{1}H NMR spectra of solutions containing: (*a*) $C_8H_{17}SO_3H$ (0.005 M); (*b*) $C_8H_{17}SO_3H$ and H_2O (0.005 M each); (*c*) $C_8H_{17}SO_3H$ and D_2O (0.05 M each) in a mixture of Freons at 90 K. (Golubev [200].)

Golubev [200] has studied the ^{1}H NMR of strong acid hydrates in Freon mixtures ($CDF_3/CDF_2Cl/CDFCl_2$ freezing point <85 K) at 85–100 K. Monomeric water molecules were found to give a signal at 0.2 ppm in this solvent. The results for octanesulphonic acid/water, at about 0.005 M each, were particularly interesting (figure 1.11): The acid proton of octanesulphonic acid in this solvent occurred as a single line at 12.3 ppm (relatively independent of concentration and temperature in the range 85–130 K), which was interpreted as being present in a cyclic hydrogen-bonded dimer. When water was added this signal broadened and disappeared and two new signals appeared at 6.1 and 21.3 ppm; these were assigned to the peripheral and the central, strongly hydrogen-bonded protons of an $H_5O_2^+$ ion. (This would be in reasonable agreement with the shifts observed for peripheral protons in solids and theoretical expectations of the relative shifts of the two kinds of proton in $H_5O_2^+$. [84, 115]) Golubev has perhaps not considered the possibility of the presence of other strongly hydrogen-bonded species (e.g. the hydrogen dioctanesulphonate ion), and, contrary to his statement, there is some indication of the presence of a small amount of H_3O^+ in the spectra shown (i.e. a shoulder at about 10–11 ppm). Nevertheless, considering the possible combinations of species which might be produced for the concentration ratios of acid and water indicated, and the signal intensities, one must conclude that $H_5O_2^+$ is a major contributor to the spectrum. A second important result was that when D_2O was used instead of H_2O the residual protons showed a preference for the bridging hydrogen-bond position, i.e. a strong isotope effect. (This has also been found in the diffraction studies of crystalline yttrium acid oxalate trihydrate. [201])

In a later paper Golubev [202] reported a study of the same system but with antimony pentafluoride added, which apparently produced much sharper lines. When octanesulphonic acid, antimony pentafluoride and water were dissolved in the Freon mixture in equimolar amounts (0.01 M) the ^{1}H NMR lines which had been present before the water was added disappeared, and were replaced by a doublet at 11.1 ppm and a triplet at 8.3 ppm with J(H–H) = 2.8 ± 0.2 Hz. These lines were interpreted as representative of an unsymmetrical H_3O^+ involved in an ion pair with two of the O–H groups hydrogen bonded to the anion (the doublet) and one O–H free (the triplet). This is a quite plausible, though unproven, model. When the quantity of water was doubled, the H_3O^+ lines reduced considerably in intensity and new lines appeared at 21.3 and 6.0 ppm (intensity ratio about 1:4), which by analogy with the earlier experiments were assigned to $H_5O_2^+$. The new lines were only 0.5 Hz wide and did not show any J splittings. This suggested that the four peripheral hydrogen atoms of the $H_5O_2^+$ were chemically equivalent. Golubev's work is perhaps the most convincing to date showing the existence of $H_5O_2^+$ in a non-aqueous solution.

Mateescu & Benedikt [103] observed a 1:3:3:1 quartet in the ^{17}O NMR spectrum of a 1.5 M water (^{17}O enriched) solution in liquid sulphur dioxide with an excess of HF/SbF_5 (1:1 mole ratio) at 258 K (see figure 1.12). The $J(^{17}O–^1H)$ coupling was measured to be 106 ± 1.5 Hz and the shift 9 ± 0.2 ppm downfield from external $^{17}OH_2$. In later work [203] Mateescu, Benedikt & Kelly found only a small isotopic shift of −0.9 ppm for $^{17}OD_3^+$. With proton decoupling the quartet reduced to a singlet. These results clearly showed that there were three hydrogen atoms attached to the ^{17}O. The authors argued that the magnitude of the J coupling was consistent with sp^2 hybridisation and hence a planar configuration. This was, however, refuted by Symons [204] who argued to the contrary, and more convincingly, that in fact the J couplings do fit well with a pyramidal configuration and an H–O–H angle of 111.3°. (Of course, all the other evidence presented in this review also strongly supports this.) Olah, Berrier & Prakash [205] reexamined this system at 253 K and measured $J(^{17}O–^1H)$ = 103.5 Hz for the quartet and a shift of 10.2 ppm for the ^{1}H decoupled singlet.

Olah *et al.* [89] very recently reported ^{17}O NMR results concerning solutions of D_2O (20% ^{17}O) in HF:SbF_5/SO_2. Shortly after preparation of the solution the ^{17}O spectrum at 258 K was a doublet, indicating D_2HO^+, but one hour later the doublet had developed into a triplet, indicating DH_2O^+, and after about four hours this became a quartet, indicating H_3O^+. Similarly for H_2O (40% ^{17}O) dissolved in DF:SbF_5/SO_2, the spectrum initially showed a triplet (DH_2O^+) which developed with time first into a doublet (D_2HO^+) and finally a singlet (D_3O^+). (These observations depend on the fact that the $^{17}O–^2H$ couplings are not resolved.) They suggested two possible mechanisms for this slow isotopic exchange:

(*a*) Deprotonation (dedeuteriation) of oxonium followed by deuteriation (protonation) of the water by the excess acid protons of the medium.

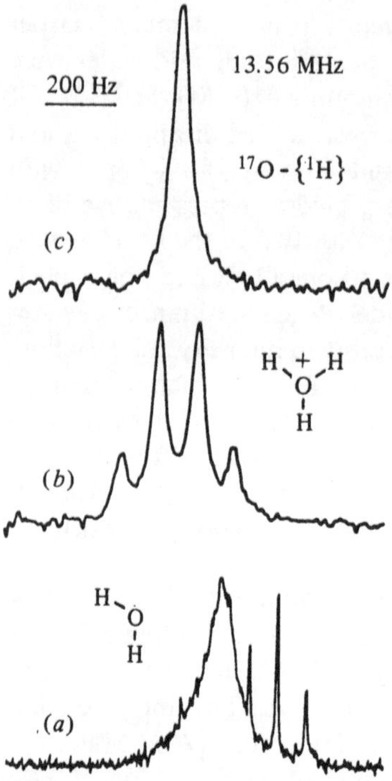

Fig. 1.12. ^{17}O NMR spectra of (*a*) water in carbon tetrachloride (the broad line is due to bulk water, the sharp triplet due to isolated H_2O molecules); (*b*) H_3O^+ in liquid sulphur dioxide at 258 K, showing a quartet due to ^{17}O–^1H spin–spin (*J*) coupling; (*c*) H_3O^+ in liquid sulphur dioxide with proton decoupling. (Mateescu & Benedikt [103].)

(*b*) Formation of the protonated or deuteriated oxonium dication (e.g. HD_3O^{2+} or other isotopic mixtures) followed by subsequent dedeuteriation (deprotonation).

They argued in favour of the latter mechanism since they found that in $DF{:}2SbF_5/SO_2$ solutions, which are more strongly acid than for 1:1 $DF{:}SbF_5$, the exchange proceeded somewhat faster. The reverse would have been expected if the first mechanism was operative. By means of *ab initio* calculations they also showed that although H_4O^{2+} is thermodynamically unstable, it has a significant kinetic stability (mentioned in section 1.1.1).

So we see at the end of this short section that some very significant information has been obtained from the studies of the hydrated proton in non-aqueous solutions.

1.2.2 *The hydrated proton in aqueous solutions*

Diffraction studies Certain information concerning structure in liquids can be obtained from X-ray [206] and neutron [207] diffraction studies. The total scattering consists of three principal components: (*a*) the 'self'-scattering from individual atoms; (*b*) the scattering due to individual molecules; (*c*) the 'distinct' scattering due to further structure within the liquid itself. Of these (*a*) can be calculated from known atomic scattering functions; (*b*) can often also be calculated from known structures, though some refinement of the molecular parameters may also be possible by fitting to the experimental results. Removal of (*a*) and (*b*) from the total scattering gives the distinct structure function. Fourier transformation then yields the pair correlation function, which corresponds to a sum of all the combinations of atom–atom pair distances. (This is sometimes also expressed in a modified form as the radial distribution function.)

Very little work has so far been done on aqueous acid solutions. The most useful work to date appears to be that of Triolo & Narten [147] who studied the systems HCl/H_2O and DCl/D_2O by X-ray and neutron diffraction respectively. This work was preceded by less extensive and lower resolution X-ray works on HCl solutions,[208–10] and one high resolution study by Lee & Kaplow [211] which was essentially in agreement with Triolo & Narten.

The X-ray results are dominated by oxygen scattering (sensitive to O··O and O··Cl) whereas neutron results are dominated by deuterium scattering (sensitive to D··D and D··O). The interpretation of the correlation functions is not entirely unambiguous and relies to a certain extent on agreement between the observed and calculated structure functions. Triolo & Narten found that the molecular component of the neutron structure function results were best fit with O–D distances of 95.5 pm for D_2O and 101.7 pm for D_3O^+. In their X-ray correlation functions for $HCl/95.7H_2O$ the dominant peak was at 285 pm due to O··O interactions of the tetrahedral network of molecules of the bulk water (see figure 1.13). As the concentration of hydrogen chloride was increased, very distinct changes occurred until at $HCl/3.99H_2O$ the O··O pair distance peaked at 252 pm. This was attributed most logically to $H_3O^+ \cdot\cdot OH_2$. The main peak at the same concentration occurred at 313 pm and was assigned to O··Cl. A smaller peak assigned to Cl··Cl was observed at 361 pm. Further broader structures in the 400–500 pm and 650–750 pm ranges were also observed at all concentrations. Model fits to the observed function for the most concentrated solution were closest when there were four coordinating waters around both H_3O^+ and Cl^-. The neutron correlation function for $DCl/3.08D_2O$ showed two small peaks at 161 and 210 pm attributable to $D_2OD^+ \cdot\cdot OD_2$ (D··O) and $D_2OD^+ \cdot\cdot DOD$ (D··D) pairs respectively. (Note Triolo & Narten refer to $D_3O^+ \cdot\cdot D$ interactions, which does not seem to make sense. Our notation above, $D_2OD^+ \cdot\cdot OD_2$, seems to fit more logically in the context of what they are saying and the distance involved.)

It was not possible to determine with any certainty the position and orientation of the fourth D_2O around the D_3O^+, though a model with the D_2O in the charge-dipole configuration was compatible with the results. Recall, however, that the Monte Carlo calculations [142] indicated that the

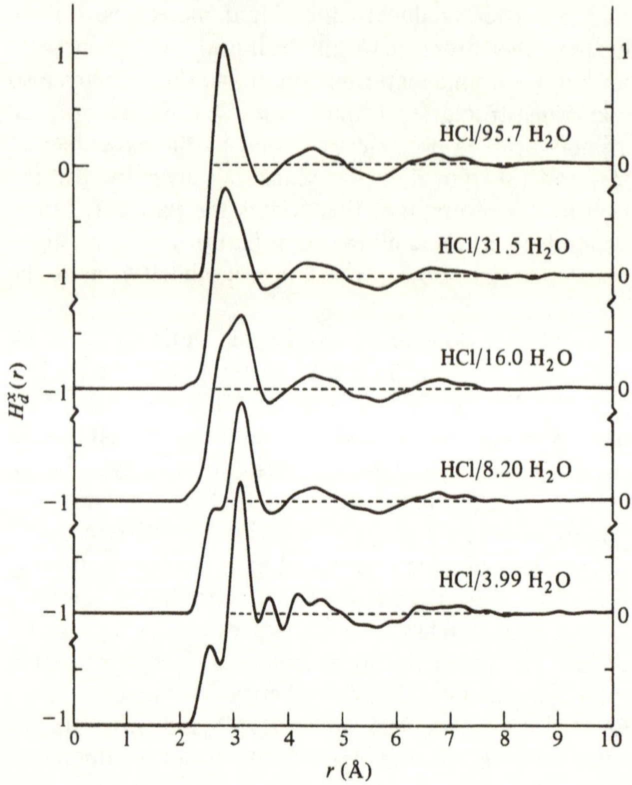

Fig. 1.13. Pair correlation functions from X-ray diffraction studies of aqueous solutions of hydrogen chloride over a wide range of concentrations. (From Triolo & Narten [147].)

fourth water does not seem to occupy the empty position above the oxygen of H_3O^+.

It should be emphasized that all these dimensions come from results at high concentration. We can compare the distances with results for the solid hydrogen chloride hydrates. Values of O··Cl for waters attached to H_3O^+ in these solids [212–14] fall in the range 301–313 pm, whereas Cl··Cl distances [212, 213] are 384–399 pm. O··O distances in the $H_9O_4^+$ unit of the hexahydrate [214] average 252.3 pm. Note that in the hexahydrate the chlorines are coordinated by six waters but the coordination decreases as the water content decreases. Except for the Cl··Cl distances the comparison between solution and solid is very fair.

In the results described above no information was obtained concerning H_3O^+ in dilute solution. This problem was addressed by Ohtomo *et al.*, [215] who studied roughly 1 M solutions of DCl and DBr in D_2O using neutron diffraction. (In such solutions D_3O^+ has a mole fraction of about $\frac{1}{50}$.) The results were analysed by means of a subtraction method which attempts to remove the contribution of the bulk water from the structure functions, leaving only the contributions of the dilute ions and their associated hydration spheres. They obtained best fits for models which had D_3O^+ tetrahedrally coordinated by four waters, with $O\cdot\cdot O = 288 \pm 5$ pm, and Cl^- and Br^- octahedrally coordinated by six waters, with $O\cdot\cdot Cl = 310 \pm 5$ pm, and $O\cdot\cdot Br = 321 \pm 5$ pm. The only surprising result here was the large $O\cdot\cdot O$ distance of 288 pm. The authors made no comment about this even though it is clearly at odds with other determinations of solutions and solids. This value is also remarkably close to the value of 285 pm observed in pure water, [205, 206] so one must wonder whether the subtraction method has actually worked.

More recently Lee, Matsumoto, Yamaguchi & Ohtaki [217] carried out X-ray studies on solutions of cobalt chloride, one of which contained concentrated hydrochloric acid and thus provided some additional results on the hydrated proton. The solution involved was roughly $HCl/3{\cdot}3H_2O$. This time after subtraction of contributions due to $Co\cdot\cdot Cl$ and $Cl\cdot\cdot Cl$ the residual scattering function was fit with a model based on H_3O^+ coordinated by four waters with $O\cdot\cdot O = 244$ pm for three waters and 290 pm for the fourth. The fourth water was modelled with a hydrogen pointing at the vacant side of the H_3O^+ oxygen, which, as was discussed in the theoretical section,[32] is probably energetically unfavourable. Besides this question about the model itself the value of 244 pm for the other distances seems remarkably small (comparable to very strong hydrogen-bond distances only found in $H_5O_2^+$ salts). Again one must wonder whether the subtraction process involved is accurate enough to give reliable results for the very dilute species.

NMR studies The NMR spectroscopy of the hydrated proton in aqueous solution presents a quite different aspect from that discussed earlier for non-aqueous solutions. In water the acid protons are able to exchange rapidly with the protons of the water and the parent acid (ionogen) under practically all attainable conditions of study (cf. non-aqueous solutions which usually do not exhibit exchange with the solvent and exchange with the acid molecules is reduced or stopped by lowering the temperature). The consequent absence both of *J* couplings and of separate peaks for the different chemical species means that structural information cannot be obtained from the NMR of aqueous solutions. Most of the NMR work has concerned attempts to deduce the degree of dissociation of numerous acids in water and mixed water/organic solvent systems. In other NMR studies which we will briefly discuss attempts to extract information about the proton exchange process in water (involving H_3O^+) are reported.

Gutowsky & Saika [218] were the first to obtain ^1H NMR spectra of acid solutions. They realised that the single, concentration-dependent resonance observed was due to averaging of the chemical shifts of the protons in the different chemical sites among which they were rapidly exchanging, i.e. among water, undissociated acid and hydrated proton species. In the fast exchange limit the observed line should occur at the population-weighted average of the resonance frequencies which the nuclei would show if they were not exchanging. This model could then explain the concentration dependence of the shifts in terms of the changing proportions of the species present. Assuming that only H_2O, HA and H_3O^+ are present, the average shift, relative to the shift of liquid water, is given by

$$S = \alpha p S_1 + (1-\alpha) p S_2/3$$

where α = degree of dissociation, $p = 3X/(2-X)$, X = stoichiometric mole fraction of acid HA, S_1 = shift of H_3O^+ protons, and S_2 = shift of unionised acid HA. If the acid is completely dissociated ($\alpha = 1$), then the plot of S vs p (the fraction of protons present as H_3O^+) should be linear and S_1 is obtainable from the slope. Linear plots were observed at low concentrations for all of the strong acids. Deviations from linearity at higher concentrations were assumed to mean that α was no longer unity. Figure 1.14 is a good illustration of such a plot. Then, using the value obtained for S_1 at low concentration and assuming that S_2 is given by the shift of the pure anhydrous acid HA, it is possible to obtain values for α in the higher concentration region, and ultimately for the thermodynamic dissociation constant K.

Gutowsky & Saika [218] were at pains to point out that a number of severe assumptions had been made in their formulation, and Hood, Redlich & Reilly [219] stressed that the one parameter actually being measured (the average chemical shift) is an average over many states and is thus a colligative

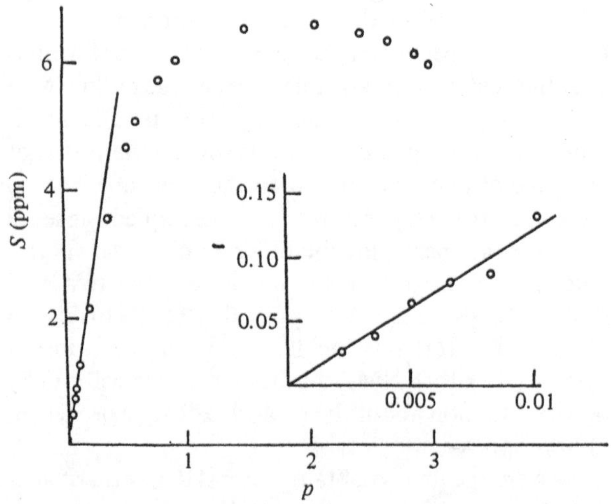

Fig. 1.14. ^{1}H NMR chemical shift (S) of the ionisable protons in aqueous solutions of CH_3SO_3H plotted against p (from Covington & Lilley [230]). The inset shows the limiting slope from which the shift (S_1) of H_3O^+ was evaluated.

property. Accordingly some of the assumptions made cannot be directly verified. Nevertheless, numerous papers have since appeared concerned with the study of the dissociation of acids using NMR. [220–43] The authors of most of these reports made use of the simple equation above, and although many seem to have ignored consideration of the inherent complications, a number have developed the equation further to try to take into account some of the other factors. Specifically the problems include the following:

(*a*) Should higher hydrates such as $H_5O_2^+$ be included as distinct species with their own characteristic shift?

(*b*) The waters involved in hydrating H_3O^+ (or $H_5O_2^+$) are likely to be quite different from bulk water.

(*c*) The waters around the anions are also different from bulk water (known from studies on salt solutions) and some anions exert a stronger 'structure breaking' effect on the water than others.

(*d*) It may not be reasonable to assume that the shift for the undissociated acid HA in solution is the same as for the anhydrous acid.

(*e*) It may not be reasonable to assume that the shifts for any of the species present remain constant over all the range of concentration studied.

(*f*) Ion pairs may also be significant species.

(*g*) Involved in most of the above is the fact that the shifts for particular species can be affected by hydrogen bonding, e.g. pure liquid water has a shift of about 4.7 ppm downfield from monomeric water in gaseous or isolated states.

Hood *et al.*, [219, 225] in their studies of nitric acid, introduced a parameter to account for $H_2O \cdot HNO_3$ as a distinct species. Duerst, in his studies of perchloric acid, [232, 233] attempted to account for the possibility of $H_5O_2^+$ and $H_9O_4^+$ by including these in the equations. The specific shifts which he derived from the results at 298 K were H_3O^+ 5.8 ppm, $H_5O_2^+$ 4.698 ppm, and $H_9O_4^+$ 3.061 ppm relative to liquid water. In the high concentration range he included H_3O^+, $H_5O_2^+$ and $HClO_4$. Akitt *et al.* [235, 238] tried to account for anionic hydration and structure breaking effects, and thus concluded that this was the major factor giving rise to the large range of derived H_3O^+ shifts, S_1.

It is not our intention, particularly given the room for scepticism introduced with all the assumptions, to go into the details of all the many determinations of the degree of dissociation, dissociation constants and hydration numbers of acids in solution. Suffice it to say that the results do give some indication of the degree of dissociation. In some cases the agreement with other determinations of α (e.g. by IR or Raman spectroscopy) is apparently good, e.g. for nitric acid, yet in others it is poor, e.g. perchloric and sulphuric acids. Apparently there were enough discrepancies and doubts for the aqueous systems of nitric, perchloric and hydrochloric acids to be investigated repeatedly over the years.

It is, however, worth taking a look at the observed ^1H shifts (i.e. the raw data) and comparing the chemical shifts for H_3O^+, derived from these, with those observed for H_3O^+ or $H_5O_2^+$ in non-aqueous solutions and solids. In the solutions of strong acids at lower concentration one can be certain of complete acid dissociation. Consequently the observed shift represents the average of the hydrated proton species and various types of water only. This observed shift is always downfield of bulk water.

Relative to the shift for bulk water the values of S_1 for H_3O^+, derived from the limiting slopes of the S vs p plots, range over 9.14–13.7 ppm. (Akitt *et al.* [235] give a comprehensive listing of values from studies carried out up to 1969, though the value of 14.5 ppm derived for *p*-toluenesulphonic acid was revised to 9.9 ppm in a later study [239].) On the TMS scale this is equivalent to a range 14.1–18.7 ppm. Also note, based on Akitt's model [238] with correction for anionic effects, that the derived specific shift for H_3O^+ is about 19.5 ppm (relative to TMS). The known values of the chemical shift for well characterised H_3O^+ in non-aqueous solutions and solids (see earlier) fall in the range 8.3–13.0 ppm (relative to TMS), with the majority of values close to the average of about 10.8 ppm. This large difference in shifts indicates either (*a*) a large error in the derived shifts for aqueous solutions due to problems with the model, or (*b*) there is a significant difference between H_3O^+ in the non-aqueous and solid systems and in water. The sceptics might choose (*a*), but there are also arguments to support (*b*). One might expect some shift to lower field due to hydrogen bonding of the H_3O^+ to other water molecules, though hydrogen bonding may also play a part in a number of the non-aqueous cases if ion pairing has occurred.

There have also been a very few studies using ^{14}N, ^{19}F and ^{35}Cl NMR chemical shifts of aqueous acid solutions. [222, 235, 241, 244] The analyses of the exchange-averaged shifts (e.g. of ^{14}N for HNO_3 exchanging with NO_3^-) to obtain α values followed similar lines to those above).

^1H NMR has been useful in the study of isotope effects in mixed hydrogen/deuterium aqueous acid systems. When deuterium is present in the system it is preferentially concentrated to a small extent in the water molecules rather than in the cations. The resulting small increase in the concentration of hydrogen in the cations relative to hydrogen in the water causes the exchange-averaged chemical shift to move to lower field. Studied as a function of stoichiometric acid concentration, and once again making numerous assumptions about the species present, this difference can be used to obtain information about the fractionation effect. Gold [245] found values of $K = 0.69$ and 0.68 for perchloric and hydrochloric acid solutions (assuming the only species were H_3O^+ and H_2O (and their isotopic variations)) for the equilibrium

$$H(+) + D(w) \rightleftarrows H(w) + D(+)$$

representing exchange of hydrogen and deuterium between an ion and a water molecule. (These results were presumably at ambient temperature.) Kresge & Allred [246], using the same NMR method, found a similar result for perchloric acid of 0.67, at about the same time.

Redlich, Duerst & Merbach [231–3] studied the effects of deuterium in the nitric and perchloric acid aqueous systems between 0–65 °C. They concluded that the hydrogen/deuterium distribution among uncharged molecules was almost uniform, and confirmed that hydrogen has a strong preference for the cation over water. DNO_3 was found to be slightly weaker than HNO_3. At high concentrations of perchloric acid, i.e. mole fractions of acid >0.75, they also found shifts to lower field on deuteriation. Since in this concentration range it is believed that all the water has been transformed into H_3O^+, the deuteriation shift in this case is explained by a preference of hydrogen to be in the cation rather than in perchloric acid.

We will now go on to consider the information which can be obtained from NMR concerning the involvement of the hydrated proton in the proton exchange process in water. Once again the earliest work on the chemical exchange-rate in acid solution was by Gutowsky & Saika. [218] They considered the effect on the lineshape of various rates of exchange among sites with different chemical shifts: for a simple two site system in the slow exchange region there will be two sharp lines. As the exchange-rate increases into a critical region these lines eventually broaden, then coalesce into a single broad line at the population-weighted, average shift. The single, broad line eventually narrows in the fast exchange limit. Gutowsky & Saika showed that the transition between slow and fast exchange between two equally populated sites is in the range of $2\tau\delta\omega = 1$–10, where $\delta\omega =$ the angular frequency separation of the two non-averaged lines and $\tau =$ the average lifetime of the species. Since the acid solutions are obviously in the fast exchange limit, the only thing one can say about the average lifetimes is that they are shorter than about 10^{-4} s.

Both acid and base catalyse the proton exchange-rate in water, so that fast exchange narrowing applies and the information is very limited, except, that is, for a small range of pH close to neutral. Consequently all subsequent NMR studies have been restricted to this range of pH (i.e. relatively low to zero concentration of dissociated acid or base). Meiboom *et al.* [247, 248] were the first to observe that, starting at neutral pH, the addition of acid or base causes the ^1H or ^{17}O NMR lines to narrow, indicating that at neutral pH the lineshape is not in the fast exchange limit. Meiboom [248] also found that the ^1H spin–spin relaxation rate $1/T_2$ depended on the concentration of ^{17}O in the water, which indicated that the frequency splitting involved in the line narrowing was, in fact, not a chemical shift difference in this case but the ^{17}O–^1H spin–spin interaction.

The theory involved in extracting exchange rates from the observed NMR parameters is rather too involved to describe here. Suffice it to say that from measurements of ^1H NMR spin–spin (T_2), rotating frame spin-lattice ($T_{1\rho}$) and low frequency spin-lattice (T_1) relaxation times, and of the ^{17}O NMR linewidths, as a function of pH and temperature, various authors have obtained rate constants and approximate activation energies for the assumed proton transfer reactions.

$$H_2O + H_3O^+ \xrightarrow{k_1} H_3O^+ + H_2O \tag{1}$$

$$H_2O + OH^- \xrightarrow{k_2} OH^- + H_2O \tag{2}$$

(or alternatively the average exchange time τ_e, equivalent to the mean lifetime of a water molecule). Several papers have appeared on this topic over the years, [247–58] the most recent of which by Halle & Karlstrom [257] gives a summary of all the determinations [248–52, 257] of the rate constants corrected to 28 °C. The results quoted fall within a small range $(4.1–7.1) \times 10^{-9}\,l\,mol^{-1}\,s^{-1}$ for k^+ $(=2k_1/3)$ and $(2.2–4.3) \times 10^{-9}\,l\,mol^{-1}\,s^{-1}$ for k^- $(=k_2)$. Halle & Karlstrom's results were 7.1×10^{-9} and $3.4 \times 10^{-9}\,l\,mol^{-1}\,s^{-1}$ respectively. They also found the acid catalysed process to be more efficient than the base catalysed process by a factor of 2 in H_2O and 3.5 in D_2O, and the kinetic isotope effect (H/D) was 1.6 for the acid and 2.7 for the base catalysed processes. Of the several previous determinations of the activation energies, Halle & Karlstrom appear to favour the results of Luz & Meiboom [250] of 10.0 kJ mol^{-1} for reaction (1) and 8.8 kJ mol^{-1} for reaction (2). They also found that salt effects begin to occur, even at quite low concentrations of potassium chloride. Hertz & Klute [253] had previously reported that at high concentrations of potassium iodide (7 m) and sodium perchlorate in pH 6.5 solutions at 25 °C the ^{17}O NMR lineshape was almost resolved into a triplet, indicating much slower proton exchange.

With the assumption that equation (1) above is truly applicable these results for the rate constants are very significant since they permit one to calculate the lifetime of H_3O^+ in aqueous solution. Giguère [164] has calculated an average lifetime for H_3O^+ of 2.2 ps (4 ps for OH^-) from some of the earlier results. [251, 252] However, Halle & Karlstrom [257, 258] have pointed out that really the lifetime calculated depends to some extent on the model chosen to describe the proton transfer mechanism. They considered different models, [258] and, in particular, tried to account for reencounters and concerted proton transfer chains, but still concluded that the lifetimes were of the order 0.22–1.49 ps for H_3O^+ and 0.36–2.45 ps for D_3O^+ (and for comparison, 0.46–3.15 ps for OH^- and 1.3–8.58 ps for OD^-) depending on the specifics of the model. Whatever the mechanism chosen, however, one can see that the lifetimes are in the picosecond range and that OH^- has a lifetime roughly twice that of H_3O^+. That is, as was pointed out by Giguère, [164] several times longer than the period of even the lowest frequency vibration of H_3O^+, and thus it should be possible to observe a discrete vibrational spectrum for H_3O^+ in solution if indeed this is the species present. One should also emphasise that the results strictly apply to the middle pH range and one would expect the lifetime to increase as the concentration of acid is increased.

Having said all this, we must now strongly emphasise that in most of the work described above the authors have generally *assumed* that the species involved in the proton transfer was H_3O^+. Only a few authors [254–6] have not tried to take their results much beyond what is actually being measured, i.e. the rate of exchange of protons on the water molecule. There is no definitive proof in any of this work for the presence of discrete H_3O^+, and, in fact, one could derive rate constants and lifetimes of the same magnitude for alternative hydrated proton species such as $H_5O_2^+$.

Vibrational spectral studies Information about the nature of the hydrated proton in aqueous solutions has also been gleaned from the study of IR and Raman spectra. Early experiments were hampered by the breadth of the bands. IR spectroscopy was favoured over Raman spectroscopy because of the polar nature of H_3O^+, [259] but no marked differences were observed between the spectra of $HNO_3 \cdot H_2O$ in the liquid and solid states. [259] Bauer [260] remarked that the lifetime of H_3O^+ could be expected to be comparable to the period of vibration and thus IR bands would be expected to be diffuse and thus not detectable. The short lifetime was attributed to the exchange of the proton between two water molecules (also see [261–4]).

The first successful experiments are attributed to Falk & Giguère. [265] They measured the IR absorption spectra of aqueous solutions of mineral acids (HCl, HBr, HNO_3, $HClO_4$, H_2SO_4, H_3PO_4) and some of their acid salts of various concentrations. In all cases three broad bands were present at 1205, 1750, and 2900 cm^{-1}. On deuteriation the bands occurred at 960, 1400, and 2170 cm^{-1}. The band positions agreed within some 60 cm^{-1} with those for H_3O^+ and D_3O^+ ions in crystals and were thus interpreted in terms of a symmetric, pyramidal oxonium ion: 1205 cm^{-1}, $\nu_2(A_1)$ symmetric bending mode; 1750 cm^{-1}, $\nu_4(E)$ antisymmetric bending mode; 2900 ± 150 cm^{-1}, $\nu_3(E)$ O–H stretching mode; $\nu_1(A_1)$, weak and masked by $\nu_3(E)$. It should be noted that the concentrations in all cases were high: thus the mixtures contained eight or less moles of water per mole acid. It should also be noted that since that time the vibrational spectra of many more solids have been reported and band positions vary considerably, dependent on the environment. (See Table 5, Part One.) Although the conclusion is plausible, other structures are not unequivocally ruled out by these IR spectra alone.

New spectral measurements made with the latest technology were reported by Giguère & Turrell [266] in 1976, in order to rebut arguments of Ackermann [267] and others. [268] Their results for hydrofluoric, hydrochloric, hydrobromic and hydriodic acids are presented in figure 1.15. The $\nu_3(E)$ band in the region centred around 2900 cm^{-1} is partially obscured by bands of water. The $\nu_2(A_1)$ band at 1200 ± 50 cm^{-1} stands out clearly and is considered 'characteristic' of H_3O^+. On deuteriation it occurs at 900 ± 10 cm^{-1}. The $\nu_4(E)$ band at 1730 ± 10 cm^{-1} is also clear. The bands are broader, and shifted from those observed for $H_3O^+SbCl_6^-$ dissolved in methylene chloride [177, 178] (cf. figures 1.9 and 1.15). For hydriodic acid the measurement was made down to a concentration of 12 mol water per mol

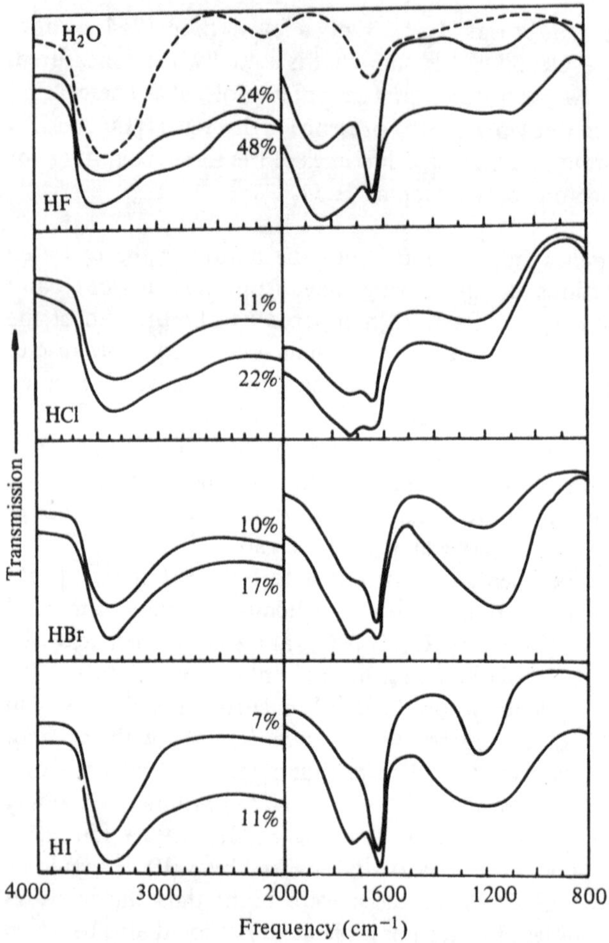

Fig. 1.15. IR absorption spectra of aqueous solutions of HF, HCl, HBr and HI at the compositions (mol %) indicated, from Giguère & Turrell [266]. The ordinate scale is expanded five times below 2000 cm^{-1}.

acid. Because of the approximately 500 cm^{-1} lower position of the O–H stretch than that observed for neat water and the higher position ($\approx$ 85 cm^{-1}) and breadth of the deformation, the authors inferred that the hydrogen bonds formed by H_3O^+ are stronger than those formed by H_2O. Giguère & Turrell were unable to reproduce the spectra of sodium hydroxide of Ackermann [267] and refute his proposal that the spectra reflect a perturbation of the water structure equally strongly caused by H^+ and OH^-. Ackermann later revised his conclusion in light of additional results of Downing & Williams. [269] They similarly rebut Zundel's proposal of $H_5O_2^+$ groupings (see below). Their case has been further presented by Giguère. [164, 270] Rhine, Williams, Hale & Querry [271] measured the spectral reflectance of aqueous solutions of hydrochloric acid, sodium hydroxide and potassium hydroxide in the spectral range 350–5000 cm^{-1} and obtained results consistent with those of Falk & Giguère. They suggested the broad absorptions from hydrochloric acid 'doubtless include contributions from both H_3O^+ and $H_5O_2^+$' but did not attempt a quantitative division between these two.

A different viewpoint has been advocated by Zundel and coworkers. [268, 272] A very recent review of their thesis has been presented by Zundel & Fritsch. [273] They argue that in aqueous solutions of strong acids, the proton is present as $H_5O_2^+$ groupings, which are embedded in a network of other water molecules. The hydrogen bond in these groupings is easily polarisable [108, 110, 111, 152–7, 273] (see section 1.1.2). In view of the extensive reviews already available [156, 157, 273] it is appropriate to present here only a cursory summary.

The authors interpret their results in terms of proton-boundary structures:

$$\underset{\text{I}}{B_1H\cdots B_2} \rightleftarrows \underset{\text{II}}{B_1^-\cdots H^+B_2}$$

where B_1 may or may not be the same proton acceptor as B_2. When $B_1 \neq B_2$ a double-minimum energy surface is present. [108, 157] A strong, continuous absorption in the IR spectrum is attributed to the 'easily polarisable hydrogen bonds' that are formed. Zundel and coworkers have reported a very large number of spectra over a range of compositions and conditions. Examples are given in figure 1.16.

For a number of strong acids the solution composition at which the acid molecule (ionogen) population is essentially zero could be estimated from the value of the water:acid mole ratio at which characteristic bands of the ionogen vanish. [274] Thus for trifluoromethanesulfonic acid $n = 1.6$ (9.3 M); H_2SO_4 (first step), $n = 1.6$ (13.6 M); $HClO_4$, $n = 1.8$ (12.9 M); H_2SeO_4 (first step), $n \approx 2.0$; $C_6H_5SO_3H$, $n = 5$ (4.9 M); and HNO_3, $n = 12$ (4.0 M). The results for $HClO_4$ and HNO_3 are consistent with recent Raman results. [275, 276] In this series more and more water molecules are required for the removal of the protons from the anions. Leuchs & Zundel [274] point out that the proton will be present both as proton-limiting structure II and as $H_5O_2^+$.

For acid mixtures with $n < 1$ the absorbance of the continuum (at 1900 cm^{-1}) increases in proportion to the molarity of water. Simultaneously the O–H stretching vibration of the acid ($n = 0$) vanishes. (The O–H of water contributes substantial intensity in this region for $n > 0$). The concentration of protons removed from the anion is proportional to the concentration of water. For $n < 1$ the number of hydrogen bonds formed is the same as the number of water molecules present and the concentration of proton-limiting structure II is equal to the concentration of protons removed. Thus from the slopes of plots of the concentration of protons removed from the anion against the concentration of water the weights of the proton-boundary structures were inferred: [277] % II: CF_3SO_3H, 85%; H_2SO_4, 85%; $HClO_4$, 75%;

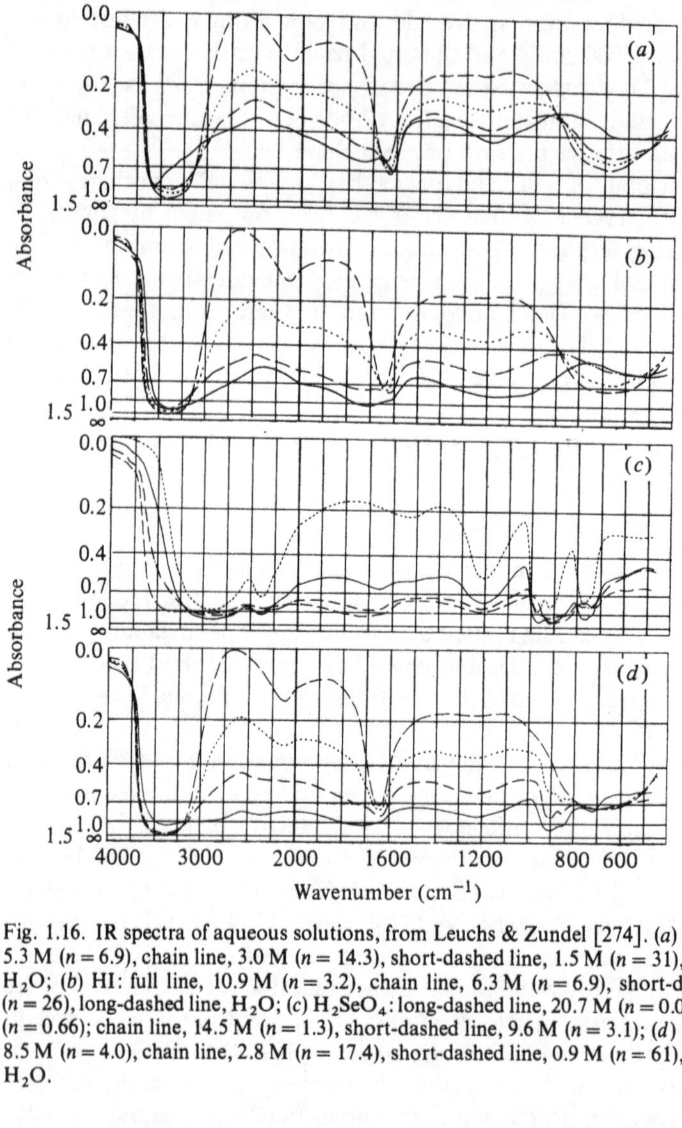

Fig. 1.16. IR spectra of aqueous solutions, from Leuchs & Zundel [274]. (*a*) $HAuCl_4$: full line, 5.3 M ($n = 6.9$), chain line, 3.0 M ($n = 14.3$), short-dashed line, 1.5 M ($n = 31$), long-dashed line, H_2O; (*b*) HI: full line, 10.9 M ($n = 3.2$), chain line, 6.3 M ($n = 6.9$), short-dashed line, 2.0 M ($n = 26$), long-dashed line, H_2O; (*c*) H_2SeO_4: long-dashed line, 20.7 M ($n = 0.0$), full line, 17.3 M ($n = 0.66$); chain line, 14.5 M ($n = 1.3$), short-dashed line, 9.6 M ($n = 3.1$); (*d*) H_2SeO_4: full line, 8.5 M ($n = 4.0$), chain line, 2.8 M ($n = 17.4$), short-dashed line, 0.9 M ($n = 61$), long-dashed line, H_2O.

$C_6H_5SO_3H$, 55%; HNO_3, 10%. [274] This series indicates that the depth of the double-minimum energy surface changes (for one water molecule per acid molecule), favouring the water molecule at the beginning of the series and the anion at the end (B_2 is water; B_1 is anion).

The addition of more water ($n > 1$) further increases the weight of proton-boundary structure II. This interpretation correlates with an increase in the intensity of the O–H stretching vibration in the region 3000–2600 cm^{-1} to a maximum near $n = 1.2$, followed by a decrease (with the exception of nitric acid). The loss of this intensity indicates that the protons have transferred from the acid–water hydrogen bonds into hydrogen bonds *between* water molecules, i.e. the excess proton is now considered to be present in $H_5O_2^+$ groupings which are embedded in the water structure network. The deformation at 1700 cm^{-1} ($n = 1$) from H^+OH_2 shifts to 1740 cm^{-1} ($n = 2$), another change attributed to $H_5O_2^+$. The observation of only one water bending mode, for $n = 2$, was also taken to indicate that the proton was distributed between two similar water molecules. Thus these authors conclude that H_3O^+ groups do not vibrate as an entity when sufficient water molecules are present to form $H_5O_2^+$ groupings. [275] They argue that H_3O^+ vibrations can only be observed when $n < 2$, and then they are apparently stabilised as an ion pair $B^- \cdots H^+OH_2$. Thus we see the basis for the entrenched difference of opinion. [164, 266 vs 273, 274, 278] Does H_3O^+ exist as an entity in water ($n \geqslant 2$) or should the proton be considered as $H_2O \cdot\cdot H^+ \cdot\cdot OH_2$ or some such other species? Irrespective of the answer it is clear that Zundel and coworkers have measured more spectra for more acids at more concentrations than any other group and have attempted the most quantitative comparison of these spectra. Thus they might be expected to have uncovered significant and subtle changes with water content. Their interpretation also is supported by results of Denisov & Golubev [179] who studied $HA/H_2O/CCl_4$ mentioned in section 1.2.1.

Further support for the presence of $H_5O_2^+$ comes from the IR work of Librovich, Sakun & Sokolov. [279] They used germanium prisms and the attenuated total reflection method to record the IR spectra of hydrochloric acid, and thus avoid problems of cell path length. They separated the spectrum of the $O \cdot\cdot H \cdot\cdot O$ fragment from the rest of the spectrum. They assigned three bands to this fragment: the antisymmetric stretching vibration (proton oscillation) at 1170 cm^{-1} (weak or absent in the Raman); the bending vibration, 1250–1400 cm^{-1}; and the continuous absorption band, 1000–3400 cm^{-1}. A band at 1710 cm^{-1} was attributed to the H–O–H bending of the water molecules in H_2OHOH_2 and a band at 2900 cm^{-1} was assigned to the O–H stretching of these water molecules. The continuum was explained in terms of a theory of proton–phonon coupling. For comparison the two different assignments are aligned in table 1.5.

The interpretation was opposed by Giguère [280] and defended by Librovich, Sakun & Sokolov. [281] $H_5O_2^+$ is the favoured form of other Soviet investigators, a conclusion reached on the basis of spectral, kinetic, and other physico-chemical investigations and cited by Librovich *et al.* [279]

As noted above the Raman spectrum of the hydrated proton is expected to have less intense bands than the IR spectrum because of the polar nature of the species. Thus, although many Raman spectral studies of acids have been made, the focus has often been on the degree of dissociation or the vibrational pattern of the ionogens and anions, rather than on the hydrated proton. (For a review of such studies see Irish & Brooker [282] and Brooker [283]; also [275, 276].) A broadening on the low-wavenumber side of the O–H bands of

Table 1.5. *Alternative assignments of the IR bands of the hydrated proton in aqueous solution*

H_3O^+ [266]	$\bar{\nu}$ (cm^{-1})	$H_5O_2^+$ [272, 279]
	1170	$\nu_{as}(O\cdots H^+\cdots O)^b$
$\nu_2(A)$	1200 ± 50	$\delta(O\cdots H^+\cdots O)$
$\nu_4(E)$	1730 ± 10	δ(H–O–H)
	1000–3000	$(O\cdots H^+\cdots O)$ continuum
$\nu_3(E)$	2900 br	ν_{as} and ν_s of H_2O in the
$\nu_1(A)$	2900^a	$H_5O_2^+$ group

[a] Weak in the IR and masked by H_2O.
[b] Assignment of [279].

water, has been detected [284–6] (see figure 1.17) corresponding to the IR continuum but much weaker. Busing & Hornig [286] considered this intensity, observed at 3025 cm^{-1} with a width of 680 cm^{-1} for hydrochloric acid solutions ranging from 3 M to 11.4 M, to arise from the 'H_3O^+ ion or some more highly polymerized species of protonated water'. Zarakhani, Maiorov & Librovich [287] inferred, from an intensity analysis, that the water bending mode (1656 cm^{-1}) and the stretching mode (3450 cm^{-1}) consist of at least two components: contributions from water and hydrated proton, the latter having the larger molar intensity scattering coefficient. Busing & Hornig [286] also reported increased intensity of these bands in the presence of potassium bromide and hydrochloric acid but not potassium hydroxide. These intensity enhancements are primarily ascribed to preresonance Raman effects arising from the charge transfer states of the hydrogen bonds between water molecules and halide ions. [288, 289]

The symmetric bending mode of H_3O^+ was absent in Zarakhani *et al.*'s spectra [287] although it was observed by Ochs *et al.* in 1940 in supersaturated hydrochloric acid. [290, 291] Giguère & Guillot [290] reported the $\nu_2(A_1)$ Raman band at 1220 cm^{-1} for HBr/H_2O solutions of $n = 4$. For $n = 2$ it occurred as a doublet at 1110 and 1060 cm^{-1}, for $n = 1.5$ it occurred at 1090 cm^{-1} and at $n = 0.8$ it occurred at 1050 cm^{-1}. Its intensity was approximately only $\frac{1}{6}$ that of the 1625 cm^{-1} bending mode of water and its FWHM (full width at half peak height) was ≈ 200 cm^{-1}. The shift to lower frequencies was attributed to a change in the environment from water molecules (strong hydrogen bonds) to bromide ions (weaker hydrogen bonds). Pernoll, Maier, Janoschek & Zundel [292] carried out similar experiments in 1975 but failed to detect this band. (But note it may be present in their Fig. 1, 6.0 M hydrochloric acid spectrum.) These workers focussed their attention on the continuous scattering in the region below 3000 cm^{-1} and on the changes below 1000 cm^{-1}, which were interpreted in terms of the large proton polarisability of the hydrogen bond in the $H_5O_2^+$ grouping.

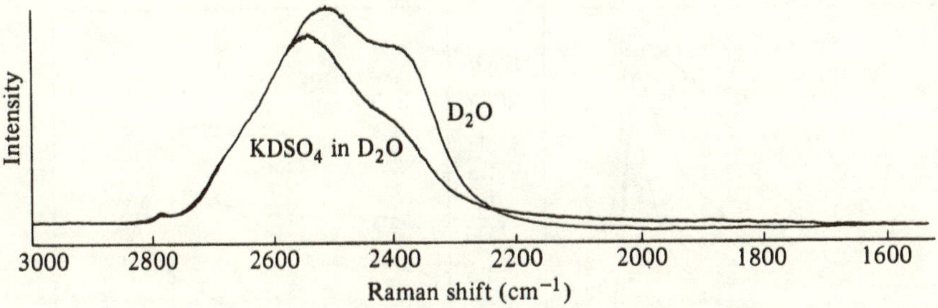

Fig. 1.17. Raman spectra of D_2O and $KDSO_4$(2.4 M)–D_2O solutions, obtained under identical instrument conditions, from [285], to illustrate the Raman continuum.

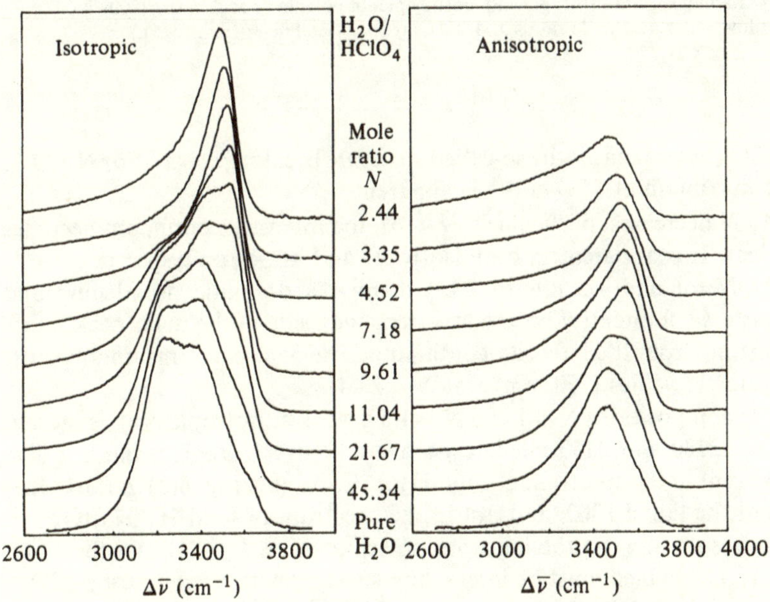

Fig. 1.18. Raman spectra (isotropic and anisotropic) of perchloric acid in the 2600–4000 cm^{-1} region at the mole ratios indicated and 25 °C. (From Ratcliffe & Irish [275].)

The isotropic and anisotropic Raman spectra in the region 2000–4000 cm^{-1} for solutions of perchloric acid in water have been reported by Ratcliffe & Irish. [275] The spectra are shown in figures 1.18 and 1.19. As the mole ratio of water to acid (N) decreases the following spectral changes can be seen:

(*a*) Intensity decreases in the 3250 cm^{-1} region and increases in the 3400–3600 cm^{-1} region. These effects are similar to those observed for perchlorate salt solutions; [293–302] the shift of intensity is usually taken to imply that hydrogen bonding of the water surrounding the

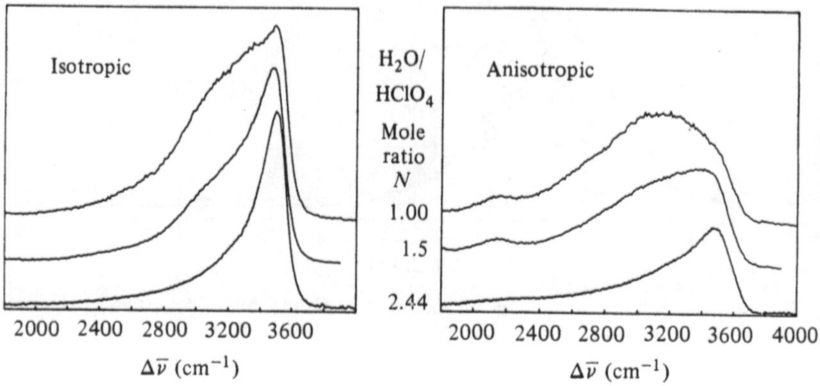

Fig. 1.19. Raman spectra (isotropic and anisotropic) of perchloric acid in the 1800–4000 cm^{-1} region at low water:acid mole ratios and 25 °C. (From Ratcliffe & Irish [275].)

ClO_4^- is weakened (the so-called structure breaking effect). For $N = 9.61$ a maximum at 3537 cm^{-1} is apparent.

(*b*) As N decreases, particularly $N < 10$, the intensity continuum becomes clear. It is apparent in both isotropic and anisotropic spectra.

(*c*) At $N = 1$ and 1.5 (figure 1.19) there is a dramatic broadening and *increase* of intensity on the low frequency side of the main peak, quite distinct from the intensity continuum. The 3537 cm^{-1} maximum shifts to lower values (3502 cm^{-1} at $N = 2.44$).

(*d*) For compositions $N = 1$ and $N = 1.5$ a weak anisotropic peak is present at $\approx$2180 and 2150 cm^{-1} respectively (figure 1.19). This band is also present in the spectrum of solid $H_3O^+ClO_4^-$ (see Fig. 6(b) in Part One) and the liquid [303] but not in the spectrum of solid $H_5O_2^+ClO_4^-$.

These results suggest the following interpretation. For $N \approx 1.5$ there is a species H_3O^+, which must be interacting strongly with neighbouring ClO_4^-, and is possibly stabilised by them, and which is sufficiently long-lived to be identifiable. This species can be described as proton boundary structure II if one wishes, *viz.* $ClO_4^- \cdots H_3O^+$. On addition of more water its environment changes to $H_2O \cdots H_3O^+$, and its lifetime is markedly shortened, giving rise to line breadth and the Raman continuum.

The aqueous perchloric acid solutions [275] generated only one band at all values of N in the 1600 cm^{-1} region. This ranged from 1638 cm^{-1} for pure water to 1602 cm^{-1} for the $N = 1$ liquid. In IR spectra there are two bands for high N; thus a band at 1620–1645 cm^{-1} (low concentration) is accompanied by a 1700–1720 cm^{-1} band at high concentration, which eventually dominates the spectrum. [277, 304]

For the aqueous perchloric acid solutions, bands from ClO_4^- or $HClO_4$ obscure the 800–1200 cm^{-1} region and thus make it virtually impossible to detect, if present, the $\nu_2(A_1)$ band of H_3O^+. For details see figures 1, 2, 4 and 5 of [275] and the discussion there.

Kanno & Hiraishi [305] have clearly identified this band in glassy aqueous hydrochloric and hydrobromic acid solutions ($N = 5.1$) at liquid nitrogen temperature, figure 1.20. Frequencies are $1240 \pm 10\,\text{cm}^{-1}$ for hydrochloric acid and $1230 \pm 10\,\text{cm}^{-1}$ for hydrobromic. For fewer water molecules per mole hydrochloric acid ($N = 3.8$), the position is $1210\,\text{cm}^{-1}$, indicating that the strength of hydrogen bonding to the environment is decreased when Cl^- replaces water in the near surroundings. The intensity is larger than that from the liquid state, an observation attributed to a longer relaxation time for the exchange of a proton between water molecules. This conclusion was supported by the observation that the continuum (3000–$2200\,\text{cm}^{-1}$) shown in figure 1.21 decreases considerably in intensity for the glassy solution. For the glass a weak band at $\approx 2800\,\text{cm}^{-1}$ was attributed to ν_1.

Manifestations of the state of the hydrated proton in aqueous solutions are also present in the Rayleigh scattering. Depolarized light scattering is evident in the Rayleigh wing and is caused by time-dependent changes of the anisotropy of the polarisability. For a review see Clarke. [306] Zundel and coworkers have measured the Rayleigh scattering from HCl and DCl solutions. [292, 307] The scattering intensity increases with acid

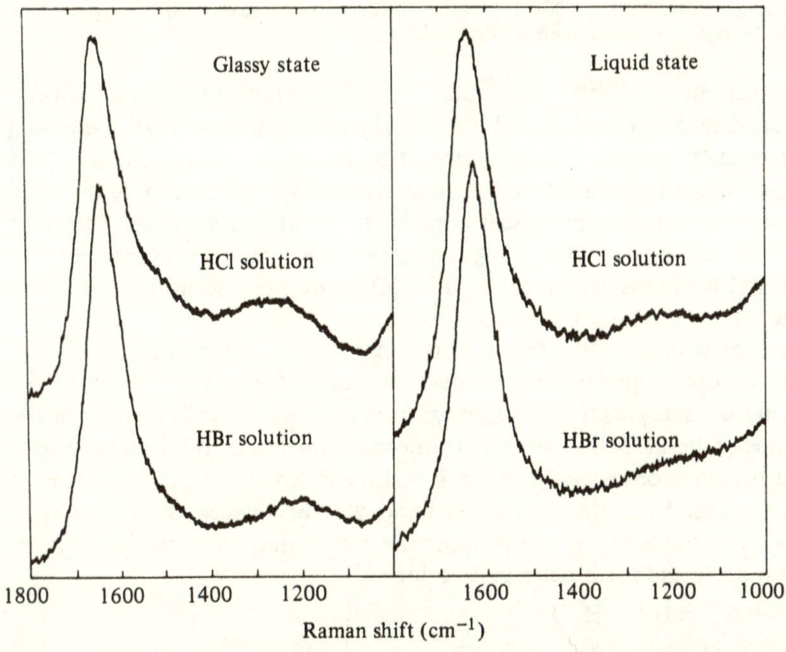

Fig. 1.20. Raman spectra of aqueous hydrochloric and hydrobromic acid solutions ($N = 5.1$) in the glassy state and the liquid state, from 1000 to 1800 cm^{-1}. (From Kanno & Hiraishi [305].)

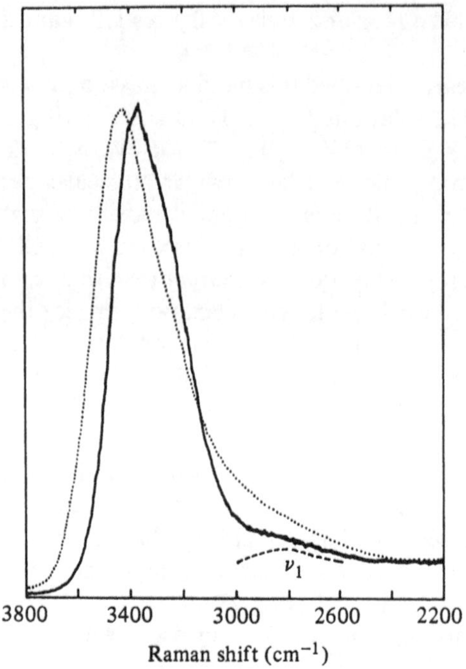

Fig. 1.21. Raman spectra of aqueous (dotted curve) and glassy (full curve) HCl ($N = 5.1$) in the frequency range from 2200 to 3800 cm^{-1}. The dashed curve shows the estimated contour of the ν_1 band of the H_3O^+. (From Kanno & Hiraishi [305].)

concentration, an effect attributed to the large anisotropic proton polarisability of the hydrogen bonds in $H_5O_2^+$ groupings. With increasing acid concentration the mean lifetime of hydrogen bonds in acid solutions increases, and the halfwidth of the depolarized Rayleigh scattering becomes narrower. The intensity of the scattering in D_2O acid solutions is less than that in hydrochloric acid solutions, suggesting that the anisotropy of the deuteron polarisabilities of the deuteron bond in $D_5O_2^+$ groupings is a little less than the analogous proton polarisabilities.

The shapes of IR and Raman vibrational lines also provide information about the excess proton in aqueous solutions. Kreevoy & Mead [308], following up a suggestion of Eigen, pioneered this field with a study of the lineshape of the 1433 cm^{-1} band of trifluoroacetate ion in trifluoroacetic acid. In the presence of excess base the FWHM, 15 cm^{-1}, was independent of concentration. In acidic media the band was broadened by as much as 7.2 cm^{-1}. A simple model was proposed, in which the proton was considered to jump back and forth between two forms

$$H_3O^+ + A^- \rightleftarrows HA + H_2O$$

with a pseudo first-order rate constant, $k_1 = \tau^{-1}$, given by

$$\tau_1^{-1} = \tfrac{1}{2}(\Delta\nu_{\frac{1}{2}} - \Delta\nu_{\frac{1}{2}}^{\circ})$$

where $\Delta\nu_{\frac{1}{2}}$ is the FWHM of the broadened line and $\Delta\nu_{\frac{1}{2}}^{\circ}$ is the FWHM of the unbroadened line (in neutral or basic solution). A value of $k_1 = 4.7 \times 10^{11}\ s^{-1}$ for a solution 1.5 M in acid and conjugate base was obtained. From the result the shape of the doublet centred at 830 cm^{-1} (843 cm^{-1} for the anion and 816 cm^{-1} for the acid) could be reproduced. The procedure is analogous to that employed in NMR spectroscopy. [309, 310] The period of a vibration of 3000 cm^{-1} is 1×10^{-14} s. Thus vibrational spectra can be expected to sense much faster processes than NMR.

At the Faraday Discussion in 1965 Kreevoy & Mead [311] presented a more thorough study of this system. Line broadenings as large as 45 cm^{-1} were reported. (Covington attributed some of this broadening, for perchloric acid/trifluoroacetic acid mixtures, to the formation of undissociated acid, which possesses a band at 1455 cm^{-1}; [312] Kreevoy rebutted this, in part, because no ionogen would be present in some solutions studied where the broadening was increased by addition of perchloric acid. [313]) A mechanism consisting of an equilibrium between the ionogen, HA, and A^- in one of three forms: (1) separated A^-; (2) A^- within 5–10 Å of H^+ but with unfavourable orientation, $H^+ \cdots A^-$, and (3) anions within one or more water molecules of A^-, H^+A^-, was suggested.

$$\underset{}{HA} \rightleftarrows \underset{(1)}{H^+A^-} \rightleftarrows \underset{(2)}{H^+\cdots A^-} \rightleftarrows \underset{(3)}{H^+ + A^-}$$

Step (3) involves translational diffusion; the two forms of A^- in this step are considered to have identical 1435 cm^{-1} Raman bands. Rotational orientation is the only thing that distinguishes A^- in step (2). It was postulated that step (1) determines the spectroscopic lifetime of A^-.

At the same Faraday Discussion Covington, Tate & Wynne-Jones [314] reported a Raman linewidth study of perchloric acid. Line broadenings from 0.3 (4.03 M) to 3.3 cm^{-1} (11.44 M) were reported. Note that the concentration of the ionogen is essentially zero for all but the highest concentration. They suggested that 'the appearance of perchloric acid molecules in concentrated solutions only occurs when there are insufficient water molecules present to solvate the proton fully as $H_9O_4^+$ as in dilute solutions.' [315] Their results were presented as a second-order rate constant k_2

$$k_2 = \pi c(\Delta\bar{\nu}_{\frac{1}{2}} - \Delta\bar{\nu}_{\frac{1}{2}}^{\circ})/[H_3O^+].$$

Values of $k_2\ [H_3O^+]$ increased from 0.6×10^{11} to $6.2 \times 10^{11}\ s^{-1}$ as the concentration increased. They noted that the FWHM of the corresponding conjugate base does not increase by more than 1 or 2% (e.g. NO_3^-, see Vollmar [316]). At this same discussion Weston [317] pointed out anomalies in the kinetic argument and cited consistent, unpublished data. Covington, Freeman & Lilley [318] later presented data for trifluoroacetic acid. Two reviewers presented aspects of the theory in this period. [319, 320]

Chen & Irish [285, 321, 322] studied the dissociation of the HSO_4^- ion (for both ammonium bisulphate and sulphuric acid) by Raman spectroscopy and presented both the degree of dissociation and the Raman line broadening of the 981 cm^{-1} line of SO_4^{2-} and the 1050 cm^{-1} line of HSO_4^-. In a separate study it was shown that an increase of the viscosity by a factor of 500 does not broaden the line. [323] A new fact emerged from this study: from mixtures containing various amounts of hydrochloric acid and ammonium sulphate or sulphuric acid and sodium hydroxide it was found that the broadening of the 981 cm^{-1} line of SO_4^{2-} is directly proportional to the $[H^+(aq)]$ and is independent of the SO_4^{2-} concentration (up to about $[H^+(aq)] = 2$ M). From the slope an overall second-order rate constant of 5.5×10^{11} M^{-1} s^{-1} was inferred for the recombination step. The broadening of the 1050 cm^{-1} line of HSO_4^- was consistent with the model. The change in the slope of line broadening against hydronium ion concentration at higher concentrations (>2 M) was attributed to a decrease in the water activity, i.e. the solvation of H_3O^+ would be diminished. Eigen, Kurtze & Tamm [324] reported a value of 1.0×10^{11} M^{-1} s^{-1} in solutions of 0.1 M ionic strength for the proton transfer rate constant, measured by the ultrasonic absorption relaxation technique. The agreement is good. At the low concentration at which they worked structural diffusion would be a more important contributing step. For DSO_4^- in D_2O the apparent second-order rate constant for D^+ transfer was twice as large as for HSO_4^- in H_2O.

Ikawa, Yamada & Kimura [325] repeated the measurements using laser excitation and computer procedures for resolving the components into Voigt functions. They confirmed the linear dependence of the FWHM of the 980 cm^{-1} band of SO_4^{2-} on hydronium ion concentration and obtained a rate constant of 4.5×10^{11} M^{-1} s^{-1}. They made allowance for an additional band at 1038 cm^{-1}; Chen made allowance for a 1024 cm^{-1} component. These have little effect on the FWHM of the 980 cm^{-1} band. The agreement with the earlier work is good.

Recently Cohen & Weiss [326] studied the IR active $\nu_3(F_2)$ mode of SO_4^{2-} in solutions of H_2SO_4/H_2O and D_2SO_4/D_2O, both in the presence and absence of added sodium chloride and at four different temperatures (for the former). They also observed direct proportionality between the FWHM and the concentration of $H^+(aq)$ and inferred a rate constant of 17×10^{11} M^{-1} s^{-1} at 23 °C (i.e. about three times larger than the Raman values). Their values for the concentration quotient of HSO_4^- dissociation are consistent with those of Chen & Irish. [322] Because the rate constant was not affected by the addition of sodium chloride, these authors concluded that the proton transfer was between closely spaced H_3O^+ and SO_4^{2-}; sodium chloride was expected to affect water structure and thus change the observation if the proton was transferred some distance over hydrogen-bond bridges. These authors reported a rate constant for H_2SO_4/H_2O about twice as large as that of D_2SO_4/D_2O. They inferred an activation energy for the process of $7.5 \pm 2.1\,kJ\,mol^{-1}$ ($10\,kJ\,mol^{-1}$ for D_2SO_4). The use of the IR band can be criticised for several reasons: it is a triply degenerate mode and ion pair formation (e.g. with H_3O^+) could cause lifting of the degeneracy and hence broadening; it is also overlapped by the $1170\,cm^{-1}$ band of H^+(aq) and the 1050 and $1195\,cm^{-1}$ bands of HSO_4^-.

Direct proportionality between the line broadening and the concentration of H^+(aq) was also observed by Puzic for nitric acid solutions. [327–9] The $1050\,cm^{-1}$ line of NO_3^- was studied. By addition of hydrochloric acid it was found that the broadening was independent of the conjugate base concentration, since it is the same, for example, for 3 M HNO_3, 2 M HNO_3 + 1 M HCl, and 1 M HNO_3 + 2 M HCl. See figure 1.22. A rate constant of $0.59 \times 10^{11}\,M^{-1}\,s^{-1}$ was calculated from the slope. Correlations were made with other acids: trifluoroacetic (TFA) and trichloroacetic (TCA). [318] For these acids as the concentration of hydronium ion increases the broadening departs markedly from the direct proportionality with [H^+(aq)] (more so than in figure 1.22). A rationalisation in terms of specific acid catalysis was suggested. [327] It was also noted that the stronger the conjugate base strength, the larger the apparent rate constant ($k \times 10^{-11}$ was 0.59, 0.71, 1.41, and $5.50\,M^{-1}\,s^{-1}$ for the bases NO_3^-, TFA^-, TCA^-, and SO_4^{2-} respectively).

Strehlow, Wagner & Hildebrandt [330] have presented a theory for the 'fast case' of chemical exchange (in contrast to the 'slow case' developed by Kreevoy & Mead and applied above) and compared the results with the NMR

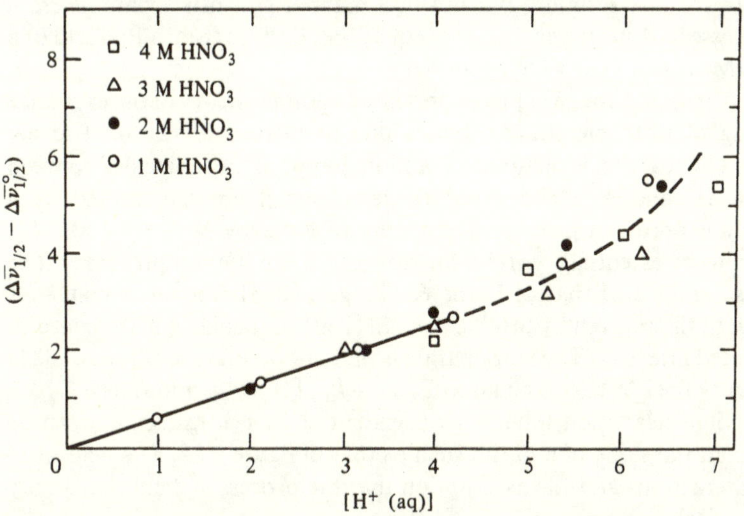

Fig. 1.22. The dependence of the increase of the FWHM of the $1048\,cm^{-1}$ line of NO_3^- on the concentration of H^+(aq). Note that solutions with different concentrations of the conjugate base, but the same concentration of H^+ (aq) (by adding hydrochloric acid), exhibit the same FWHM. (From [324, 326].)

case. They showed that if the lifetimes broaden the Raman lines the integrated line intensity is not proportional to concentration, but rather becomes smaller (i.e. it shows a kinetic damping). They also made allowance for a band from an ion pair close to the band of the ion in the two-step reaction

$$H_3O^+ + NO_3^- \underset{k_{1'}}{\overset{k_1}{\rightleftarrows}} H_3O^+NO_3^- \underset{k_{2'}}{\overset{k_2}{\rightleftarrows}} H_2O + HNO_3$$

They illustrated the theory by application to ion pair formation, $Na^+ \cdot NO_3^-$, and proton transfer in nitric acid. They obtained an effective bimolecular rate constant ($k'_{AB} = k_1/[H_3O^+]$) of order 2×10^{11} (dependent on acid concentration). The strong increase of the rate constant above the diffusion controlled value at high concentrations of acid was explained by the fact that, at high concentrations of nitric acid, translational diffusion is not required since the average separation of the recombining ions is sufficiently small and therefore the recombination reaction itself is rate determining. Based on very few data, the temperature independence of the result suggested a proton transfer by the quantum mechanical tunnel effect.

MacPhail & Strauss [331] have presented a critique of the use of Bloch equations to describe the vibrational spectra of a reacting molecule. They conclude that the simple Bloch equations will only be valid if a number of stringent conditions apply; but on the vibrational time scale the requirements of high barrier, rapid transit time and rapid reaction time are incompatible. The extracted lifetime is thus not considered to be a meaningful quantity. "The width of the spectral lines or the collapse of a multiplet is thus 'due' to the presence of the fast reaction, but not in a manner which is separable from other processes." A more general set of equations, derived from a Redfield or a Zwanzig–Mori analysis, were proposed.

Bratos, Tarjus & Viot [332] have presented another theory of the exchange broadening of isotropic Raman bands due to ultrarapid proton transfer processes. The reaction is pictured as random jumps of the molecules between the forms AH and A^-. The variables describing it are represented by a dichotomic Markovian process. The theory of Kreevoy & Mead [308–11] was found to be essentially correct for slow exchange Raman processes. The theory also confirmed that of Berne & Giniger. [333] But the conclusions disagree with those given by Strehlow *et al.*, [330] particularly with regard to the integrated intensity. The work parallels that of MacPhail & Strauss. [331] In this theory the FWHM is equal to $T_{2,A^-}^{-1} + k_{12}[H_3O^+]$, and thus if T_{2,A^-}^{-1}, the vibrational relaxation time, is independent of concentration, k_{12} can be inferred from the slope of a graph such as that of figure 1.22. The neglect of A^-/A^- interactions and the assumption that the process is dichotomic may affect the conclusions.

From the above it is clear that the application of Raman (and IR) spectroscopy to the study of proton transfer processes is an evolving field. Essentially no experimental work has been done with the new theories in mind. The experimental work that has been done clearly shows that the proton transfer processes are manifest in the vibrational lineshapes of a normal mode localised in an unmodified fragment of the conjugate base but perturbed by the reaction; further theoretical and experimental work can be anticipated.

In the above discussion mention has been made of ion pairs. An ionogen, HA, will form ions only by chemical reaction with solvent molecules. The reaction of the ionogen with water can be considered to occur in two steps: [318, 328, 334] an ionisation step results from transfer of a proton to a water molecule in the hydration sphere and yields a contact ion pair $H_3O^+ \cdot A^-$; the ion pair is then dispersed by the bulk solvent to yield the equilibrium populations of solvated ions. The processes and the corresponding equilibrium constants are as follows: [328]

$$\mathrm{HA} + \mathrm{H_2O} \underset{K_a}{\rightleftharpoons} \mathrm{H^+(aq)} + \mathrm{A^-} \qquad \mathrm{HA} + \mathrm{H_2O} \underset{K_i}{\rightleftharpoons} \mathrm{H_3O^+ \cdot A^-}$$

$$\mathrm{H^+(aq)} + \mathrm{A^-} \underset{K_d}{\rightleftharpoons} \mathrm{H_3O^+ \cdot A^-}$$

$$K_i = \frac{\{H_3O^+ \cdot A^-\}}{\{HA\}\{H_2O\}}$$

$$K_d = \frac{\{H^+(aq)\}\{A^-\}}{\{H_3O^+ \cdot A^-\}}$$

$$K_a = K_i K_d$$

It is clear that the traditional thermodynamic acidity constant K_a has been defined in such a way that it is a product of the ionisation constant, K_i, and the dissociation constant, K_d.

The ionisation step corresponds to the formation of Zundel's proton-boundary structure II [273] and to step (1) of the Kreevoy & Mead mechanism for proton transfer, [311] as discussed earlier. The question is can the $H_3O^+ \cdot A^-$ species be detected?

The vibration spectrum has provided evidence for such a species, although the identity as $H_3O^+ \cdot A^-$, as opposed to some other form of the hydrated proton, must be considered conjecture. New bands in the spectrum of a polyatomic anion suggest the formation of new bonds or a significant perturbation of the anion resulting from the juxtaposition of a strongly interacting second species. Raman spectroscopy is particularly sensitive to first nearest-neighbour interactions. Thus for many nitrate salts, the formation of an ion pair (or complex ion) results in a second band close to the intense symmetric stretching band at 1048 cm^{-1} [282, 328] and similarly for sulphate salts and salts of other oxyanions.

In the case of nitric acid a band 12 cm^{-1} lower in frequency than the NO_3^- symmetric stretching vibration (1048 cm^{-1}) has been attributed to $H_3O^+ \cdot NO_3^-$. [328] The species has been detected by IR spectroscopy in argon matrices containing >3% water. [335] The evidence here is the lifting of the $\nu_3(E)$ degeneracy (the antisymmetric stretching mode of NO_3^-). For less than 3% water the IR spectra are understood in terms of the monohydrate $H_2O \cdot HNO_3$. As the water content increases from 3 to 100% the line splitting ($\Delta\nu_3$) decreases from ≈150 cm^{-1} (indicating a very strong and specific interaction between H_3O^+ and NO_3^-) to 65 cm^{-1}; this splitting is assigned to the $H_3O^+NO_3^-$ existing as the completely hydrated but contacting ion pair. It is interesting to note that both proton-boundary structures I and II are found in the matrix and II appears to give way to a fully hydrated form of ion pair; cf. [338]. Mention was made [335] that broad bands at 2800, 1750 and 1200 cm^{-1}, attributed to H_3O^+, were observed for concentrated nitric acid in H_2O matrices. Other evidence for this ion pair can be found in the work of Leuchs & Zundel, [336] Strehlow *et al.* [330] and Herzog-Cance, Potier & Potier. [337]

A weak 984 cm^{-1} line, near the 981 cm^{-1} symmetric stretching band of SO_4^{2-}, was assigned to $H_3O^+ \cdot SO_4^{2-}$ by Irish & Chen, [321] although it was not confirmed by recent studies. [338] For H_2O:$HClO_4$ mole ratios of 2.44 a high frequency shoulder on the Raman ν_1 stretching band of ClO_4^- (isotropic spectrum) and a low frequency shoulder of ν_3 (anisotropic spectrum) have been interpreted as evidence for an ion pair $H^+(aq)ClO_4^-$. [275, 339] Hydrofluoric acid, noted for its weakness among the hydrohalic acids, has a vibrational spectrum that has been rationalised in terms of the $H_3O^+F^-$ complex as the predominant species in solution. [340]

Although all of the above observations relate to interactions of the hydrated proton with an anion, none of them unequivocally provide information about the stoichiometry and structure of that species.

Other studies The hydrated proton has been the subject of study by many other techniques and is the active participant in the kinetics and mechanisms of processes in chemical, biochemical and physical phenomena. These range from the measurement of pH with a glass electrode to proton transport in membranes and in ice. We make no attempt here to be exhaustive in coverage. Our focus is on stoichiometry and geometry. The role of the proton in organic chemistry has recently been reviewed, [341] following on the authoritative work of Bell. [342]

We must cite the classic studies of Eigen and coworkers [343] however. In his 1964 review Eigen summarises the case for the H_3O^+ ion, the $H_9O_4^+$ ion (secondary hydration), structural diffusion resulting from the directed formation and breaking of hydrogen bridges on the periphery (cf. Zundel [344, 345]) and the mechanism of proton transfer. Three Faraday Discussions have addressed this theme since 1964. [346–8] (Also see [349].) The papers that were presented address the proton mobility, proton transfer reactions, acid catalysis, etc. in a variety of situations ranging from biological to gas phase. The structure of the hydrated proton was not particularly addressed; most often it was (conveniently) described as H_3O^+ (or $H_3O^+(H_2O)_x$ where x varies with concentration). [348]

Picosecond spectroscopic techniques have recently been applied to the study of the hydration of the proton [350] and an extensive review has been written. [351] Lee, Griffin & Robinson [350] investigated the proton transfer process of 2-naphthol in its first excited singlet state, in water/methanol mixtures at different temperatures. The proton transfer rate was found to increase as the temperature was raised and to decrease as the methanol concentration was raised. Using a Markov random walk theory it was shown that a water cluster containing 4 ± 1 members is the proton acceptor (i.e. $H_9O_4^+$). The observed activation energy, 14.4 kJ mol^{-1} in pure water, was attributed to the energy required for rearrangement of the hydrogen bonding in the normal water structure to form the proton accepting cluster. The authors claim that this is the first direct determination of the actual existence of an entity that in all probability is $H_9O_4^+$. (The intuitive reasoning for this structure from thermochemical, mobility data and theory was given by Conway. [352]) Lee *et al.* conclude that a cluster size of four is actually a critical size for caging a proton. In the liquid state a proton in a cluster of three water molecules or less is not energetically favoured. Water molecules in the second coordination shell play a secondary role insofar as the structure is concerned; thus methanol molecules are able to serve just as well.

Huppert *et al.* [353] have recently measured the pressure dependence of the proton transfer rate of 8-hydroxy-1,3,6-pyrene trisulphonate (HPTS), using time resolved picosecond fluorescence techniques. The proton transfer rates exhibited a large linear increase with pressure from 8×10^9 s^{-1} at 1 atm and 294 K to 2.5×10^{10} s^{-1} at the liquid → ice VI transition point of 9 kbar and 294 K. In D_2O the deuteron transfer rate was three times slower in the entire pressure range. The change in rate was attributed to the relatively large changes of the water structure caused by pressure. The following mechanism was visualised: a proton is first transferred from HPTS to an adjacent, properly aligned, water molecule forming H_3O^+. The formation of this species causes the water in the local environment to rearrange and form a cluster. The proton transfers through the cluster to distance itself from the excited anion. Pressure facilitates this by the reduction of the free volume of water and the resulting compactness of the water structure.

Natzle & Moore [354] have photoionised water with a short laser pulse and have measured the recombination rate of H^+ and OH^- from the transient conductivity, as a function of temperature and isotopic composition. The relaxation times ranged between 233 and 14 μs. A recombination distance of 580 ± 50 pm, independent of temperature and isotopic composition, is consistent with the four-molecule ion pair intermediate of Eigen.

Marcus has proposed a set of consistent aqueous ionic radii for 35 ions, including H_3O^+, for which he gives 113 ± 5 pm. [355] Swaddle & Mak [356] have measured partial molar volumes of a number of ions and correlated these through an equation involving the number of water molecules in the unit $M(H_2O)_n^{z+}$. H^+ (aq) fits the correlation quite well with $n = 2$. Laforgue *et al.* [163] have presented a new model of the conduction mechanism in an acid; $H_5O_2^+$ was considered to be the essential structure responsible for conduction. These authors make no reference to the theory put forward by Halle & Karlström [258] to account for the complementary information obtained from nuclear spin relaxation, which monitors proton dynamics, and electric conductivity data, which probe charge transport. (See p. 40.) These authors argue that the prototropic charge migration (pcm) and the diffusive contributions are not independent: there is a significant contribution from cross correlations, in fact of the same order of magnitude as the normal diffusive contribution. A concerted pcm model was presented. The authors conclude with a critique of the oft-cited Conway–Bockris–Linton model [352, 357–9] and of conclusions of Gagliardi. [360, 361] These mechanistic studies address 'structure' in that an initial state and a 'transition state' are implied or assumed.

1.3 Concluding remarks

In the solid state, the gaseous state and some non-aqueous environments the species H_3O^+ is well characterized. In solids and non-aqueous environments $H_5O_2^+$ also is readily distinguishable; in the gas phase the isolated species $H_3O^+(H_2O)_x$, $x = 1$, 2, and 3, can be prepared as we have seen (Part One, p. 188 [1]) and the last member (formally $H_9O_4^+$) is a favoured form compared to those with $x > 3$. Also the observation of a basic cluster of four waters with the excess proton has now been claimed in liquid water [350]; but the structural data for these hydrates are lacking. Larger molecular clusters (e.g. $(H_2O)_nH^+$ where $n = 21$ and 28) are slightly more stable than other clusters with large hydration numbers. We have also seen that theoretical studies have contributed to our understanding of the structure and stability of the hydrated proton.

It has been remarked that the liquid state 'has the reputation of a kind of purgatory between solid and gas . . .'. [362] This is particularly true for liquid water. When one visualises either H_3O^+ or $H_5O_2^+$ hydrogen bonded into the network of bulk water, which itself is not fully understood, it becomes evident that there is not very much difference between the two entities. The particular distinguishing feature of $H_5O_2^+$ would be a short O···O distance (less than 248 pm; Part One, p. 155). If one takes into account the dynamic nature of the liquid system and the fact that a distribution of hydrogen bond lengths and angles exist, one can anticipate a distribution of O···O distances in a constant state of change. Some configurations would correspond to $H_5O_2^+$ and others to H_3O^+. (Although the diffraction results for HCl/3.99 H_2O [147] give a peak with a maximum corresponding to an O···O distance of 252 pm it is possible that the peak arises from a distribution of lengths; in this particular case it would not be possible to determine the width of the distribution, since it is probably less than the resolution.) Zundel would argue that the excess proton moves so rapidly between two water molecules that the lifetime of the H_3O^+ group is too short to allow it to vibrate as an entity. [363] This highly polarisable hydrogen bond model explains the breadth and intensity of the spectral continua. Giguère prefers to call $H_5O_2^+$ a transition state and argues for a less rapid proton transfer time. [280]

At this point one is reminded of some thoughtful words of Frank on the topic of a related phenomenon – the structure of liquid water. [364]

> That experiments should be undertaken either to prove or to disprove a point violates one of the widely held popular views of what a sound scientific attitude ought to be. According to that view scientists are, in their scientific work, 'without passion' – are, on the contrary, indifferent to the implications of any given experimental result, so long only as it has been properly arrived at. In particular, they are supposed to be ready to abandon any 'preconceived' idea as soon as an experimental result is found to be inconsistent with it. .. In point of fact, scientists as a group are strongly motivated men and women, and would not be able to put forth the effort involved in performing a careful experiment or in puzzling through the complexities of a line of theoretical reasoning if they were not able to visualize at least the possibility of a concrete outcome. They do not work at random. And between visualizing a particular outcome and desiring it, the boundary is very nebulous. Why, then, should such outcomes not be discredited as reflecting the 'will to believe' of the worker rather than the objective 'truth'? Because in addition to such a will to believe the true scientist also possesses a penetrating critical faculty, along with a certain ruthlessness which enables him, so to speak, to slaughter his (intellectual) offspring without pity should they be found to be defective. It is only when he permits his critical faculties to be lulled into inattentiveness that the scientist is in danger from his wishful thinking. So long as the critical faculties are active and functioning without impairment, the capacity for wishful thinking is not only permissible – it is actually indispensable if work of the highest quality is to be done. Moreover, . . . , the existence of competing (wishful) interpretations can, at its best, result in sounder and more rapid progress than might otherwise be possible.

The 'structure' adopted for the hydrated proton in water depends on the time scale and, with or without 'wishful thinking', in due time with the new picosecond technologies the flickering picture can be expected to clarify.

Appendix

Since the preparation of Part One of this review a number of new papers concerning the hydrated proton in the solid and gaseous phases have appeared and we will briefly review these here.

Solids

The new works on solids (and a few older ones which have also come to our attention since Part One) are catalogued in Table A1.1, which takes the form of an extension to the appendix of Part One. A few new X-ray diffraction structural studies have been reported; [A2, A10, A12, A14, A15, A19, A20, A56] these include two 18-crown-6-acid complexes. [A19, A20] Several studies, which were mentioned in Part One, have now been published in more detail. In particular these include the clathrate hydrate structures of acids [A15, A16, A17] determined by Mootz, Wiebcke and Oellers, in which the excess acid proton is incorporated into the hydrogen-bonded water network which forms the cage structures. Hencke's structure of $HSbCl_6 4\frac{1}{2}H_2O$ [A12] shows this to be another example of a solid containing an $H_{14}O_6^{2+}$ grouping, consisting of $(H_5O_2^+)_2(H_2O)_2$ subunits in a centrosymmetric ring, with further H_2O molecules linking the rings. $H_{14}O_6^{2+}$ is also present in $HSbBr_6 3H_2O$ [A13] whose structure has been determined by neutron diffraction. The other results include vibrational spectroscopic studies of proton conducting materials [A5, A6] and crown ether complexes [A18] and ^1H NMR studies of the presence and motions of H_3O^+. [A1, A3, A9] In addition to these there is continuing interest in the involvement of H_3O^+ in defects in ice caused by hydrochloric acid doping [A21, A22] and UV photolysis. [A23]

Gas phase

New works concerning the inversion mode include: (*a*) The calculation of the forbidden $\Delta k = \pm 3$ transitions in the ν_3 inversion band for H_3O^+ and D_3O^+ [A24] (such forbidden bands have been detected in the spectrum of NH_3); (*b*) the calculation of rotation-inversion energies for H_2DO^+ and HD_2O^+ [A25] to aid in the search for such transitions; and (*c*) observation of the $1- \leftarrow 1+$ inversion–rotation spectrum of H_3O^+. [A26] Okumura, Yeh, Myers & Lee [A27] have obtained vibrational-predissociation IR spectra of $H_7O_3^+ \cdot H_2$ and $H_9O_4^+ \cdot$cluster ions. These spectra show lines in the 3000–4000 cm^{-1} region at much higher resolution than the older work of Schwarz [148] on $(H_3O^+)(H_2O)_n$ ($n = 3$–5). Both ions show two doublets centred at 3640 and 3730 cm^{-1} (cf. free water at 3657 and 3756 cm^{-1}) assigned to the ν O–H symmetric and antisymmetric stretches of the H_2O units attached to H_3O^+.

The $H_7O_3^+ \cdot H_2$ also gave a line at 3590 cm^{-1}, assigned to ν OH of the free OH of the H_3O^+ unit. These values must represent a perturbation from the isolated ion case by the weakly bound hydrogen molecule, though this is thought to be small. Begemann & Saykally [A28] obtained high resolution IR vibration–rotation spectra of the ν_3 band of H_3O^+ and found 3535.974 and 3519.397 cm^{-1} for the s–s and a–a inversion components of ν_3 respectively. The results of Stahn, Solka, Adams & Urban [A57] were in close agreement. H_3O^+ was detected in the coma of comet Giacobini–Zinner during a spacecraft flyby in September 1985. [A29] This result is hardly surprising considering the 'icy' make-up of cometary nuclei.

Results on the electron impact dissociation of H_3O^+ suggest $e^- + H_3O^+ \rightarrow e^- + (H_3O^+)^* \rightarrow HO^+ + O^+ + 2e^-$ as the most probable reaction. [A30] $(H_2O)_n^+$ and $(H_2O)_nH^+$ ions have been observed in the mass spectra of UV photoionised water clusters. [A31] The $(H_2O)_nH^+$ ions are produced by ionisation of neutral water clusters followed by intracluster proton transfer. The first and second steps in the hydration of H_3O^+ have recently been studied by pulsed electron beam mass spectrometry. [A32].

For $H_3O^+ + H_2O \rightleftarrows H_5O_2^+$

$\Delta H° = -148.5$ kJ mol^{-1}, $\Delta S° = -126$ J K^{-1} mole^{-1}

and $H_5O_2^+ + H_2O \rightleftarrows H_7O_3^+$

$\Delta H° = -84.5$ kJ mol^{-1}, $\Delta S° = -98.7$ J K^{-1} mole^{-1}.

Meot-Ner & Speller [A33] have very recently obtained high pressure mass spectra concerning successive hydration of $H_3O^+(H_2O)_n$, and confirm a distinct filling of the first hydration shell at $n = 3$. Similarly a study by Christiansen, Tsong & Lin [A58] of the mass distribution of hydrated proton species, desorbed from ice following ion bombardment, also showed a higher yield of $H_3O^+(H_2O)_3$ than other species (all relative to the number of H_3O^+ ions). They did not observe an enhanced yield of $(H_2O)_{21}H^+$. Sharma & Kebarle [A59] have studied 18-crown-6, 15-crown-5 and 12-crown-4 complexes of H_3O^+ using mass spectrometry.

In Part One we did not cover the area of gas phase reactions involving the hydrated proton with molecules other than water, but it has become clear that this is a subject attracting considerable attention, especially in terms of *ab initio* calculations (discussed earlier). We will list here a number of experimental studies of such reactions to give the interested reader an introduction, which is by no means exhaustive. The following include reactions of H_3O^+ or higher hydrates with hydrogen sulphide, [A34] ammonia, [A35, A36] deuterium, [A37] carbon dioxide, [A38, A39] sulphur dioxide, [A39] alcohols, [A40–2] ketones and aldehydes, [A42, A43] crown ethers, [A44] and numerous other organic molecules. [A42, A45–9]

Ion–cyclotron resonance has been used to study the fractionation of

Table A1.1. *Studies of the hydrated proton in solids*

	Claimed species	X-ray structure	Neutron structure	Vibrational spectroscopy	NMR	Other
$H_3PW_{12}O_{40} \cdot nH_2O$	H_3O^+				[A1]	[A1]
$HFe_3(HPO_4)_2(H_2PO_4)_6 \cdot 5H_2O$	H_3O^+	[A2]				
WO_3 amorphous films	H_3O^+				[A3]	[A3]
$Zr(HASO_4)(HPO_4) \cdot H_2O$	H_3O^+(?)					[A4]
$HUO_2PO_4 \cdot 4H_2O$(HUP)	$H_5O_2^+$			[A5]		
$H_2Sb_4O_{11} \cdot 3H_2O$	H_3O^+			[A6]		
Hydrogen β-alumina	H_3O^+			[A5]		[A7]
Hydrogen β''-alumina	H_3O^+					[A7]
Zeolites	H_3O^+					[A8]
$HClO_4 \cdot H_2O$	H_3O^+				[A9]	
$H_2SO_4 \cdot 6\frac{1}{2}H_2O$	$H_7O_3^+/H_5O_2^+$	[A60]				
$H_2SO_4 \cdot 8H_2O$	$H_5O_2^+$	[A60]				
$HFeCl_4 \cdot 4H_2O$	$H_9O_4^+$	[A61, A62]				
$HFeCl_4 \cdot 6H_2O$	H_{aq}^+ in water layer	[A62]				
$H_2PtCl_6 \cdot 2H_2O$	H_3O^+				[A9]	
$(HClO_4)_2(NO_2ClO_4)_9 \cdot 2H_2O$	H_3O^+				[A9]	
$HCl_4NO_2ClO_4 \cdot H_2O$	H_3O^+	[A10]			[A9]	
$HSbCl_6 \cdot 2H_2O$	$H_5O_2^+$			[A11]		
$HSbCl_6 \cdot 4\frac{1}{2}H_2O$	$H_{14}O_6^{2+}$	[A12]				
$HSbBr_6 \cdot 3H_2O$	$H_{14}O_6^{2+}$		[A13]			

$HSbF_6 \cdot H_2O$	H_3O^+	[A14, A15]	
$HAsF_6 \cdot H_2O$	H_3O^+	[A14, A15]	
$H_2SiF_6 \cdot 4H_2O$	$H_5O_2^+$	[A56]	
$H_2SiF_6 \cdot 6H_2O$	$H_5O_2^+$	[A56]	
$H_2SiF_6 \cdot 9\frac{1}{2}H_2O$	$H_5O_2^+/H_7O_3^+$	[A56]	
Clathrate hydrate structures:			
$HPF_6 \cdot 7\frac{2}{3}H_2O$	(H_3O^+)	[A15, A16]	
$HBF_4 \cdot 5\frac{3}{4}H_2O$	(H_3O^+)	[A15, A16]	
$HClO_4 \cdot 5\frac{1}{2}H_2O$	(H_3O^+)	[A15, A16]	
$HPF_6 \cdot 5H_2OHF$	(H_3O^+)	[A15, A17]	
$HAsF_6 \cdot 5H_2OHF$	(H_3O^+)	[A15, A17]	
$HSbF_6 \cdot 5H_2OHF$	(H_3O^+)	[A15, A17]	
$HAsF_66H_2O$	(H_3O^+)	[A15]	
$HSbF_6 \cdot H_2O$	(H_3O^+)	[A15]	
18-crown-6-acid complexes:			
$HBF_4$18-C-6	H_3O^+		[A18]
HF18-C-6	H_3O^+		[A18]
$HClO_4$18-C-6	H_3O^+		[A18]
$[C_6(NO_2)_3Cl_2OH \cdot H_2O]_2$18-C-6 (dichloropicric acid)	H_3O^+	[A19]	
$H_2Mo_6O_{19} \cdot 2H_2O(18\text{-C-}6)_2$	H_3O^+	[A20]	

(?) indicates possible H_3O^+ not suggested by original authors.
(H_3O^+), these parentheses show that the hydrated proton forms part of a cage structure.

deuterium among water and oxonium ions in the gas phase. [A50] A fractionation factor of 0.79 was obtained for

$$\tfrac{1}{3}H_3O^+ (g) + \tfrac{1}{2}D_2O(l) \rightleftarrows \tfrac{1}{3}D_3O^+(g) + \tfrac{1}{2}H_2O(l)$$

Other experiments on isotope exchange reactions are reported in [A51, A52].

A new compilation of thermochemical data on gas phase clustering and hydration reactions by Keesee & Castleman Jr. [A53] includes much information on $\Delta H°$, $\Delta G°$ and $\Delta S°$ for hydrated proton hydration reactions.

Very recently Okumura & Lee [A54] have obtained IR spectra of ions $H_3O^+(H_2O)_n$, by detecting their vibrational predissociation. Their strategy is to attach a weakly bound 'messenger' to the hydrate cluster, such that the vibrations of the hydrate itself are only slightly perturbed. In this case the hydrogen molecule was attached. When the O–H stretch of the hydrate is excited the messenger atom is detached by vibrational predissociation. The messenger thus acts as a mass label, allowing vibrational predissociation to be used to probe the low resolution IR spectra of a large variety of previously unobserved ions. 'Like a spy its (the hydrogen molecule) role is to gather information as unobtrusively as possible.'

The process ($n = 0$, 1, 2, and 3)

$$H_3O^+(H_2O)_n(H_2) \rightarrow H_3O^+(H_2O)_n + H_2$$

was studied by searching for spectra over the frequency range 2300–4200 cm^{-1}. Spectra of the clusters $H_3O^+(H_2)_m$, with $m = 1$, 2 and 3, over the frequency range from 3100 to 4200 cm^{-1}, from their vibrational predissociation

$$H_3O^+(H_2)_m \rightarrow H_3O^+(H_2)_{(m-m')} + m'H_2$$

were also measured. Fundamental bands of the O–H and H–H stretches, as well as several weak combination and overtone bands, were observed for hydronium ion solvated by hydrogen molecules. For $H_3O^+H_2$, the bond of the hydrogen molecule is roughly perpendicular to the hydrogen bond, $H_2OH^+\cdots H_2$. The H–H stretch (4046 cm^{-1}) is red-shifted from the H_2 monomer value of 4161 cm^{-1}, the free O–H stretches occur near the stretching frequencies of H_3O^+, at 3480 cm^{-1} and $\approx$3560 cm^{-1}, and the hydrogen-bonded $OH\cdots H_2$ intramolecular stretch is red-shifted, intensified and broadened at 3110 cm^{-1}. The H–H stretch shows clear resolution of the P, Q, and R branches and partially resolved rotational lines in the P and R branches. This is the first report of rotational structure in the spectra of weakly bound cluster ions. On addition of two hydrogen molecules the free O–H stretch occurred at 3550 cm^{-1}; on addition of three hydrogen molecules no strong peaks were observed in this region, indicating that all three O–H bonds of H_3O^+ are involved in hydrogen bonding. The hydrogen-bonded O–H stretches lie at 3200 cm^{-1} and 3290 cm^{-1} for this case. The former is tentatively assigned to the A_1 symmetric stretch and the latter to the E type, antisymmetric stretch.

For the higher hydrates $H_3O^+(H_2O)_nH_2$, hydrogen is expected to bind less strongly than water (12 kJ mol^{-1} compared with 75 kJ mol^{-1}), and to locate on a given cluster at the same location as the next additional H_2O would be expected to bind. Thus for $H_5O_2^+$, the hydrogen molecule appears to distort the structure to the asymmetric $H_3O^+ \cdot (H_2O)$. Since the symmetric structure (figure 1.2) is more stable by only about 0.8 kJ mol^{-1}, this is not surprising. The free O–H stretches occur at 3662 cm^{-1} and 3528 cm^{-1}. Bands at 3617 cm^{-1} and 3693 cm^{-1} are assigned to red-shifted water symmetric and antisymmetric stretches. (See figure 1.23.) The spectrum of the symmetrical $H_5O_2^+$ has been recently observed in the same laboratory, [A55] and is strikingly different. Two major bands occur, at 3610 cm^{-1} and 3697 cm^{-1}.

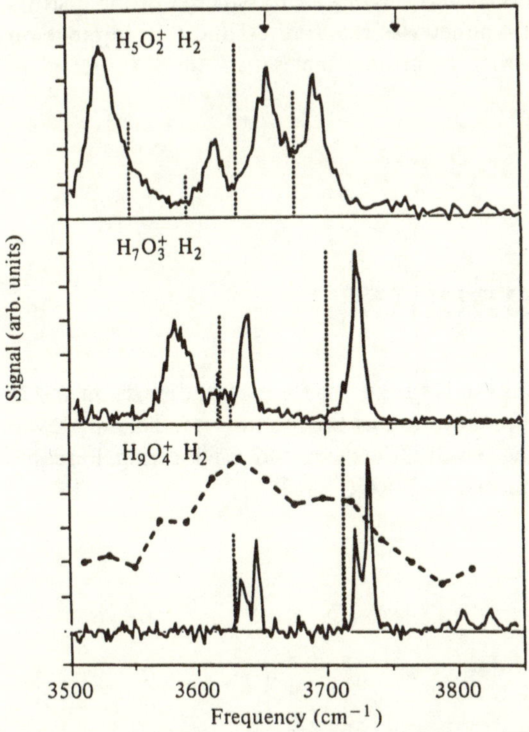

Fig. 1.23. Infrared spectra of $H_5O_2^+H_2$, $H_7O_3^+H_2$, and $H_9O_4^+(H_2)$. The laser linewidth was 1.2 cm^{-1}. The arrows at the top indicate the band origins of transitions to the symmetric (3657 cm^{-1}) and antisymmetric (3756 cm^{-1}) vibrations for the H_2O monomer. The dashed stick spectra are *ab initio* predictions of the vibrational frequencies and intensities of the respective $H_3O^+(H_2O)_n$ clusters, from the unpublished work of Remington & Schaefer at the DZ + P SCF level. The theoretical calculations for the $H_5O_2^+$ ion are at the asymmetric C_s geometry, the minimum energy configuration predicted at this level. The dashed line on the lower frame is the spectrum of Schwarz [148] with a resolution of 40 cm^{-1}.

For $H_3O^+(H_2O)_2$ two bands of H_2O occur at 3644 and 3725 cm^{-1}; notice that as the hydration number increases, the perturbation of the H_2O groups decreases. A band at 3589 cm^{-1} is assigned to the O–H stretch of H_3O^+; the hydrogen messenger will bind to this OH group, lowering its frequency from a calculated value of 3640 cm^{-1} for a free OH on $H_7O_3^+$.

For $H_9O_4^+(H_2)$ two strong doublet bands at 3648 cm^{-1} and 3733 cm^{-1} are assigned to the symmetric and antisymmetric stretches of the solvent H_2O groups. The doublet structure is attributed to the hydrogen messenger. A broad band at 2670 cm^{-1}, attributed to the donor OH, supports Schwarz's assignment. [148] This band is intrinsic to the $H_9O_4^+$ cluster.

These studies have significantly advanced the understanding of the small, gas phase clusters. Larger clusters, with other probes and without probes, are under investigation. As cluster size increases phenomena such as intracluster proton transfer and cyclic hydrogen bonds may be elucidated. Rotational structure of $H_5O_2^+$ will be studied and the degree to which a probe distorts $H_5O_2^+$ will give insight to the symmetry of this species, and thus impinge on the controversy discussed earlier regarding the nature of this species in aqueous solutions.

Acknowledgments

The authors express their thanks to Dr John Tse for helpful discussion and to Mrs Lisa Dowsett for careful preparation of the manuscript. This work was supported by a grant from the Natural Sciences and Engineering Research Council of Canada. It is published as NRCC 28411.

References

1. C. I. Ratcliffe & D. E. Irish, in *Water Science Reviews 2* (ed. F. Franks). Cambridge University Press: Cambridge, 1986, p. 149.
2. J. Almlof & U. Wahlgren. *Theoret. Chim. Acta* **28** (1973), 161.
3. M. Fournier, M. Allavena & A. Potier. *Theoret. Chim. Acta* **42** (1976), 145.
4. J. Angyan, M. Allavena, M. Picard, A. Potier & O. Tapia. *J. Chem. Phys.* **77** (1982), 4723.

5. F. Hund. *Zeits. Physik.* **31** (1925), 81.
6. F. Hund. *Zeits. Physik.* **32** (1925), 1.
7. H. Hartmann & G. Gliemann. *Z. Phys. Chem. N.F.* **19** (1959), 29.
8. R. Grahn. *Arkiv Fysik* **19** (1961), 147.
9. R. Gaspar, I. Tamassy-Lentei & Y. Kruglyak. *J. Chem. Phys.* **36** (1962), 740.
10. R. Grahn. *Arkiv Fysik* **21** (1962), 81.
11. R. Grahn, *Arkiv Fysik* **21** (1962), 1.
12. J. W. Moskowitz & M. C. Harrison. *J. Chem. Phys.* **43** (1965), 3550.
13. D. M. Bishop. *J. Chem. Phys.* **43** (1965), 4453.
14. B. D. Joshi. *J. Chem. Phys.* **47** (1967), 2793.
15. A. C. Hopkinson, N. K. Holbrook, K. Yates & I. G. Csizmadia. *J. Chem. Phys.* **49** (1968), 3596.
16. P. A. Kollman & L. C. Allen. *J. Am. Chem. Soc.* **92** (1970), 6101.
17. M. DePaz, S. Ehrenson & L. Friedman. *J. Chem. Phys.* **52** (1970), 3362.
18. M. D. Newton & S. Ehrenson. *J. Am. Chem. Soc.* **93** (1971), 4971.
19. W. A. Lathan, W. J. Hehre, L. A. Curtiss & J. A. Pople. *J. Am. Chem. Soc.* **93** (1971), 6377.
20. G. H. F. Diercksen & W. P. Kraemer. *Theor. Chim. Acta* **23** (1972), 387.
21. P. A. Kollman & C. F. Bender. *Chem. Phys. Lett.* **21** (1973), 271.
22. H. Lischka & V. Dyczmons. *Chem. Phys. Lett.* **23** (1973), 167.
23. H. Lischka. *Theor. Chim. Acta* **31** (1973), 39.
24. G. Alagona, R. Cimiraglia & U. Lamanna. *Theor. Chim. Acta* **29** (1973), 93.
25. M. Allavena & E. LeClec'h. *J. Mol. Struct.* **22** (1974), 265.
26. W. A. Lathan, L. A. Curtiss, W. J. Hehre, J. B. Lisle & J. A. Pople. *Prog. Phys. Org. Chem.* **11** (1974), 175.
27. M. Dixon, R. E. Overill & T. A. Claxton. *J. Magn. Reson.* **15** (1974), 477.
28. P. C. Hariharan & J. A. Pople. *Mol. Phys.* **27** (1974), 209.
29. J.-J. Delpuech, G. Serratrice, A. Strich & A. Veillard. *Mol. Phys.* **29** (1975), 849.
30. G. H. F. Diercksen, W. P. Kraemer & B. O. Roos. *Theor. Chim. Acta* **36** (1975), 249.
31. R. Ahlrichs, F. Driessler, H. Lischka & V. Staemmler. *J. Chem. Phys.* **62** (1975), 1235.
32. M. D. Newton. *J. Chem. Phys.* **67** (1977), 5535.
33. R. E. Kari & I. G. Csizmadia. *J. Am. Chem. Soc.* **99** (1977), 4539.
34. R. C. Raffenetti, H. J. T. Preston & J. J. Kaufman. *Chem. Phys. Lett.* **46** (1977), 513.
35. M. Urban, V. Kello & P. Carsky. *Theor. Chim. Acta* **45** (1977), 205.
36. F. H. Stillinger & C. W. David. *J. Chem. Phys.* **69** (1978), 1473.
37. V. Kello, M. Urban, I. Hubac & P. Carsky. *Chem. Phys. Lett.* **58** (1978), 83.
38. V. Kello, M. Urban, P. Carsky & Z. Slanina. *Chem. Phys. Lett.* **53** (1978), 555.
39. G. Alagona, E. Scrocco & J. Tomasi. *Theor. Chim. Acta* **47** (1978), 133.
40. W. I. Ferguson & N. C. Handy. *Chem. Phys. Lett.* **71** (1980), 95.
41. P. J. Desmeules & L. C. Allen. *J. Chem. Phys.* **72** (1980), 4731.
42. M. J. Frisch, J. E. Del Bene, K. Raghavachari & J. A. Pople. *Chem. Phys. Lett.* **83** (1981), 240.
43. K. Ishida, S. Kadowaki & T. Yonezawa. *Bull. Chem. Soc. Japan* **54** (1981), 967.
44. W. R. Rodwell & L. Radom. *J. Am. Chem. Soc.* **103** (1981), 2865.
45. I. J. F. Poplett. *J. Magn. Reson.* **44** (1981), 488.

46. S. Scheiner & L. B. Harding. *Chem. Phys. Lett.* **79** (1981), 39.
47. V. Spirko & P. R. Bunker. *J. Mol. Spec.* **95** (1982), 226.
48. J. J. Kaufman. *Int. J. Quant. Chem.: Quant. Chem. Symposia* **16** (1982), 649.
49. J. E. Del Bene, M. J. Frisch, K. Raghavachari & J. A. Pople. *J. Phys. Chem.* **86** (1982), 1529.
50. S. Scheiner. *Chem. Phys. Lett.* **93** (1982), 540.
51. S. Scheiner. *J. Chem. Phys.* **77** (1982), 4039.
52. H. Fukui, K. Miura & F. Tada. *J. Chem. Phys.* **79** (1983), 6112.
53. P. R. Bunker, W. P. Kraemer & V. Spirko. *J. Mol. Spec.* **101** (1983), 180.
54. H. Chojnacki & Z. Laskowski. *Bull. Pol. Acad. Sci. Chem.* **31** (1983), 67.
55. P. Botschwina, P. Rosmus & E.-A. Reinsch. *Chem. Phys. Lett.* **102** (1983), 299.
56. J. E. Del Bene, H. D. Mettee, M. J. Frisch, B. T. Luke & J. A. Pople. *J. Phys. Chem.* **87** (1983), 3279.
57. M. L. Hendewerk, R. Frey & D. A. Dixon. *J. Phys. Chem.* **87** (1983), 2026.
58. W. A. Sokalski & R. A. Poirier. *Chem. Phys. Lett.* **98** (1983), 86.
59. H. Z. Cao, M. Allavena, O. Tapia & E. M. Evleth. *Chem. Phys. Lett.* **96** (1983), 458.
60. P. S. Martin, K. Yates & I. G. Csizmadia. *Theor. Chim. Acta* **64** (1983), 117.
61. T. R. Knowles. *J. Electroanal. Chem.* **150** (1983), 365.
62. A. M. Kuznetsov. *J. Electroanal. Chem.* **159** (1983), 241.
63. V. Galasso. *J. Mol. Struct. (Theochem)* **93** (1983), 201.
64. S. Tomoda & K. Kimura. *Chem. Phys.* **82** (1983), 215.
65. M. E. Colvin, G. P. Raine, H. F. Schaefer III & M. Dupuis. *J. Chem. Phys.* **79** (1983), 1551.
66. S. Scheiner & L. B. Harding. *J. Phys. Chem.* **87** (1983), 1145.
67. K. Sato, S. Tomoda & K. Kimura. *Chem. Phys. Lett.* **95** (1983), 579.
68. N. Shida, K. Tanaka & K. Ohno. *Chem. Phys. Lett.* **104** (1984), 575.
69. T.-K. Ha & M. T. Nguyen. *J. Phys. Chem.* **88** (1984), 4295.
70. C. M. Cook, K. Haydock, R. H. Lee & L. C. Allen. *J. Phys. Chem.* **88** (1984), 4875.
71. C. Reynolds & C. Thomson. *Int. J. Quant. Chem.: Quant. Biol. Symposia* **11** (1984), 167.
72. Z. Latajka & S. Scheiner. *Chem. Phys. Lett.* **105** (1984), 435.
73. J. Donnella & J. R. Murdoch. *J. Am. Chem. Soc.* **106** (1984), 4724.
74. V. Kello, M. Urban, J. Noga & G. H. F. Diercksen. *J. Am. Chem. Soc.* **106** (1984), 5864.
75. E. Magnusson. *J. Am. Chem. Soc.* **106** (1984), 1177.
76. E. Magnusson. *J. Am. Chem. Soc.* **106** (1984), 1185.
77. L. Pataki, A. Mady, R. D. Venter, R. A. Poirier, M. R. Peterson & I. G. Csizmadia. *Chem. Phys. Lett.* **109** (1984), 198.
78. L. Pataki, A. Mady, R. D. Venter, R. A. Poirier & I. G. Csizmadia. *J. Mol. Struct.* **110** (1984), 229.
79. M. A. Lopez Bote & S. Montero. *J. Raman Spec.* **15** (1984), 4.
80. W. H. Jones, R. D. Mariani & M. L. Lively. *Chem. Phys. Lett.* **108** (1984), 602.
81. J. Fritsch, G. Zundel, A. Hayd & M. Maurer. *Chem. Phys. Lett.* **107** (1984), 65.
82. C. N. R. Rao, G. V. Kulkarni, A. Muralikrishna Rao & U. Chandra Singh. *J. Mol. Struct.* **108** (1984), 113.

83. H. Z. Cao, M. Allavena, O. Tapia & E. M. Evleth. *J. Phys. Chem.* **89** (1985), 1581.
84. C. I. Ratcliffe, J. A. Ripmeester & J. S. Tse. *Chem. Phys. Lett.* **120** (1985), 427.
85. G. Fronzoni & V. Galasso. *J. Mol. Struct.* (*Theochem*) **122** (1985), 327.
86. D. J. DeFrees & A. D. McLean. *J. Chem. Phys.* **82** (1985), 333.
87. H. Fukui, K. Miura, H. Yamazaki & T. Nosaka. *J. Chem. Phys.* **82** (1985), 1410.
88. F. Bernardi, A. Bottoni, M. Olivucci & G. Tonachini. *J. Mol. Struct.* (*Theochem*) **133** (1985), 243.
89. G. A. Olah, G. K. S. Prakash, M. Barzaghi, K. Lammertsma, P. von R. Schleyer & J. A. Pople. *J. Am. Chem. Soc.* **108** (1986), 1032.
90. P. Botschwina. *J. Chem. Phys.* **84** (1986), 6523.
91. W. G. Richards, P. R. Scott, V. Sackwild & S. A. Robins, in *Oxford Science Research Papers*, 'A Bibliography of Ab Initio Molecular Wave Functions', Supplement for 1978–80, Clarendon, Oxford, 1981, and previous supplements.
92. K. Ohno & K. Morokuma. 'Quantum Chemistry Literature Database. Bibliography of *Ab Initio* Calculations, 1978–1980', *Physical Sciences Data 12*, Elsevier Scientific, NY, 1982.
93. K. Ohno & K. Morokuma, Annual supplements to 'Bibliography of Ab Initio Calculations' appear in *Theochem* (*J. Mol. Struct.*) **8** (1982), **15** (1983), **20** (1984), **27** (1985), **33** (1986).
94. P. R. Bunker, T. Amano & V. Spirko. *J. Mol. Spec.* **107** (1984), 208.
95. T. J. Sears, P. R. Bunker, P. B. Davies, S. A. Johnson & V. Spirko. *J. Chem. Phys.* **83** (1985), 2676.
96. M. H. Begemann & R. J. Saykally. *J. Chem. Phys.* **82** (1985), 3570.
97. M. H. Begemann, C. S. Gudeman, J. Pfaff & R. J. Saykally. *Phys. Rev. Lett.* **51** (1983), 554.
98. T. B. McMahon & P. Kebarle. *J. Am. Chem. Soc.* **107** (1985), 2612.
99. D. K. Bohme & G. I. Mackay. *J. Am. Chem. Soc.* **103** (1981), 2173.
100. S. M. Collyer & T. B. McMahon. *J. Phys. Chem.* **87** (1983), 909.
101. S. G. Lias, J. F. Liebman & R. D. Levin. *J. Phys. Chem. Ref. Data* **13** (1984), 695.
102. J. F. Wolf, R. H. Staley, I. Koppel, M. Taagepera, R. T. McIver, Jr., J. L. Beauchamp & R. W. Taft. *J. Am. Chem. Soc.* **99** (1977), 5417.
103. G. D. Mateescu & G. M. Benedikt. *J. Am. Chem. Soc.* **101** (1979), 3959.
104. A. F. Beecham, A. C. Hurley, M. F. Mackay, V. W. Maslen & A. McL. Mathieson. *J. Chem. Phys.* **49** (1968), 3312.
105. M. C. Flanigan & J. R. de la Vega. *Chem. Phys. Lett.* **21** (1973), 521.
106. M. A. Muniz, J. Bertran, J. L. Andres, M. Duran & A. Lledos. *J. Chem. Soc. Faraday Trans. 1* **81** (1985), 1547.
107. W. P. Kraemer & G. H. F. Diercksen. *Chem. Phys. Lett.* **5** (1970), 463.
108. R. Janoschek, E. G. Weidemann, H. Pfeiffer & G. Zundel. *J. Am. Chem. Soc.* **94** (1972), 2387.
109. W. Meyer, W. Jakubetz & P. Schuster. *Chem. Phys. Lett.* **21** (1973), 97.
110. R. Janoschek, E. G. Weidemann & G. Zundel. *J. Chem. Soc. Faraday Trans. 2* **69** (1973), 505.
111. R. Janoschek, in *The Hydrogen Bond* (ed. P. Schuster, G. Zundel & C. Sandorfy). North-Holland, NY, 1976, chap. 3.
112. S. Scheiner. *Int. J. Quant. Chem.: Quant. Biol. Symposia* **7** (1980), 199.
113. S. Scheiner. *J. Chem. Phys.* **75** (1981), 5791.

114. S. Scheiner. *J. Am. Chem. Soc.* **103** (1981), 315.
115. C. McMichael Rohlfing, L. C. Allen & R. Ditchfield. *Chem. Phys. Lett.* **86** (1982), 380.
116. S. Scheiner. *J. Phys. Chem.* **86** (1982), 376.
117. S. Scheiner, M. M. Szczesniak & L. D. Bigham. *Int. J. Quant. Chem.* **23** (1983), 739.
118. S. Yamabe, T. Minato & K. Hirao. *J. Chem. Phys.* **80** (1984), 1576.
119. A. Potier, J. M. Leclercq & M. Allavena. *J. Phys. Chem.* **88** (1984), 1125.
120. E. Kochanski. *Nouv. J. Chim.* **8** (1984), 605.
121. E. A. Hillenbrand & S. Scheiner. *J. Am. Chem. Soc.* **106** (1984), 6266.
122. J. E. Del Bene, M. J. Frisch & J. A. Pople. *J. Phys. Chem.* **89** (1985), 3669.
123. M. M. Szczesniak & S. Scheiner. *J. Phys. Chem.* **89** (1985), 1835.
124. S. Scheiner & L. D. Bigham. *J. Chem. Phys.* **82** (1985), 3316.
125. S. Scheiner, P. Redfern & M. M. Szczesniak. *J. Phys. Chem.* **89** (1985), 262.
126. M. J. Frisch, J. E. Del Bene, J. S. Binkley & H. F. Schaefer III. *J. Chem. Phys.* **84** (1986), 2279.
127. G. R. J. Williams. *J. Mol. Struct.* (*Theochem*) **138** (1986), 333.
128. W. Joswig, H. Fuess & G. Ferraris. *Acta Cryst.* **B38** (1982), 2798.
129. M. Ishikawa. *Acta Cryst.* **B34** (1978), 2074.
130. H. Savage, in *Water Science Reviews 2* (ed. F. Franks). Cambridge University Press: Cambridge, 1986, p. 67.
131. A. J. Cunningham, J. D. Payzant & P. Kebarle. *J. Am. Chem. Soc.* **94** (1972), 7627.
132. M. Meot-Ner & F. H. Field. *J. Am. Chem. Soc.* **99** (1977), 998.
133. M. C. Flanigan & J. R. de la Vega. *J. Chem. Phys.* **61** (1974), 1882.
134. J. H. Busch & J. R. de la Vega. *J. Am. Chem. Soc.* **99** (1977), 2397.
135. G. V. Yukhnevich, E. G. Kokhanova, A. I. Pavlyuchko & V. V. Volkov. *J. Mol. Struct.* **122** (1985), 1.
136. L. A. Curtiss & J. A. Pople. *J. Mol. Spec.* **55** (1975), 1.
137. R. Grahn. *Arkiv Fysik* **21** (1962), 13.
138. N. Salaj. *Acta Chem. Scand.* **23** (1969), 1534.
139. S. H. Suck, J. L. Kassner Jr., R. E. Thurman, P. C. Yue & R. A. Anderson. *J. Atmos. Sci.* **38** (1981), 1272.
140. E. E. Ferguson & F. C. Fehsenfeld. *J. Geophys. Res., Space Res.* **74** (1969), 5743.
141. E. Kochanski. *J. Am. Chem. Soc.* **107** (1985), 7869.
142. S. L. Fornili, M. Migliore & M. A. Palazzo. *Chem. Phys. Lett.* **125** (1986), 419.
143. P. M. Holland & A. W. Castleman, Jr. *J. Chem. Phys.* **72** (1980), 5984.
144. U. Nagashima, H. Shinohara, N. Nishi & H. Tanaka. *J. Chem. Phys.* **84** (1986), 209.
145. I. P. Buffey & W. B. Brown. *Chem. Phys. Lett.* **109** (1984), 59.
146. J. K. Fang, K. Godzik & G. L. Hofacker. *Ber. Bunsenges. Phys. Chem.* **77** (1973), 980.
147. R. Triolo & A. H. Narten. *J. Chem. Phys.* **63** (1975), 3624.
148. H. A. Schwarz. *J. Chem. Phys.* **67** (1977), 5525.
149. O. Matsuoka, E. Clementi & M. Yoshimine. *J. Chem. Phys.* **64** (1976), 1351.
150. V. Hermann, B. D. Kay & A. W. Castleman, Jr. *Chem. Phys.* **72** (1982), 185.
151. G. A. Jeffrey, in *Inclusion Compounds* (ed. J. L. Atwood, J. E. D. Davies & D. D. MacNicol). Academic Press: London, 1984, Vol. 1, chap. 5.

152. E. G. Weidemann & G. Zundel. *Zeits. Physik* **198** (1967), 288.
153. E. G. Weidemann & G. Zundel. *Zeits. Naturforsch.* **25a** (1970), 627.
154. R. Janoschek, A. Hayd, E. G. Weidemann, M. Leuchs & G. Zundel. *J. Chem. Soc. Faraday Trans. 2* **74** (1978), 1238.
155. A. Hayd, E. G. Weidemann & G. Zundel. *J. Chem. Phys.* **70** (1979), 86.
156. G. Zundel. *Hydration and Intermolecular Interaction*. Academic Press: NY, 1969.
157. G. Zundel, in *The Hydrogen Bond* (ed. P. Schuster, G. Zundel & C. Sandorfy). North-Holland: NY, 1976, Vol. 2, chap. 15.
158. N. B. Librovich, V. P. Sakun & N. D. Sokolov. *Chem. Phys.* **39** (1979), 351.
159. N. B. Librovich, V. P. Sakun & N. D. Sokolov. *Teor. Eksp. Khim.* **14** (1978), 435.
160. G. Zundel & E. G. Weidemann. *Chem. Phys.* **44** (1979), 427.
161. N. B. Librovich, V. P. Sakun & N. D. Sokolov. *Chem. Phys.* **44** (1979), 429.
162. H. Romanowski & L. Sobczyk. *Chem. Phys.* **19** (1977), 361.
163. A. Laforgue, C. Brucena-Grimbert, D. Laforgue-Kantzer, G. Del Re & V. Barone. *J. Phys. Chem.* **86** (1982), 4436.
164. P. A. Giguère. *J. Chem. Educ.* **56** (1979), 571.
165. Y. K. Lau, S. Ikuta & P. Kebarle. *J. Am. Chem. Soc.* **104** (1982), 1462.
166. S. A. Pope, I. H. Hillier & M. F. Guest. *Faraday Symp. Chem. Soc.* **19** (1984), 109.
167. S. Scheiner & E. A. Hillenbrand. *J. Phys. Chem.* **89** (1985), 3053.
168. S. Scheiner. *J. Chem. Phys.* **80** (1984), 1982.
169. T. A. Weber & F. H. Stillinger. *J. Phys. Chem.* **86** (1982), 1314.
170. L. S. Bagster & G. Cooling. *J. Chem. Soc.* **117** (1920), 693.
171. L. S. Bagster & B. D. Steele. *Trans. Faraday Soc.* **8** (1912), 51.
172. N. N. Greenwood & R. L. Martin. *J. Chem. Soc.* (1951) 1915.
173. N. N. Greenwood & R. L. Martin. *J. Chem. Soc.* (1953), 1427.
174. N. N. Greenwood & R. L. Martin. *Quart. Revs.* **8** (1954), 1.
175. M. Schneider & P.-A. Giguère. *Compt. Rendu. Acad. Sci. Paris* **B267** (1968), 551.
176. P. A. Giguère & C. Madec. *Chem. Phys. Lett.* **37** (1976), 569.
177. P. V. Huong & B. Desbat. *J. Raman Spec.* **2** (1974), 373.
178. B. Desbat & P. V. Huong. *Spectrochim. Acta* **31A** (1975), 1109.
179. G. S. Denisov & N. S. Golubev, referenced by G. Zundel and J. Fritsch in 'The Chemical Physics of Solvation', Part B (ed. R. R. Dogonadze, E. Kalman, A. A. Kornyshev & J. Ulstrup). *Studies in Physical and Theoretical Chemistry*. Elsevier: New York, 1986, Vol. 38, chap. 2.
180. W. H. Baldwin, C. E. Higgins & B. A. Soldano. *J. Phys. Chem.* **63** (1959), 118.
181. D. G. Tuck & R. M. Diamond. *J. Phys. Chem.* **65** (1961), 193.
182. D. C. Whitney & R. M. Diamond. *J. Phys. Chem.* **67** (1963), 209.
183. Y. Marcus. *Chem. Revs.* **63** (1963), 139.
184. J. Axelrod & E. H. Swift. *J. Am. Chem. Soc.* **62** (1940), 33.
185. R. J. Myers, D. E. Metzler & E. H. Swift. *J. Am. Chem. Soc.* **72** (1950), 3767.
186. H. L. Friedman. *J. Am. Chem. Soc.* **74** (1952), 5.
187. A. H. Laurene, D. E. Campbell, S. E. Wiberley & H. M. Clark. *J. Phys. Chem.* **60** (1956), 901.
188. I. M. Kolthoff & M. K. Chantooni Jr. *J. Amer. Chem. Soc.* **90** (1968), 3320.
189. M. K. Chantooni Jr. & I. M. Kolthoff. *J. Amer. Chem. Soc.* **92** (1970), 2236.
190. V. Skarda, J. Rais & M. Kyrs. *J. Inorg. Nucl. Chem.* **41** (1979), 1443.
191. J. Talarmin, M. L'Her, A. Laouenan & J. Courtot-Coupez. *J. Electroanal. Chem.* **103** (1979), 203.

192. J. Talarmin, M. L'Her, A. Laouenan & J. Courtot-Coupez. *J. Electroanal. Chem.* **106** (1980), 347.
193. J. L. Kurz, M. T. Myers & K. M. Ratcliff. *J. Am. Chem. Soc.* **106** (1984), 5631.
194. C. MacLean & E. L. Mackor. *J. Chem. Phys.* **34** (1961), 2207.
195. C. MacLean & E. L. Mackor. *Disc. Faraday Soc.* **34** (1962), 165.
196. A. Commeyras & G. A. Olah. *J. Amer. Chem. Soc.* **91** (1969), 2929.
197. V. Gold, J. L. Grant & K. P. Morris. *J. Chem. Soc. Chem. Comm.* (1976), 397.
198. J. Emsley, V. Gold & M. J. B. Jais. *J. Chem. Soc. Chem. Comm.* (1979), 961.
199. P. Rimmelin, S. Schwartz & J. Sommer. *Org. Magn. Reson.* **16** (1981), 160.
200. N. S. Golubev. *Khim. Fiz.* **1** (1983) 42; English version, *Sov. J. Chem. Phys.* **2** (1985), 63.
201. G. D. Brunton & C. K. Johnson. *J. Chem. Phys.* **62** (1975), 3797.
202. N. S. Golubev. *Khim. Fiz.* **3** (1984), 772; English version, *Sov. J. Chem. Phys.* **3** (1985), 1182.
203. G. D. Mateescu, G. M. Benedikt & M. P. Kelly. *Synth. Appl. Isot. Labeled Compnd., Proc. Int. Symp. 1982* (ed. W. O. Duncan & A. B. Susan). Elsevier: Amsterdam, 1983, p. 483.
204. M. C. R. Symons. *J. Am. Chem. Soc.* **102** (1980), 3982.
205. G. A. Olah, A. L. Berrier & G. K. S. Prakash. *J. Am. Chem. Soc.* **104** (1982), 2373.
206. A. H. Narten & H. A. Levy, in *Water: a Comprehensive Treatise* (ed. F. Franks). Plenum Press: New York, 1972, Vol. 1, p. 311.
207. J. G. Powles. *Adv. Phys.* **22** (1973), 1.
208. G. Licheri, G. Piccaluga & G. Pinna. *Chem. Phys. Lett.* **12** (1971), 425.
209. D. L. Wertz. *J. Solution Chem.* **1** (1972), 489.
210. D. S. Terekhova. *J. Struct. Chem.* **11** (1970), 483.
211. S. C. Lee & R. Kaplow. *Science* **169** (1970), 477.
212. J.-O. Lundgren & I. Olovsson. *Acta Cryst.* **23** (1967), 966.
213. J.-O. Lundgren & I. Olovsson. *Acta Cryst.* **23** (1967), 971.
214. I. Taesler & J.-O. Lundgren. *Acta Cryst.* **B34** (1978), 2424.
215. N. Ohtomo, K. Arakawa, M. Takeuchi, T. Yamaguchi & H. Ohtaki. *Bull. Chem. Soc. Japan* **54** (1981), 1314.
216. N. Ohtomo & K. Arakawa. *Bull. Chem. Soc. Japan* **51** (1978), 1649.
217. H.-G. Lee, Y. Matsumoto, T. Yamaguchi & H. Ohtaki. *Bull. Chem. Soc. Japan* **56** (1983), 443.
218. H. S. Gutowsky & A. Saika. *J. Chem. Phys.* **21** (1953), 1688.
219. G. C. Hood, O. Redlich & C. A. Reilly. *J. Chem. Phys.* **22** (1954), 2067.
220. M. G. Morin, G. Paulett & M. E. Hobbs. *J. Phys. Chem.* **60** (1956), 1594.
221. G. C. Hood & C. A. Reilly. *J. Chem. Phys.* **27** (1957), 1126.
222. O. Redlich & G. C. Hood. *Disc. Faraday Soc.* **24** (1957), 87.
223. G. C. Hood, A. C. Jones & C. A. Reilly. *J. Phys. Chem.* **63** (1959), 101.
224. R. J. Gillespie & R. F. M. White. *Can. J. Chem.* **38** (1960), 1371.
225. G. C. Hood & C. A. Reilly. *J. Chem. Phys.* **32** (1960), 127.
226. L. Kotin & M. Nagasawa. *J. Am. Chem. Soc.* **83** (1961), 1026.
227. J. C. Hindman. *J. Chem. Phys.* **36** (1962), 1000.
228. R. H. Dinius & G. R. Choppin. *J. Phys. Chem.* **66** (1962), 268.
229. J. W. Akitt, A. K. Covington, J. G. Freeman & T. H. Lilley. *J. Chem. Soc. Chem. Comm.* (1965), 349.

230. (*a*) A. K. Covington & T. H. Lilley. *Trans. Faraday Soc.* **63** (1967), 1749; (*b*) A. K. Covington, J. G. Freeman & T. H. Lilley. *Trans. Faraday Soc.* **65** (1969), 3136; (*c*) A. K. Covington & R. Thompson. *J. Solution Chem.* **3** (1974), 603.

231. A. Merbach. *J. Chem. Phys.* **46** (1967), 3450.

232. R. W. Duerst. *J. Chem. Phys.* **48** (1968), 2275.

233. O. Redlich, R. W. Duerst & A. Merbach. *J. Chem. Phys.* **49** (1968), 2986.

234. P. S. Knapp, R. O. Waite & E. R. Malinowski. *J. Chem. Phys.* **49** (1968), 5459.

235. J. W. Akitt, A. K. Covington, J. G. Freeman & T. H. Lilley. *Trans. Faraday Soc.* **65** (1969), 2701.

236. R. Radeglia & A. Weber. *Zeits. Chem.* **11** (1971), 236.

237. F. J. Vogrin, P. S. Knapp, W. L. Flint, A. Anton, G. Highberger & E. R. Malinowski. *J. Chem. Phys.* **54** (1971), 178.

238. J. W. Akitt. *J. Chem. Soc. Dalton Trans.* (1973), 49.

239. C. S. Baker & H. A. Strobel. *J. Phys. Chem.* **83** (1979), 728.

240. N. Soffer, Y. Marcus & J. Shamir. *J. Chem. Soc. Faraday Trans. 1* **76** (1980), 2347.

241. N. N. Shapte'ko, Yu. S. Bogachev, V. G. Khutsishvili & N. P. Bulatova. *Zh. Obshch. Khom.* **52** (1982), 740.

242. P. Mirti & V. Zalano. *Nouv. J. Chim.* **7** (1983), 381.

243. P. Mirti & V. Zelano. *J. Chem. Soc. Faraday Trans. 1* **81** (1985), 2365.

244. Y. Masuda & T. Kanda. *J. Phys. Soc. Japan* **8** (1953), 432.

245. V. Gold. *Proc. Chem. Soc. (London)* (1963), 141.

246. A. J. Kresge & A. L. Allred. *J. Am. Chem. Soc.* **85** (1963), 1541.

247. S. Meiboom, Z. Luz & D. Gill. *J. Chem. Phys.* **27** (1957), 1411.

248. S. Meiboom. *J. Chem. Phys.* **34** (1961), 375.

249. A. Loewenstein & A. Szoke. *J. Am. Chem. Soc.* **84** (1962), 1151.

250. Z. Luz & S. Meiboom. *J. Am. Chem. Soc.* **86** (1964), 4768.

251. R. E. Glick & K. C. Tewari. *J. Chem. Phys.* **44** (1966), 546.

252. S. W. Rabideau & H. G. Hecht. *J. Chem. Phys.* **47** (1967), 544.

253. H. G. Hertz & R. Klute. *Z. Phys. Chem. (NF)* **69** (1970), 101.

254. R. R. Knispel & M. M. Pintar. *Chem. Phys. Lett.* **32** (1975), 238.

255. V. Graf, F. Noack & G. J. Bene. *J. Chem. Phys.* **72** (1980), 861.

256. W. J. Lamb, D. R. Brown & J. Jonas. *J. Phys. Chem.* **85** (1981), 3883.

257. B. Halle & G. Karlstrom. *J. Chem. Soc. Faraday Trans. 2* **79** (1983), 1031.

258. B. Halle & G. Karlstrom. *J. Chem. Soc. Faraday Trans. 2* **79** (1983), 1047.

259. D. E. Bethell & N. Sheppard. *J. Chim. Phys.* **50** (1953), C72, C118.

260. E. Bauer. *J. Chim. Phys.* **45** (1948), 242.

261. E. Darmois. *J. Phys. Radium* **11** (1950), 577.

262. R. Suhrmann & I. Wiederisch. *Z. Elektrochem.* **57** (1953), 93.

263. R. Suhrmann & I. Wiederisch. *Z. Anorg. Chem.* **273** (1953), 166.

264. E. Wicke, M. Eigen & T. Ackermann. *Z. Physik Chem.* **1** (1954), 342.

265. M. Falk & P. A. Giguère. *Can. J. Chem.* **35** (1957), 1195.

266. P. A. Giguère & S. Turrell. *Can. J. Chem.* **54** (1976), 3477.

267. T. Ackermann. *Z. Phys. Chem. Frankfurt* **27** (1961), 253.

268. Reference 156, pp. 169–72.

269. H. D. Downing & D. Williams. *J. Phys. Chem.* **80** (1976), 1640 (See Editor's note).

270. P. A. Giguère. *Can. J. Chem.* **61** (1983), 588.
271. P. Rhine, D. Williams, G. M. Hale & M. R. Querry. *J. Phys. Chem.* **78** (1974), 1405.
272. G. Zundel & H. Metzger. *Z. Physik. Chem.* **58** (1968), 225.
273. G. Zundel & J. Fritsch, in *The Chemical Physics of Solvation, Part B, Spectroscopy of Solvation* (ed. R. R. Dogonadze, E. Kalman, A. A. Kornyshev & J. Ulstrup). Elsevier: Amsterdam, 1986, chap. 2, p. 21. Particularly see pp. 74 ff.
274. M. Leuchs & G. Zundel. *Can. J. Chem.* **58** (1980), 311.
275. C. I. Ratcliffe & D. E. Irish. *Can. J. Chem.* **62** (1984), 1134.
276. C. I. Ratcliffe & D. E. Irish. *Can. J. Chem.* **63** (1985), 3521.
277. M. Leuchs & G. Zundel. *J. Chem. Soc. Faraday Trans. 2* **74** (1978), 2256.
278. G. C. Pimentel & A. L. McClellan. *Ann. Rev. Phys. Chem.* **22** (1971), 357.
279. N. B. Librovich, V. P. Sakun & N. D. Sokolov. *Chem. Phys.* **39** (1979), 351.
280. P. A. Giguère. *Chem. Phys.* **60** (1981), 421.
281. N. B. Librovich, V. P. Sakun & N. D. Sokolov. *Chem. Phys.* **60** (1981), 425.
282. D. E. Irish & M. H. Brooker, in *Advances in Infrared and Raman Spectroscopy* (ed. R. J. H. Clark & R. E. Hester). Heyden: London, 1976, Vol. 2, chap. 6, p. 263.
283. M. H. Brooker, in *The Chemical Physics of Solvation, Part B, Spectroscopy of Solvation* (ed. R. R. Dogonadze, E. Kalman, A. A. Kornyshev & J. Ulstrup). Elsevier: Amsterdam, 1986, chap. 4, p. 119. Particularly see pp. 146 ff.
284. N. G. Zarakhani & M. I. Vinnik. *Zh. Fiz. Khim.* **37** (1963), 503.
285. H. Chen, Ph.D. Thesis, University of Waterloo, Waterloo, Ontario, Canada, 1970.
286. W. R. Busing & D. F. Hornig. *J. Phys. Chem.* **65** (1961), 284.
287. N. G. Zarakhani, V. D. Maiorov & N. B. Librovich. *J. Structural Chem.* **14** (1973), 332.
288. H. Kanno & J. Hiraishi. *Chem. Phys. Lett.* **62** (1979), 82; **72** (1980), 541; *J. Phys. Chem.* **87** (1983), 3664.
289. N. Abe & M. Ito. *J. Raman. Spec.* **7** (1978), 161.
290. P. A. Giguère & J. G. Guillot. *J. Phys. Chem.* **86** (1982), 3231.
291. L. Ochs, J. Guéron & M. Magat. *J. Phys. Radium* (1940), 85.
292. I. Pernoll, U. Maier, R. Janoschek & G. Zundel. *J. Chem. Soc. Faraday Trans. 2* **71** (1975), 201.
293. G. E. Walrafen. *J. Chem. Phys.* **52** (1970), 4176.
294. G. Brink & M. Falk. *Can. J. Chem.* **48** (1970), 3019.
295. D. M. Adams, M. J. Blandamer, M. C. R. Symons & D. Waddington. *Trans. Faraday Soc.* **67** (1971), 611.
296. L. J. Bellamy, M. J. Blandamer, M. C. R. Symons & D. Waddington. *Trans. Faraday Soc.* **67** (1971), 611.
297. P. Dryjanski & Z. Kecki. *J. Mol. Struct.* **12** (1972), 219.
298. M. C. R. Symons & D. Waddington. *J. Chem. Soc. Faraday Trans. 2* **71** (1975), 22.
299. T. J. V. Findlay & M. C. R. Symons. *J. Chem. Soc. Faraday Trans. 2* **72** (1976), 820.
300. Z. Kecki. *J. Mol. Struct.* **45** (1978), 23.
301. D. Schioberg. *Ber. Bunsenges. Phys. Chem.* **85** (1981), 513.
302. A. Sokolowska & Z. Kecki. *J. Mol. Struct.* **101** (1983), 113.

303. M. P. Thi, M.-H. Herzog-Cance, A. Potier & J. Potier. *J. Raman Spec.* **12** (1982), 238.

304. L. V. Chernykh, L. A. Myund, N. E. Rumyantseva & D. N. Glebovskii. *Vestn. Lening. Univ. Fiz. Khim.* **4** (1979), 56.

305. H. Kanno & J. Hiraishi. *Chem. Phys. Lett.* **107** (1984), 438.

306. J. H. R. Clarke, in *Advances in Infrared & Raman Spectroscopy* (ed. R. J. H. Clark & R. E. Hester). Heyden: London, 1978, Vol. 4, p. 109.

307. W. Denninger & G. Zundel. *J. Chem. Phys.* **74** (1981), 2769.

308. M. M. Kreevoy & C. A. Mead. *J. Am. Chem. Soc.* **84** (1962), 4596.

309. J. A. Pople, W. G. Schneider & H. J. Bernstein. *High-Resolution Nuclear Magnetic Resonance.* McGraw-Hill: NY, 1959, chap. 10.

310. J. I. Kaplan & G. Fraenkel. *NMR of Chemically Exchanging Systems.* Academic Press: NY, 1980.

311. M. M. Kreevoy & C. A. Mead. *Disc. Faraday Soc.* **39** (1965), 166.

312. A. K. Covington. *Disc. Faraday Soc.* **39** (1965), 176.

313. M. M. Kreevoy. *Disc. Faraday Soc.* **39** (1965), 180.

314. A. K. Covington, M. J. Tait & Lord Wynne-Jones. *Disc. Faraday Soc.* **39** (1965), 172.

315. A. K. Covington, M. J. Tait & W. F. K. Wynne-Jones. *Proc. Royal Soc.* (*London*) **A286** (1965), 235.

316. P. Vollmar. *J. Chem. Phys.* **39** (1963), 2236.

317. R. E. Weston, Jr. *Disc. Faraday Soc.* **39** (1965), 178.

318. A. K. Covington, J. G. Freeman & T. H. Lilley. *J. Phys. Chem.* **74** (1970), 3773.

319. E. Grunwald. *Progr. Phys. Org. Chem.* **3** (1965), 317.

320. W. J. Albery. *Progr. React. Kinetics* **4** (1967), 353.

321. D. E. Irish & H. Chen. *J. Phys. Chem.* **74** (1970), 3796.

322. H. Chen & D. E. Irish. *J. Phys. Chem.* **75** (1971), 2672.

323. D. E. Irish & R. C. Meatherall. *J. Phys. Chem.* **75** (1971), 2684.

324. M. Eigen, G. Kurtze & K. Tamm. *Z. Elektrochem.* **57** (1953), 103.

325. S. Ikawa, M. Yamada & M. Kimura. *J. Raman Spec.* **6** (1977), 89.

326. B. Cohen & S. Weiss. *J. Phys. Chem.* **90** (1986), 6275.

327. O. Puzic. MSc. Thesis, University of Waterloo, Waterloo, Ontario, Canada, 1976.

328. D. E. Irish & O. Puzic. *J. Solution Chem.* **10** (1981), 377.

329. D. E. Irish. *Proc. 6th Int. Conf. Raman Spectrosc.*, Volume 1. Heyden: London, 1978, p. 245.

330. H. Strehlow, I. Wagner & P. Hildebrandt. *Ber. Bunsenges. Phys. Chem.* **87** (1983), 516.

331. R. A. MacPhail & H. L. Strauss. *J. Chem. Phys.* **82** (1985), 1156.

332. S. Bratos, G. Tarjus & P. Viot. *J. Chem. Phys.* **85** (1986), 803; P. Viot, G. Tarjus & S. Bratos. *J. Mol. Liquids* **36** (1987), 185.

333. B. J. Berne & R. Giniger. *Biopolym.* **12** (1973), 1161.

334. J. E. Prue. *Proc. 3rd Symposium on Coordination Chemistry*, Vol. 1 (ed. M. T. Beck). Debrecen: Hungary, 1970, p. 25.

335. G. Ritzhaupt & J. P. Devlin. *J. Phys. Chem.* **81** (1977), 521.

336. M. Leuchs & G. Zundel. *J. Phys. Chem.* **82** (1978), 1632.

337. M. H. Herzog-Cance, A. Potier & J. Potier. *Can. J. Chem.* **63** (1985), 1492.

338. B. S. W. Dawson, D. E. Irish & G. E. Toogood. *J. Phys. Chem.* **90** (1986), 334.
339. M. P. Thi, M.-H. Herzog-Cance, A. Potier & J. Potier. *J. Raman Spec.* **12** (1982), 238.
340. P. A. Giguère & S. Turrell. *J. Am. Chem. Soc.* **102** (1980), 5473.
341. R. Ross. *The Proton: Applications to Organic Chemistry*. Academic Press: Orlando, USA, 1985.
342. R. P. Bell. *The Proton in Chemistry*, 2nd edn. Chapman and Hall: London, 1973.
343. M. Eigen. *Angew. Chem. Internat. Edit.* **3** (1964), 1.
344. E. G. Weidemann & G. Zundel. *Z. Naturforsch.* **25A** (1970), 627.
345. G. Zundel. *J. Membrane Science* **11** (1982), 249.
346. The kinetics of proton transfer processes. *Disc. Faraday Soc.* **39** (1965).
347. Proton transfer. *Faraday Symposia of the Chem. Soc.* **10** (1975).
348. Electron and proton transfer. *Faraday Disc. of the Chem. Soc.* **74** (1982).
349. A. J. Kresge. *Accts. Chem. Res.* **8** (1975), 345.
350. J. Lee, R. D. Griffin & G. W. Robinson. *J. Chem. Phys.* **82** (1985), 4920.
351. G. W. Robinson, P. J. Thistlethwaite & J. Lee. *J. Phys. Chem.* **90** (1986), 4224.
352. B. E. Conway, in *Mod. Aspects Electrochem.* (ed. J. O'M. Bockris & B. E. Conway). Butterworths: London, 1964, No. 3, chap. 2, p. 43.
353. D. H. Huppert, A. Jayaraman, R. G. Maines, Sr., D. W. Steyert & P. M. Rentzepis. *J. Chem. Phys.* **81** (1984), 5596.
354. W. C. Natzle & C. B. Moore. *J. Phys. Chem.* **89** (1985), 2605.
355. Y. Marcus. *J. Solution Chem.* **12** (1983), 271.
356. T. W. Swaddle & M. K. S. Mak. *Can. J. Chem.* **61** (1983), 473.
357. B. E. Conway, J. O'M. Bockris & H. Linton. *J. Chem. Phys.* **24** (1956), 834.
358. T. Erdey-Gruz & S. Lengyel, in *Mod. Aspects Electrochem.* (ed. J. O'M. Bockris & B. E. Conway). Plenum: NY, 1977, No. 12, chap. 1.
359. J. O'M. Bockris & S. U. M. Khan. *Quantum Electrochemistry*. Plenum: New York, 1979, chap. 9.
360. L. J. Gagliardi. *J. Chem. Phys.* **58** (1973), 2193.
361. L. J. Gagliardi. *J. Chem. Phys.* **61** (1974), 5465.
362. C. A. Croxton. *Endeavour* **34** (1975), 79.
363. G. Zundel. *Allgemeine und Praktische Chemie* **21** (1970), 329.
364. H. S. Frank, in *Structure of Water and Aqueous Solutions* (ed. W. A. P. Luck). Verlag Chemie, Physik Verlag: Weinheim, 1974, p. 24.
365. E. Kochanski. *Chem. Phys. Lett.* **133** (1987), 143.
366. T. Clark. *A Handbook of Computational Chemistry*. J. Wiley & Sons: New York, 1985.
367. W. J. Hehre, L. Radom, P. von R. Schleyer & J. A. Pople. *Ab Initio Molecular Orbital Theory*. J. Wiley & Sons: New York, 1986.
A1. E. A. Ukshe, L. S. Leonova, L. O. Atovmyan, A. I. Korosteleva, L. N. Erofeev, V. P. Tarasov & V. G. Shteinberg. *Dokl. Akad. Nauk SSSR* **285** (1985), 1157.
A2. W. P. Bosman, P. T. Beurskens, J. M. M. Smits, H. Behm, J. Mintjens, W. Meisel & J. C. Fuggle. *Acta Cryst.* **C42** (1986), 525.
A3. P. Molinie & S. Paoli. *C.R. Acad. Sci. Paris, Ser. II* **299** (1984), 1243.
A4. M. L. Berardelli, P. Galli, A. La Ginestra, M. A. Massucci & K. G. Varshney. *J. Chem. Soc. Dalton Trans.* (1985), 1737.
A5. A. Novak, M. Pham-Thi & Ph. Colomban. *J. Mol. Struct.* **141** (1986), 211.

A6. M. H. Herzog-Cance, J. F. Herzog, A. Potier, J. Potier, H. Arribart, C. Doremieux-Morin & Y. Piffard. *J. Mol. Struct.* **143** (1986), 67.

A7. P. S. Nicholson, M. Nagai, K. Yamashita, M. Sayer & M. F. Bell. *Sol. Stat. Ionics* **15** (1985), 317.

A8. P. Chu & F. G. Dwyer. *Zeolites* **3** (1983), 72.

A9. M.-H. Herzog-Cance, M. P. Thi & A. Potier, in *Solid State Protonic Conduct. Fuel Cells Sens., 3rd Eur. Workshop, 1984* (ed. J. B. Goodenough, J. Jensen & A. Potier). Odense University Press: Odense, 1985, p. 129.

A10. M.-H. Herzog-Cance, C. Belin & J.-F. Herzog. *C.R. Acad. Sci. Paris, Ser. II* **298** (1984), 531.

A11. G. Picotin & J. Rozier. *Compt. Rend. Chim. Phys.* **69** (1972), 372.

A12. H. Henke. *Zeits. Kristallog.* **172** (1985), 263.

A13. H. Henke & W. F. Kuhs. preliminary report, 1986.

A14. D. Mootz & M. Wiebcke. *Inorg. Chem.* **25** (1986), 3095.

A15. M. Wiebcke, Dissertation, University of Dusseldorf, 1986.

A16. D. Mootz, E. J. Oellers & M. Wiebcke. *J. Am. Chem. Soc.* **109** (1987), 1200.

A17. M. Wiebcke & D. Mootz. *Zeits. Kristallog.* **177** (1986), 291.

A18. W. P. McKenna & E. M. Eyring. *Appl. Spec.* **40** (1986), 20.

A19. D. Britton, M. K. Chantooni Jr., W. J. Wang & I. M. Kolthoff. *Acta Cryst.* **C40** (1984), 1584.

A20. C. Brink Shoemaker, L. V. McAfee, D. P. Shoemaker & C. W. DeKock. *Acta Cryst.* **C42** (1986), 1310.

A21. I. Takei & N. Maeno. *J. Chem. Phys.* **81** (1984), 6186.

A22. I. Takei & N. Maeno. *Abstracts VIIth Symposium on the Physics and Chemistry of Ice, Grenoble* (1986), 65.

A23. P. J. Wooldridge & J. P. Devlin. *J. Chem. Phys.* **84** (1986), 4112.

A24. D. Papousek, S. Urban, V. Spirko & K. Narahari Rao. *J. Mol. Struct.* **141** (1986), 361.

A25. V. Danielis & V. Spirko. *J. Mol. Spec.* **117** (1986), 175.

A26. D.-J. Liu, T. Oka & T. J. Sears. *J. Chem. Phys.* **84** (1986), 1312.

A27. M. Okumura, L. I. Yeh, J. D. Myers & Y. T. Lee. *J. Chem. Phys.* **85** (1986), 2328.

A28. M. H. Begemann & R. J. Saykally. *J. Chem. Phys.* **82** (1985), 3570.

A29. K. W. Ogilvie, M. A. Coplan, P. Bochsler & J. Geiss. *Science* **232** (1986), 374.

A30. P. A. Schulz, D. C. Gregory, F. W. Meyer & R. A. Phaneuf. *J. Chem. Phys.* **85** (1986), 3386.

A31. H. Shinohara, N. Nishi & N. Washida. *J. Chem. Phys.* **84** (1986), 5561.

A32. K. Hiraoka, H. Takimoto & S. Yamabe. *J. Phys. Chem.* **90** (1986), 5910.

A33. M. Meot-Ner & C. V. Speller. *J. Phys. Chem.* **90** (1986), 6616.

A34. J. M. Hopkins & L. I. Bone. *J. Chem. Phys.* **58** (1973), 1473.

A35. J. F. Wolf, R. H. Staley, I. Koppel, M. Taagepera, R. T. McIver Jr., J. L. Beauchamp & R. W. Taft. *J. Am. Chem. Soc.* **99** (1977), 5417.

A36. R. S. Hemsworth, J. D. Payzant, H. I. Schiff & D. K. Bohme. *Chem. Phys. Lett.* **26** (1974), 417.

A37. R. J. Cotter & W. S. Koski. *J. Chem. Phys.* **59** (1973), 784.

A38. K. Hiraoka, T. Shoda, K. Morise, S. Yamabe, E. Kawai & K. Hirao. *J. Chem. Phys.* **84** (1986), 2091.

A39. C. M. Banic & J. V. Iribarne. *J. Chem. Phys.* **83** (1985), 6432.
A40. A. J. Stace & C. Moore. *J. Am. Chem. Soc.* **105** (1983), 1814.
A41. J. E. Moryl, W. R. Creasy & J. M. Farrar. *J. Chem. Phys.* **82** (1985), 2244.
A42. G. I. Mackay, S. D. Tanner, A. C. Hopkinson & D. K. Bohme. *Can. J. Chem.* **57** (1979), 1518.
A43. W. R. Creasy & J. M. Farrar. *J. Phys. Chem.* **88** (1984), 6162.
A44. R. B. Sharma & P. Kebarle. *J. Am. Chem. Soc.* **106** (1984), 3913.
A45. A. J. Stace & C. Moore. *J. Phys. Chem.* **86** (1982), 3681.
A46. A. C. Hopkinson, G. I. Mackay & D. K. Bohme. *Can. J. Chem.* **57** (1979), 2996.
A47. A. B. Raksit & D. K. Bohme. *Can. J. Chem.* **63** (1985), 854.
A48. S. D. Tanner, G. I. Mackay, A. C. Hopkinson & D. K. Bohme. *Int. J. Mass Spec. Ion Phys.* **29** (1979), 153.
A49. K. Hiraoka. *Int. J. Mass Spec. Ion Phys.* **27** (1978), 139.
A50. J. W. Larson & T. B. McMahon. *J. Am. Chem. Soc.* **108** (1986), 1719.
A51. N. G. Adams, D. Smith & M. J. Henchman. *Int. J. Mass Spec. Ion Phys.* **42** (1982), 11.
A52. P. W. Ryan, C. R. Blakley, M. L. Vestal & J. H. Futrell. *J. Phys. Chem.* **84** (1980), 561.
A53. R. G. Keesee & A. W. Castleman Jr. *J. Phys. Chem. Ref. Data* **15** (1986), 1011.
A54. M. Okumura. Ph.D. Thesis, Lawrence Berkeley Laboratory, University of California, 1986.
A55. L. I. Yeh & J. D. Myers, unpublished results.
A56. D. Mootz & E.-J. Oellers, submitted for publication, 1987.
A57. A. Stahn, H. Solka, H. Adams & W. Urban. *Mol. Phys.* **60** (1987), 121.
A58. J. W. Christiansen, I. S. T. Tsong & S. H. Lin. *J. Chem. Phys.* **86** (1987), 4701.
A59. R. B. Sharma & P. Kebarle. *J. Am. Chem. Soc.* **106** (1984), 3913.
A60. D. Mootz & A. Merschenz-Quack. *Z. Naturforsch.* **42b** (1987), 1231.
A61. T. Gustafsson. *Acta Cryst.* **C43** (1987), 816.
A62. T. Gustafsson. Ph.D. Thesis, Uppsala, Sweden (1987).

Water as a plasticizer: physico-chemical aspects of low-moisture polymeric systems

HARRY LEVINE AND LOUISE SLADE

General Foods Corporation, Technical Center T22-1, 555 South Broadway, Tarrytown, New York 10591, *USA*
(Present address: Nabisco Brands Inc., Corporate Technology Group, East Hanover, NJ 07936, USA)

2.1 Introduction

'Water is the most ubiquitous plasticizer in our world.' [1] It has become well established that plasticization by water affects the glass-to-rubber transition temperatures (T_g) of many synthetic and natural amorphous polymers (particularly at low moisture contents), and that T_g depression can be advantageous or disadvantageous to material properties, processing, and stability. [2] Eisenberg [3] has stated that 'the glass transition is perhaps the most important single parameter which one needs to know before one can decide on the application of the many non-crystalline (synthetic) polymers that are now available.' Karel has noted that 'water is the most important ... plasticizer for hydrophilic food components.' [4] The physico-chemical effect of water, as a plasticizer, on the T_g of starch and other amorphous or partially-crystalline (PC) polymeric food materials has been increasingly discussed in several recent reviews and reports [5–10], dating back to the pioneering doctoral research of van den Berg. [11, 12]

The critical role of water as a plasticizer of amorphous materials (both water-soluble and water-sensitive ones) has been a focal point of our research, and has developed into a central theme during six years of an active industrial program in food polymer science. Recently reported studies from our laboratories were based on thermal and thermomechanical analysis methods used to illustrate and characterize the polymer physico-chemical properties of various food ingredients and products (e.g. starch and rice [13–22]; gelatin [13–15, 23, 24]; gluten [16, 21]; frozen aqueous solutions of small sugars, derivatized sugars, polyols, and starch hydrolysis products (SHPs) [14, 18, 20, 22, 25–9]; and 'intermediate moisture food' (IMF) carbohydrate systems [14, 18, 20, 22, 28]), all of which were described as systems of amorphous or PC polymers, oligomers, and/or monomers, soluble in and/or plasticized by water.

In this review, we shall not try to survey exhaustively the literature on the broad spectrum of synthetic and natural polymers plasticised by water. Rather, we shall attempt a critical analysis and present some new perspectives on the subject, by concentrating on a few major cases from our experiences with low-moisture polymeric food systems. It will be left to the reader to recognize that food 'polymers' (a term we always use generically to include oligomers and monomers as well) serve in this review only as excellent examples of many other water-soluble and -sensitive polymers. The fundamental structure–property principles illustrated by their behavior can be generalized to polymeric systems of interest in many other industries, including plastics, rubbers and resins, pharmaceuticals, textiles, paper conversion, adhesives, paints, and tobacco. [2] In fact, our polymer science approach to understanding the structure–property relations of food polymers at low moisture emphasizes the generic similarities between synthetic polymers and food molecules. From a theoretical basis of established structural principles from the field of synthetic polymer science, the functional properties of food materials during processing and storage can be explained and often predicted.

This food polymer science approach has developed to unify structural aspects of food materials (conceptualized as amorphous or PC polymer systems, the latter typically based on the classic 'fringed micelle' morphological model [30, 31]) with functional aspects described in terms of water dynamics and glass dynamics. [20, 22, 28] These integrated approaches focus on the non-equilibrium nature of all 'real world' food products and processes, and stress the importance to ultimate product quality and stability of the maintenance of food systems in kinetically-metastable 'states' (as opposed to equilibrium thermodynamic phases), which are always subject to potentially-detrimental plasticization by water. Through this unification, the kinetically-controlled behavior of polymeric food materials may be described by a single 'map' (which is derived from a solute–solvent 'state' diagram [32, 33]), in terms of the critical variables of moisture content, temperature, and time. [26] The map domains of moisture content and temperature, traditionally described using concepts such as 'water activity' (A_w), 'bound water', cryoprotection, water vapor sorption isotherms, and sorption hysteresis, can be treated as aspects of water dynamics. [20] The concept of water dynamics can be used to explain the moisture management and structural stabilization of IMF systems [20, 22, 28] or the cryostabilization of frozen, freezer-stored, or freeze-dried aqueous glass-forming materials [14, 18, 25–9], as will be reviewed here.

The concept of glass dynamics focuses on the temperature dependence of the relationships among composition, structure, thermomechanical properties, and functional behavior, and has been used to describe a unifying concept for interpreting 'collapse' phenomena, which govern, for example, the caking during storage of low-moisture, amorphous or PC food powders. [14, 26, 27] As will be reviewed here, collapse phenomena are regarded as diffusion-controlled consequences of a structural relaxation which depends on the occurrence of an underlying 'state' transformation at T_g. The critical effect of plasticization by water (leading to increased mobility in the dynamically-constrained amorphous solid) on T_g is a central element of the concept and the mechanism derived from it. A general physico-chemical mechanism for collapse has been described [26], based on the occurrence of a material-specific structural transition at T_g, followed by viscous flow in the rubbery liquid state. The mechanism was derived from Williams–Landel–Ferry (WLF) free volume theory for amorphous polymers [34] (see section 2.2.5), and led to a conclusion of the fundamental identity of T_g with the transition temperatures observed for structural collapse (T_c) and recrystallization (T_r). [26, 27] The non-Arrhenius kinetics of collapse and/or recrystallization in the high-viscosity (η) rubbery state, which are governed by the mobility of the water-plasticized polymer matrix, depend on the magnitude of ΔT above T_g [14], as defined by an exponential relationship derived from WLF theory. [34] Glass dynamics has also proved a useful concept for elucidating the physico-chemical mechanisms of structural changes involved in various melting and (re)crystallization phenomena which are relevant to many PC food polymers and processing/storage situations, including, for example, the gelatinization and retrogradation of starches and the gelation of gelatin. [20–2] Glass dynamics has also been used to describe the amorphous polymeric behavior of proteins such as native wheat gluten and elastic at low moisture. [14, 16, 20, 23] These topics will also be covered in this review.

The key to our new perspective on water-plasticized polymers at low moisture relates to the recognition of the fundamental importance of the dynamic map. The results of our research program in food polymer science [13–29] have demonstrated that the critical feature of the map is the identification of the glass transition as the reference surface which serves as a basis for the description of the non-equilibrium behavior of polymeric materials, in response to changes in moisture content, temperature, and time. The kinetics of all diffusion-controlled relaxation processes, which are governed by the mobility of the water-plasticized polymer matrix, vary (from Arrhenius- to WLF-type) between distinct temperature/structural domains, which are divided by this glass transition. The viscoelastic, rubbery fluid state, for which WLF kinetics apply [34], represents the most significant domain for the study of water dynamics. One particular location on the reference surface results from the behavior of water as a crystallizing plasticizer and corresponds to an invariant point on a state diagram for any particular solute. This location represents the glass with maximum moisture content as a kinetically-metastable, dynamically-constrained solid which is pivotal to characterization of the structure and function of amorphous and PC polymeric food materials. From the theoretical basis provided by the integrated concepts of water dynamics and glass dynamics, a new experimental approach has been suggested [18, 20, 22] for predicting the technological performance, product quality and stability of many polymeric food systems, and examples which illustrate the utility of this approach vs. others based on the traditional concept of A_w will be described.

In closing this Introduction, we must point out that, in the context of this review, a 38% (w/w) fructose solution at 20 °C represents an example of a 'low-moisture' situation. [20] Franks [35] has emphasized that 'low', with respect to temperatures, is a relative concept which refers to a range of some 800 degrees when the interests of both metallurgists and quantum physicists are considered. Here, it is equally necessary to emphasize that 'low', with respect to moisture contents, is also a relative concept which refers to everything which is 'not very dilute'. [18] This insight arose from the results of a groundbreaking study by Soesanto & Williams [36] on viscosities of concentrated aqueous sugar solutions. They demonstrated that, for such glass-forming liquids, in their rubbery state ($\eta > 10$ Pa s) at $20 < T < 80$°C, the WLF equation [34] characterizes $\eta(T)$ extremely well. [36] Arrhenius kinetics would not be applicable to the behavior of such 'polymers at low moisture'. [20]

2.2 Theoretical background

In the field of synthetic polymers, the classical definition of a plasticizer is 'a material incorporated in a polymer to increase the polymer's workability, flexibility, or extensibility.' [1] A plasticizer may lower a polymer's T_g, melt viscosity, or elastic modulus. Plasticization, on a molecular level, leads to increased intermolecular space or free volume, and may involve the weakening or breaking of selective interpolymer bonds. Plasticization implies intimate mixing, such that a plasticizer is dissolved in a polymer or a polymer is dissolved in a plasticizer. Accordingly, a true solvent for a polymer, in terms of a system of high thermodynamic compatibility and miscibility at all proportions, is always also a plasticizer, but a plasticizer is not always a solvent. [1]

In the context of polymers at low moisture content, the term 'water-soluble' can be confusing, because its definition can depend on the nature and concentration range of a particular polymer. Certain highly hydrophilic, polar, ionic, and/or H-bonding polymers, including both synthetic and natural ones [14], are highly water-soluble. One example is poly(vinyl pyrrolidone) (PVP). [37] In a dilute PVP solution, water, as the solvent, is the major component in which the minor PVP component, as the solute, is soluble in the sense of Flory–Huggins theory. [30] However, many so-called 'water-soluble' polymers, including PVP, are, in fact, not completely miscible with water at all proportions and temperatures, and show mesophase separation behavior at upper or lower critical solution temperatures (UCST or LCST). [36]

Most other synthetic and natural polymers, which are more hydrophobic, less polar, or lesser H-bonders, are not highly water-soluble, but are water-sensitive to some extent, especially at low moisture. For these, water is a plasticizer, but not a good solvent. [2] In such a binary mixture, the plasticizer is the minor solute component, which is soluble in the polymer solvent only at relatively low plasticizer–polymer ratios. [1] This case has been described as the Henry's law solution region, in the 'infinite dilution limit' of water dissolved in a polymer. [39] Many such polymers, at low moisture, manifest their limited thermodynamic compatibility with water by their swelling behavior. [39] One example of a water-sensitive polymer, which forms a homogeneous one-phase solution by dissolving a limited amount of water (about 4% w/w at 23 °C), is poly(vinyl acetate) (PVAc). [40]

Between the limits delineated by highly water-soluble PVP and slightly water-sensitive PVAc lie such biopolymeric materials as starch, gelatin, gluten, and elastin. These high polymers, at low moisture contents, demonstrate essentially the same physico-chemical responses to plasticization by water as do many readily-soluble monomeric and oligomeric sugars, polyols, and SHPs. In this review, we prefer to describe all such materials as 'water-compatible.' [20]

2.2.1 *Polymer structure–property principles*

For 'dry' polymers which are solids at room temperature, two possible structural forms are described as amorphous ('glassy') and partially crystalline. (The latter term, rather than semicrystalline, is preferred to describe polymers of relatively low percentage crystallinity, such as native starches (about 15–39% crystallinity [7, 9]) and gelatin (about 20% crystallinity [41]). The term semicrystalline is the preferred usage to describe polymers of about or above 50% crystallinity [30, 42], so that it may be conceptually misleading when it is used [9] to describe granular starches.) Amorphous homogeneous polymers manifest a single, 'quasi-second-order', kinetic state transition from metastable amorphous solid to unstable amorphous liquid at a characteristic T_g. PC polymers show two types of characteristic transitions, (1) a T_g for the amorphous component, and (2) at a crystalline melting temperature, T_m, which is always at a higher temperature than T_g for homopolymers, a first-order, 'equilibrium phase' transition from crystalline solid to amorphous liquid. [43] The same is true for PC oligomers and monomers. [20, 36] Both T_g and T_m are measurable as thermomechanical transitions by various instrumental methods, including differential scanning calorimetry (DSC), differential thermal analysis, thermomechanical analysis (TMA), dynamical mechanical analysis, and others. [43]

Structural models for PC polymers The 'fringed micelle' model, used classically to describe the morphology of PC synthetic polymers [30, 31, 42, 44], is illustrated in figure 2.1. It is particularly useful for conceptualizing a three-dimensional network composed of microcrystallites crosslinking amorphous regions of random-coil chain segments. [41] The model is especially applicable to polymers which crystallize from an undercooled melt or concentrated solution, to produce a metastable network of relatively low percentage crystallinity, containing small crystalline regions of only about

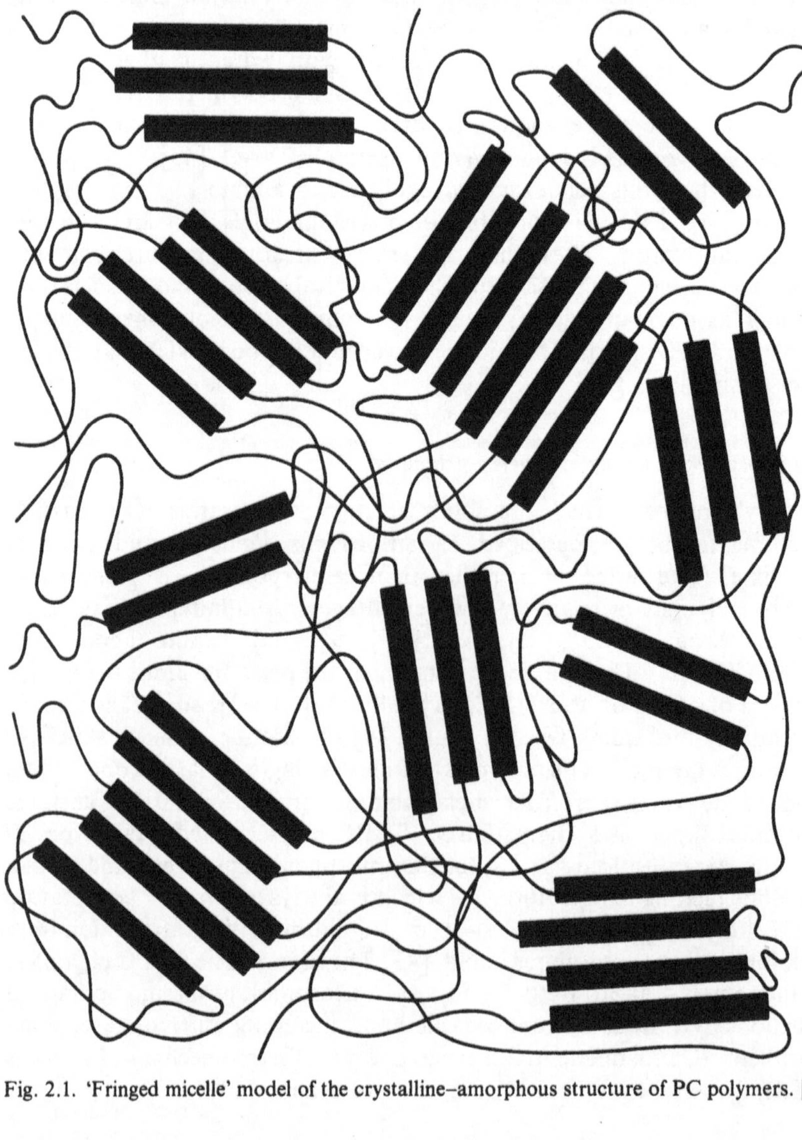

Fig. 2.1. 'Fringed micelle' model of the crystalline–amorphous structure of PC polymers. [24]

10 nm dimensions. [24, 41, 42] Thus, it has also often been used to describe the PC structure of aqueous gels of biopolymers such as gelatin [13, 15, 23, 24, 41, 42, 45–9] and starch [13–17, 21, 22, 28], in which the amorphous regions contain plasticizing water and the microcrystalline regions, which serve as physical junction zones, are crystalline hydrates. We have also used this model to describe the PC morphology of lower-molecular weight (MW) carbohydrate systems [14, 18, 20, 22, 28], and of frozen aqueous solutions of water-compatible, non-crystallizing materials. [13–15, 18, 20, 22, 25–9] In the latter case, ice crystals represent the 'micelles' dispersed in a continuous amorphous matrix of solute–unfrozen water. [35, 37] An important feature of the 'fringed micelle' model, as applied to high-MW polymers including gelatin and starch, concerns the interconnections between crystalline and amorphous regions. A single, long polymer chain can have helical (or other ordered) segments located within one or more microcrystallites and random-coil segments in one or more amorphous regions. [13, 15, 41, 42] Moreover, in the amorphous regions, chain segments may experience random intermolecular 'entanglement coupling.' [26] So, in terms of their thermomechanical behavior in response to plasticization by water and/or heat, the crystalline and amorphous regions are certainly not independent phases, as mistakenly described by Biliaderis *et al.* [9] and incorrectly attributed to us. [15]

In recent years, an extension of the simple 'fringed micelle' model (originally proposed in 1930 [44]) has been described to explain further the thermomechanical behavior of various semicrystalline synthetic homopolymers. [50–3] This 'three-microphase' model incorporates two distinct types of amorphous domains, a bulk mobile-amorphous fraction and an interfacial (i.e. intercrystalline regions within spherulites or between crystallites in stressed systems) rigid-amorphous fraction, each capable of manifesting a separate T_g, plus a third rigid-crystalline component. [54] The sequence of thermal transitions predicted by this model is T_g (mobile-amorphous) $< T_g$ (rigid-amorphous) $< T_m$ (rigid-crystalline), although the magnitude of the change in heat capacity for the middle transition may become vanishingly small due to steric restraints on chain-segmental mobility. [50–4] (However, it should still be possible to detect this T_g by dynamic mechanical measurements.) Recently, the 'three-microphase' model has also been postulated to explain the multiple thermal transitions observed during the non-equilibrium melting of native granular starch. [9] However, while the model has proven applicable to several linear homopolymers [50–3], its application to normal starches, while well-intentioned, is not necessary and may even be inappropriate. Normal starch is not a homopolymer, but a mixture of two glucose polymers, linear amylose (MW 10^5–10^6) and highly-branched amylopectin (MW 10^8–10^9). [55] (Even 'waxy' starch, which contains only amylopectin, is not best described as a homopolymer of glucose, but as a special type of block copolymer in which backbone segments and branch points exist in amorphous domains and the crystallizable branches exist in microcrystalline domains. [55–7]) In a native granule of normal starch, each of these polymers may be partially crystalline [56], and their amorphous components may each manifest a distinct T_g (at a temperature dependent on MW [31] as well as local moisture content) characteristic of a predominant mobile-amorphous domain. [21]

2.2.2 *Crystallization mechanism and kinetics for PC polymers*

A classical three-step crystallization mechanism has been widely used to describe PC synthetic polymers crystallized, from the melt or concentrated solution, by undercooling from $T > T_m$ to $T_g < T < T_m$. [14, 41, 44] A familiar example of such a process is the blow-molding of plastic soda bottles made from poly(ethylene terephthalate). [14] The mechanism is compatible with both the generic 'fringed micelle' model [58] and a specific 'three-microphase' model, where for the latter, the operative T_g would be that of the predominant mobile-amorphous phase. It involves the following sequential steps: [44] (1) nucleation – formation of critical nuclei by initiation of ordered chain segments intramolecularly; (2) propagation – growth of crystals from nuclei by intermolecular aggregation of ordered segments; and (3) maturation – crystal perfection (by annealing of metastable microcrystallites) and/or continued slow growth (via Ostwald ripening).

The thermoreversible gelation, from concentrated solution, of a number of crystallizable synthetic homopolymers (e.g. polystyrene [60], polyethylene [61]) as well as copolymers (with flexible chains of high MW, which may be linear or highly branched [59]) has recently been reported to occur by the above crystallization mechanism. [58–61] This gelation-via-crystallization process (described as a nucleation-controlled growth process [61]) produces a metastable three-dimensional network [59, 61] crosslinked by 'fringed micellar' [58] or chain-folded lamellar [61] microcrystalline junction zones composed of intermolecularly associated helical chain segments. [60] Such PC gel networks may also contain random interchain entanglements in their amorphous regions. [60, 61] The non-equilibrium nature of the process [61] is manifested by 'well-known aging phenomena' [58] (i.e. maturation, which can involve polymorphic crystalline forms [61]), attributed to time-dependent crystallization processes which occur subsequent to initial gelation. The thermoreversibility of such gels is explained in terms of a crystallization (on undercooling)/melting (on heating to $T > T_m$) process. [59, 61] Only recently has it been recognized that such synthetic homopolymer–organic diluent gels are not glasses [62] ('gelation is not the glass transition of highly-plasticized polymer' [58]) but PC rubbers [60], in which the mobility of the diluent (in terms of rotational and translational motion) is not significantly restricted by the gel structure. [62] The temperature of gelation (T_{gel}) is above T_g [62], in the rubbery fluid range up to $T_g + 100$ °C, and is related to the T_m observed in melts of PC polymers. [58, 60] The MW-dependence of T_{gel} has been identified [60] as an isoviscous state (which may include the existence of interchain entanglements) of $\eta_{gel} = 10^5$ Pa s in comparison to η_g (at T_g) $\approx 10^{12}$ Pa s. [35]

Curiously, it has been well established for much longer [61] that the same three-step polymer crystallization mechanism describes the gelation mechanism for the classical gelling system, gelatin–water. [41, 46–8, 57, 239, 241] The fact that the resulting PC aqueous gels [63] can be modeled by the 'fringed micelle' structure is also widely recognized. [13–15, 23, 24, 41, 44–9] However, while the same facts are true with regard to the aqueous gelation process for starch (called 'retrogradation', a gelation-via-crystallization process that follows the 'gelatinization' and 'pasting' of PC native granular starch–water mixtures [56, 64]), recognition of starch retrogradation as a polymer crystallization process has been more recent and less widespread. [7–10] Many of the persuasive early insights in this area resulted from the food polymer science approach of Slade and her various coworkers. [9, 10, 13–22]

Slade [13–16] was the first to stress the importance of treating gelatinization as a non-equilibrium melting process of kinetically-metastable PC native starch in the presence of plasticizing water. In this process, the melting of the microcrystallites is controlled by the prerequisite plasticization ('softening') of the random-coil chain segments in the interconnected amorphous regions of the 'fringed micelle' network. [13, 15] Slade and coworkers [13–17] recognized that previous attempts [e.g. 65, 66] to interpret the effect of water content on the observed T_m of the starch by the Flory–Huggins thermodynamic treatment [30] were inappropriate and had failed, because the Flory–Huggins theory only applies to the equilibrium melting of a PC polymer with diluent. Slade described retrogradation as a non-equilibrium recrystallization process (involving amylopectin) in the amorphous (in the case of waxy starches) starch–water melt. [13–16] She noted that amylopectin recrystallization is a nucleation-controlled process which occurs, at $T > T_g$, in the mobile, viscoelastic 'fringed micelle' network plasticized by water. [13–16] As for the aging effects observed in starch gels, Slade reported [16] that 'analysis of the results (of measurements of extent of recrystallization vs. time after gelatinization) by the classical Avrami equation may provide a convenient means to represent empirical data from retrogradation experiments, but published theoretical interpretations [e.g. 67] have been misleading.' Complications, due to the non-equilibrium nature of starch recrystallization via the three-step mechanism, limit the theoretical utility of the Avrami parameters [16], which were originally derived to define the mechanism of crystallization under conditions of thermodynamic equilibrium. [44]

It should be noted that the three-step polymer crystallization mechanism also applies to concentrated aqueous solutions and melts of oligomers and monomers such as low-MW carbohydrates [14, 18, 20, 22, 26–8] and to recrystallization processes in frozen systems of water-compatible materials. [13–15, 18, 20, 22–9]

The classical theory of crystallization kinetics developed for synthetic PC polymers [44], which is illustrated in figure 2.2 (adapted from Jolley [41]), has also been shown to describe the kinetics of gelatin gelation [41, 61, 239] and starch retrogradation. [16, 21] Figure 2.2 shows the dependence of crystallization rate on temperature within the range $T_g < T < T_m$, and emphasizes the fact that gelation-via-crystallization can only occur in the rubbery (undercooled liquid) state [49, 68, 241], between the temperature limits defined by T_g and T_m. (These limits, for gelatin (high-MW) solutions of concentrations up to about 65% (w/w) gelatin, are about −12 °C and 37 °C, respectively [241]; while for homogeneous and amorphous sols or pastes of gelatinized B-type starch containing ≳27% (w/w) water, they are about −5 °C and 60 °C, respectively. [13–15]) The rate of crystallization would be essentially zero at $T < T_g$, because nucleation is a liquid-state phenomenon which requires orientational mobility, and such mobility is virtually disallowed (i.e. over realistic times) in the solid glass of $\eta > 10^{12}$ Pa s. [37] Likewise, the rate of propagation goes essentially to zero below T_g, because propagation is a diffusion-controlled process which also requires the liquid state. At $T > T_m$, the rate of crystallization also goes to zero, because, intuitively, one realizes that crystals can neither nucleate nor propagate at any temperature at which they would be melted instantaneously.

As illustrated in figure 2.2, the mechanistic steps of nucleation and propagation each manifest an exponential dependence of rate on temperature, such that the rate of nucleation increases exponentially with decreasing temperature, down to $T = T_g$, while the rate of propagation increases exponentially with increasing temperature, up to $T = T_m$. [41, 44] The rate of

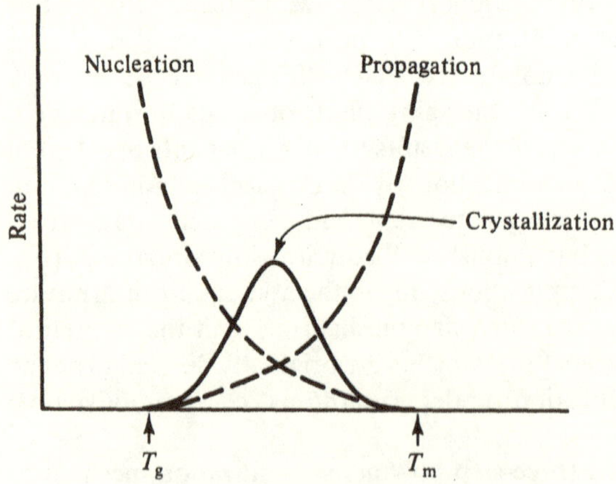

Fig. 2.2. Crystallization kinetics of PC polymers. [24]

maturation for non-equilibrium crystallization processes also increases with increasing temperature, up to the maximum T_m of the most mature crystals. [16] The overall rate of crystallization (i.e. nucleation + propagation), at a single temperature, can be maximized at a temperature about midway between T_g and T_m. However, Ferry [69] showed that the rate of gelatin gelation can be increased further, while the phenomenon of steadily increasing gel maturation over extended storage time can be eliminated, by a two-step temperature-cycling gelation protocol that capitalizes on the crystallization kinetics defined in figure 2.2. Ferry showed that a short period of nucleation at a temperature just above T_g, followed by another short period for crystal growth at a temperature just below T_m, produced a gelatin gel of maximum and unchanging gel strength in the shortest possible overall time. Recently, Slade [21] has shown that a similar temperature-cycling protocol can be used to maximize the rate of starch recrystallization in freshly-gelatinized starch–water mixtures containing at least 27 w% water.

2.2.3 *Viscoelastic properties of amorphous and PC polymers*

In the absence of plasticizer, the viscoelastic properties of glassy and PC polymers depend critically on temperature, relative to the T_g of the undiluted polymer. These properties include, for example, polymer specific volume, V, as illustrated in figure 2.3 (adapted from [34, 70, 71]) for glassy, PC, and crystalline polymers. From free volume theory, T_g is defined as the temperature at which the slope changes (due to a discontinuity in thermal expansion coefficient) in the V vs. T plot for a glass in figure 2.3 [34], while at T_m, V shows a characteristic discontinuity, typically increasing by about 15% for crystalline polymers. [57] Glassy and PC polymers also manifest η vs. T behavior as illustrated in figure 2.4 (adapted from Franks [37]), which provides a working definition of a glass as a mechanical solid capable of supporting its own weight against flow. The viscosity equals about 10^{11}–10^{14} Pa s at T_g [36, 37, 72], which represents the intersection of the curve

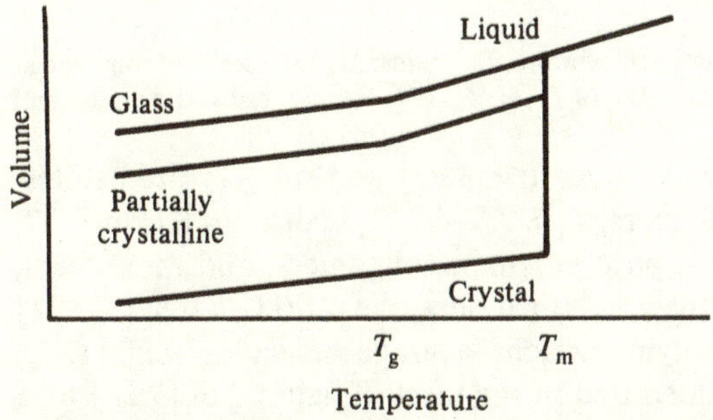

Fig. 2.3. Specific volume as a function of temperature for glassy, crystalline, and PC polymers. [14]

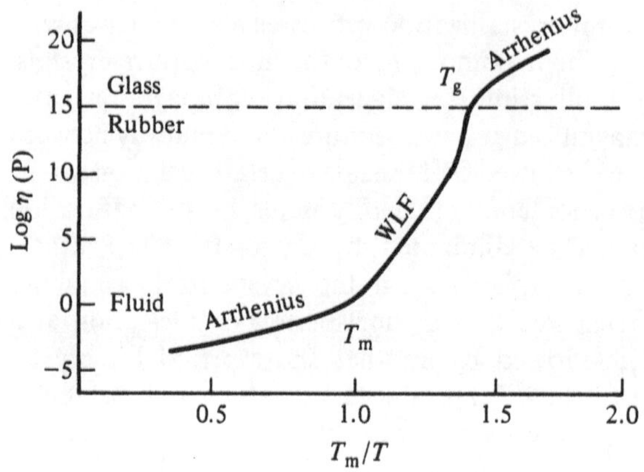

Fig. 2.4. Viscosity as a function of reduced temperature (T_m/T) for glassy and PC polymers. [18]

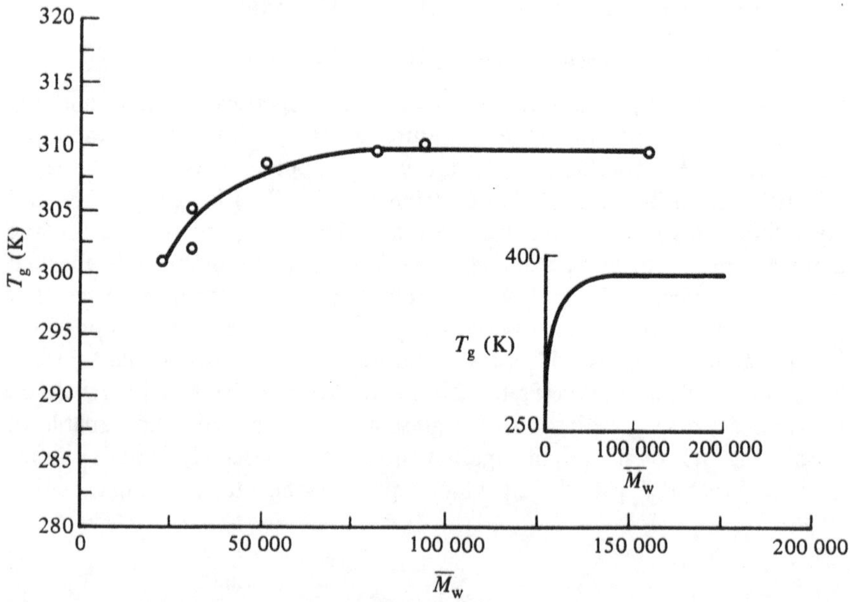

Fig. 2.5. Variation of the glass transition temperature, T_g, against $\bar{M}_w$ for a series of commercial PVAc polymers. [26] (Inset: an idealized plot of T_g vs. $\bar{M}_w$. [74] Reproduced with permission.)

in figure 2.4 with the boundary between the glassy and rubbery fluid states. For many typical synthetic polymers, $T_g \approx 0.5$–$0.8\ T_m$ in degrees Kelvin. [37, 57, 73] For highly-symmetrical pure polymers, oligomers, and monomers, $T_g/T_m < 0.5$, while for highly-unsymmetrical ones, the ratio is >0.8. [57, 74] T_g is also known to vary with polymer weight-average molecular weight ($\bar{M}_w$) in the characteristic fashion illustrated in the inset of figure 2.5. [75] For a homologous series of amorphous linear polymers (e.g. PVAc, shown in the main plot of figure 2.5 [26]), T_g increases with increasing $\bar{M}_w$, up to the plateau limit for the region of 'entanglement coupling' in rubber-like viscoelastic random networks (typically at $\bar{M}_w = 10^4$–10^5 daltons), then levels off with further increases in $\bar{M}_w$. [31, 34]

2.2.4 *Effects of water as a plasticizer on the thermomechanical properties of solid polymers*

As is well known, the effects of water as a plasticizer pertain to both the T_g and T_m of polymers. The primary plasticizing effect of increasing moisture content at constant temperature, which is equivalent to the effect of increasing temperature at constant moisture content, leads to increased segmental mobility of the chains in the amorphous regions of glassy and PC polymers, which in turn produces a primary structural relaxation transition, T_g, at decreased temperature. [77–9] The state diagrams in figure 2.6 illustrate the extents of this T_g-depressing effect of water for two particular glassy polymers. As shown in the main part of figure 2.6, PVP represents a typical water-soluble, completely-miscible, non-crystallizable polymer [37], which manifests a smooth T_g curve from about 100 °C for 'dry' PVP-44 to about −135 °C, the value commonly stated (but never yet measured) for glassy water. [37] The dramatic effect of water on T_g is seen at low moisture, such that for PVP-44, T_g decreases about 6 °C/w% water for the first 10% moisture. [26] In contrast to PVP, glassy PVAc, as shown in the inset of figure 2.6, is only somewhat water-sensitive, being plasticized (albeit as strongly, i.e. about 6 °C/w% water) only up to its compatibility limit of about 4% water. [40] Two further examples of water-plasticized, PC polymers are the polyamides, collagen and nylon 6. The plasticizing effect of up to 10% moisture on the T_g of native collagen protein exceeds that of synthetic nylon 6, but the extent of plasticization for both at low moisture falls in the range of about 5–10 °C/w% water. [73] We have found this same range to apply widely to amorphous and PC water-compatible materials, including many monomeric and oligomeric carbohydrates. [20, 22]

Mechanism of plasticization While depression of polymer T_g by plasticizing water is a very well-known and widely-reported phenomenon [2], details as to specific polymers, a specific mechanism of plasticization, quantitation of the extent of T_g depression by water, and the specific structural properties of polymers impacting thereon are still very topical subjects in the current polymer literature. [51, 80, 81] The results illustrated in figure 2.6 and described in the preceding paragraph typify those for many other synthetic and natural polymers, as reviewed elsewhere [e.g. 2, 14].

Efforts to calculate, theoretically, the extent of T_g depression by water that agrees with experimental values for glassy polymers have utilized several classical thermodynamic approaches, including, for example: (1) the Flory–Fox equation [71] for compatible plasticizer–polymer blends of low plasticizer content, which has been applied successfully to nylon and other

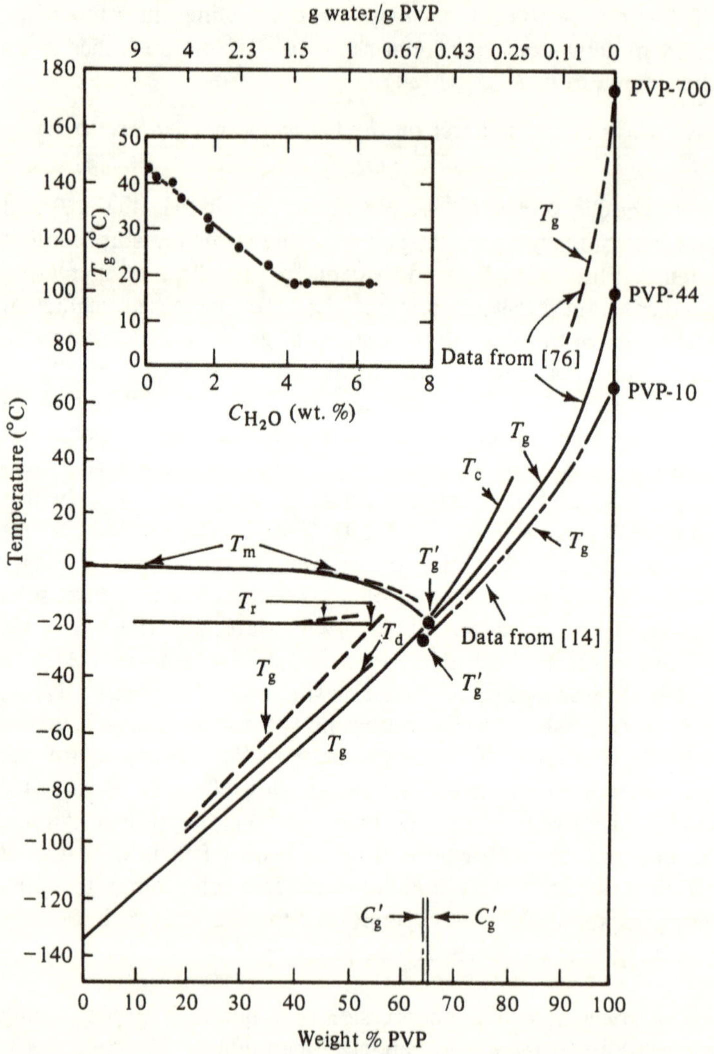

Fig. 2.6. Solid–liquid state diagram for water–PVP, showing the following transitions: T_m, T_g, T'_g, T_d, T_r, T_c. [26, 32, 145] (Inset: T_g as a function of moisture content for PVAc. [40] Reproduced with permission.)

water-sensitive fibers at low moisture (i.e. <10% water) [43]; (2) the Gibbs–Dimarzio equation derived from configurational entropy theory of glass formation, which is more refined than the Flory–Fox equation, and is more broadly applicable to systems of higher plasticizer content [43]; (3) Gordon's modification of the Gibbs–Dimarzio equation, which is useful for binary plasticizer–polymer mixtures that obey Henry's solution law, and which has been applied successfully to epoxy resin at very low moisture (i.e.

<3%) [82]; and (4) refinements [81] of the Couchman–Karasz approach for miscible polymer–diluent blends, which have proved applicable to covalently-crosslinked polymer networks such as epoxies and polystyrenes containing 1–3% moisture. [51, 81]

According to the view still prevalent in much of the polymer (synthetic and natural) literature [82, 83], the overall mechanism of plasticization of water-compatible glassy polymers by water may have two major components. The first derives from free volume theory, which provides the general concept that low-MW plasticizing diluents have large free volume, so that molecular-level interactions between water and a polymer lead to increased free volume in the diluted aqueous polymer rubber, which allows increased chain-segmental mobility of the backbone, which in turn results in decreased T_g. [1, 82–5] (In fact, it is well known that the ability of a diluent to depress T_g decreases linearly with increasing diluent MW, albeit with scatter [60], as predicted by free volume theory.) This free volume increase alone accounts completely for the plasticizing effect of water on the T_g of polystyrene. [85] However, for polymers capable of H-bonding with water, it has been claimed that the free volume contribution to the plasticization mechanism can be augmented by a contribution due to the extent of polymer–water H-bonding. [82, 83] The particular nature of such water–polymer interactions has been suggested to be site-specific H-bonding between water and the polar groups on the polymer, and the extent of this interaction has been said to increase with increasing hydrophilicity of the polymer. [73, 76, 77, 82–8] Furthermore, it has been suggested that when the glassy polymer network structure involves interpolymer H-bonds, in addition to any covalent crosslinks such as are present in, for example, epoxy resins and elastin, the plasticization mechanism may be augmented further by the breakage of these interpolymer H-bonds by water, and their replacement by labile water–polymer H-bonds, which would lead to greatly increased molecular mobility, and result in even greater T_g depression. [76, 82, 83, 86–8] However, recent publications [51, 80] have expressed strong favor for the free volume concept, contending that the effectiveness of water as a plasticizer of synthetic polymers merely reflects, in part, its low molar mass. [51] Our studies of water-compatible food materials have led us to the same conclusion. [13–25] Karasz and coworkers [51, 80] have come to discount the concepts of specific interactions, such as disruptive H-bonding in polymer H-bonded networks and plasticizing molecules becoming 'firmly bound' to polar sites along a polymer chain, in explaining water's plasticizing ability. To negate the prevailing argument for site-binding, they have cited [51] NMR results which clearly indicate that water molecules in epoxy resin have a large degree of mobility. We [14, 16, 18, 20, 22, 25–9] have followed the lead of Franks [35, 37, 89, 90] in taking a similar position and using similar evidence to try to dispel the popular myths about 'bound water' and 'water-binding capacity' in food polymers and other aqueous glass-forming materials, including sugars and polyols.

Behavior of PC polymers In PC polymers, the fact that water plasticization occurs only in the amorphous regions [11, 51, 77, 91, 92] both further complicates the effect of the crystalline regions on the thermomechanical behavior of the amorphous regions and explains how the amorphous regions cause the non-equilibrium melting behavior of the crystalline regions. In PC synthetic polymers which have anhydrous crystalline regions and a relatively low capacity for water in the amorphous regions (e.g. nylons [41, 77]), the percentage crystallinity affects T_g, such that increasing percentage crystallinity generally leads to increasing T_g [21, 51], due to two factors. One is the stiffening or 'antiplasticizing' effect of the dispersed microcrystalline crosslinks, which leads to decreased mobility of the chain segments in the interconnected amorphous regions. [51, 91] The same effect is produced by covalent crosslinks [51, 81], which, when produced by radiation, occur only in the amorphous regions. [51, 80] (Note the obvious connection between this stiffening effect in PC polymers modeled by the 'fringed micelle' (whereby T_g of the homogeneous amorphous 'fringe' is increased) and the previously-mentioned effect of the rigid-crystalline phase on the segmental mobility in the rigid-amorphous phase (resulting in a higher T_g than that of the bulk-amorphous phase) of polymers modeled by the 'three-microphase' structure.) The second factor is the hydrophobicity of the crystalline regions relative to the hydrophilicity of the glassy matrix, which is another way of saying that polymer–polymer contacts are much preferred over polymer–water contacts in the crystals. [77] (It has been pointed out that, in such polymers, only the amorphous regions are accessible to penetration and therefore plasticization by moisture. [51, 80] Even in native starches, composed of PC polymers which have hydrated crystalline regions, it is known that hydrolysis by aqueous acid ('acid etching') or enzymes, at $T < T_m$, can occur initially only in the amorphous regions. [56]) An interesting variation on this effect of crystallinity on T_g has been reported for several PC polyacrylates crystallized from the melt. [96] In these cases, the percentage crystallinity increases, but T_g decreases, with increasing isotacticity; because, with increasing tacticity, the crystal content increases, thereby forcing the 'as is' moisture contained in the material to the remaining amorphous regions, which results in decreased, rather than increased, T_g. [92] For the type of PC polymers with anhydrous crystalline regions, which have been analyzed based on the 'three-microphase' model (e.g. synthetic polyamides), it has been proposed [80] that a concentration gradient of the plasticizing water may be present. Experimental evidence has suggested that the moisture content is zero in the rigid-crystalline phase, low in the rigid-amorphous phase, and highest (i.e. the majority of the water in the polymer) in the bulk-amorphous phase. [51, 80]

Two related phenomena are observed in situations of overall low moisture content for PC polymers which have hydrated crystalline regions, such as gelatin [63, 241], collagen [73], and starch [17]: (1) atypically high [36, 57] T_g/T_m ratios much greater than 0.80 (e.g. 0.93 for collagen vs. 0.67 for nylon 6

[73]); and (2) a pronounced apparent depressing effect of water on T_m as well as on T_g, such that T_m decreases with increasing percentage moisture. [10] Previous attempts to analyze the latter phenomenon, using the Flory–Huggins thermodynamic treatment for the equilibrium melting of a PC polymer with diluent [30], had failed. [e.g. 65, 66] We recently reported [13–16], for starch and gelatin, that the correct explanation lies in the *non*-equilibrium nature of the melting process, and the indirect effect of the plasticizing water on T_m. Relative dehydration of the amorphous regions at the initially low overall moisture content leads to the kinetically-metastable condition where the effective T_g is higher than the equilibrium T_m of the crystalline regions. Consequently, the apparent T_m is elevated and observed after the softening of the amorphous regions at T_g. Added water acts directly on the continuous glassy regions, depressing their T_g, and thus allowing sufficient mobility and swelling for the interconnected microcrystallites, embedded in the 'fringed micelle' matrix (wherein the 'fringe' is an unstable rubber above T_g), to melt (by dissociation) on heating to a less kinetically-constrained T_m only slightly above the depressed T_g. In contrast to the case of limited moisture, in an excess moisture situation (e.g. a gelatin gel with >35% water), where the amorphous matrix would be fully plasticized and the temperature would be above T_g (about −12 °C for the gelatin gel), the fully-hydrated and matured crystalline junctions in such a polymer would show the actual (lower) equilibrium T_m (35–45 °C for gelatin). [10, 13–17, 23, 24, 63] As mentioned earlier, once Slade [13–16] had suggested that starch gelatinization is such a non-equilibrium melting process (for which the Flory–Huggins treatment has no theoretical basis), others (who had previously tried to treat starch melting data by the Flory–Huggins theory [66]) subsequently began to recognize and describe results for the non-equilibrium melting of granular starches and amylose–lipid crystalline complexes [9, 10], in agreement with Slade's concept.

Water plasticization of sub-T_g transitions Water plasticization, which causes the primary relaxation transition at T_g (the α transition), also causes sub-T_g (generally subzero) secondary relaxations (the β and γ viscoelastic, mechanical, or dielectric loss transitions) in many glassy and PC polymers. [77, 86, 93–7] In solid polymers (at $T < T_g$), the micro-Brownian translational motions of long main-chain segments (e.g. about 15 amide groups for nylon 66 [77]), which are associated with T_g, are 'frozen in'. [93] In the amorphous regions, only local and cooperative motions (rotations and vibrations) can occur below T_g. These include the β relaxation of polar, mobile side groups or chains [94–8] and the γ relaxation of short main-chain segments. [77, 93, 95, 96] The effects of water on the β and γ relaxations in nylon 66 [77, 95, 96] exemplify similar plasticizing effects observed for many other polymers. [86, 93, 94, 97, 98] With increasing percentage moisture, the intensity of the β transition (dynamic mechanical loss peak) increases and T_β

decreases, while the intensity of the γ transition decreases and T_γ decreases slightly. At $T_\gamma < T < T_\alpha$, increasing percentage moisture leads to increased modulus, which is the familiar 'mechanical antiplasticization' effect of water added to a polymer below its T_g. [77] This is because, as the percentage water increases, T_g decreases, while the intensity of the γ loss peak decreases, so that the modulus between the two transitions increases. [77] An important general characteristic of these cooperative, low-temperature, secondary relaxations which occur well below T_g is that they show linear Arrhenius plots for the temperature dependence of their relaxation rates and activation energies. [98]

2.2.5 *Effects of water as a plasticizer on the properties of polymers in the rubbery state – the domain of WLF kinetics*

At $T > T_g$, plasticization by water affects the viscoelastic, thermomechanical, electrical, and gas permeability properties of amorphous and PC polymers by means of the effect on T_g. [2, 14] The dependence of polymer properties on temperature, in the rubbery range above T_g (typically from T_g to $T_g + 100$ °C), is successfully predicted [71] by the WLF equation derived from free volume theory. [34, 99] The WLF equation [36, 99] is shown in equation (2.1):

$$\log\left(\frac{\eta}{\rho T}\bigg/\frac{\eta_g}{\rho_g T_g}\right) = -\frac{C_1(T - T_g)}{C_2 + (T - T_g)} \tag{2.1}$$

where η = viscosity or other diffusion-controlled relaxation process, ρ = density, and C_1 and C_2 are 'universal constants' (17.44 and 51.6, respectively, as extracted from data on numerous polymers [36, 99]). Equation (2.1) describes the kinetic nature of the glass transition [34], and is universally applicable to any glass-forming polymer, oligomer, or monomer (e.g. molten glucose [36, 99]). [34, 36] The equation defines the exponential temperature dependence of any diffusion-controlled relaxation process, occurring at a temperature T, vs. the rate of the relaxation at a reference temperature, namely T_g below T, in terms of log η proportional to ΔT, where $\Delta T = T - T_g$. The WLF equation applies in the temperature range of the rubbery or undercooled liquid state above T_g, and is based on the temperature dependence of free volume (i.e. the temperature dependence of segmental mobility), as illustrated in figure 2.3. It is not applicable much below T_g (i.e. in the glassy solid state) or in the very mobile liquid state ($\eta < 10$ Pa s [36]) more than about 100 °C above T_g, where Arrhenius-type kinetics apply. [34, 99] The WLF equation depends critically on the appropriate reference T_g for any particular glass-forming polymer (of any MW and extent of plasticization [36, 99]), where T_g is defined as an iso-free volume state of limiting free volume for the liquid, and also approximately as an isoviscosity state somewhere in the range of 10^{11}–10^{14} Pa s. [34, 36, 37, 99]

The impact of WLF-type behavior on the kinetics of diffusion-controlled relaxation processes in water-plasticized polymers at low moisture can be illustrated by several examples. For instance, relative relaxation rates vs. ΔT, calculated from equation (2.1), demonstrate the exponential relationship as follows: for ΔTs of 0, 3, 7, 11, and 21 °C, the corresponding relative rates would be 1, 10, 100, 1000, and 100000, respectively. Such rates are dramatically different from those that would be defined by the familiar Q_{10} rule of Arrhenius kinetics for dilute solutions. Another example has already been illustrated in figure 2.2. The propagation step in the mechanism of recrystallization of an amorphous but crystallizable polymer (or low-MW sugar), initially quenched from the melt to a kinetically-metastable solid state, reflects a zero rate at $T < T_g$. Due to immobility in the glass, the migratory diffusion of large main-chain segments, required for crystal growth, would be inhibited over realistic times. However, the propagation rate increases exponentially with increasing ΔT above T_g (up to T_m), due to the mobility allowed in the rubbery state. [100] Many other specific examples of WLF-governed behavior in various rubbery polymer–water systems have been reported in the recent literature, including: (1) rates of structural relaxation vs. ΔT for water-sensitive structural plastics [101]; (2) dielectric loss transition data for PVAc [102]; (3) drying rates for polymer salts [103] and water-soluble polymeric coatings [104]; (4) crystallization kinetics in miscible polymer blends [105]; (5) T_g depression in water-plasticized epoxy resins [83]; and (6) mechanical relaxation behavior of water-swollen elastin. [106]

The critical message to be distilled from the preceding material in this section is that the structure–property relations of water-compatible polymers at low moisture are dictated by a moisture–temperature–time superposition. [77, 79] Referring to the PVP–water state diagram in figure 2.6 as a conceptual 'map', one sees that the T_g curve represents a boundary between physical states in which various diffusion-controlled processes (e.g. collapse phenomena) either can (at $T > T_g$, the domain of 'water dynamics') or cannot (at $T < T_g$, the 'glass dynamics' domain) occur over realistic times. [107] The WLF equation defines the kinetics of molecular-level relaxation processes that will occur above T_g, in terms of a non-Arrhenius exponential function of ΔT above this boundary condition.

2.2.6 *Water sorption by glassy and rubbery polymers*

Water vapor absorption may be treated as a diffusion-controlled transport process involving relaxations in glassy and rubbery polymers. The kinetics of diffusion associated with adsorption leading to absorption [108] and with sorption–desorption hysteresis [109] depend, in part, on the ever-changing structural state of a polymer, *vis-a-vis* its T_g, and on the polymer's extent of plasticization by water. [107] For sorption at $T \ll T_g$, classical Fickian diffusion of low-MW plasticizing sorbate, which may appear to be time-independent [110] and to show Arrhenius-type temperature dependence, may actually be an indication of extremely slow and inconspicuous relaxation in a kinetically-metastable glassy polymer moving toward its equilibrium state.

[107, 111, 112] In the temperature range near, but below, T_g to 5–10 °C above T_g (the latter called the 'leathery' region), observations of anomalous [111], non-Fickian [113], time-dependent [114], cooperative [115] diffusion have been suggested to indicate that the glassy state is relaxing more rapidly to the rubber. [107, 116] These sorption situations also reflect the fact that water plasticization of glassy polymers leads to increasing permeability of the substrate to gases and vapors, due to increasing segmental mobility of the polymer as T_g decreases relative to the constant temperature of sorption, T_s. [76, 94, 117] In the rubbery state above T_g, diffusion and relaxation rates increase sharply, as does polymer free volume [107] (which is known to cause a dramatic increase in permeability to gases and vapors [117, 118]), as WLF-governed temperature dependence takes over [115], and Fickian diffusion behavior again applies. [91, 107]

Non-equilibrium characteristics of 'equilibrium' water vapor sorption and sorption isotherms From countless studies of equilibrium water vapor sorption and sorption isotherms for amorphous or PC, water-compatible polymers, two general characteristics have become widely acknowledged. One is that such experiments do not usually represent a true thermodynamic equilibrium situation, since the polymer substrate is changing structurally (and slowly, during sorption experiments over too short a period) due to plasticization by sorbed water. [11, 107, 109, 113, 119–21] Secondly, since T_g decreases during sorption, such experiments are not even isothermal with respect to the ΔT-governed viscoelastic properties of the polymer, because ΔT ($= T_s - T_g$) changes over the sorption time course. [107, 109] Consequently, both the extent of sorption and the mobility of sorbed molecules generally increase with increasing plasticization by water. [111, 119, 120]

Any discussion of water vapor sorption isotherms must start with an explanation of the characteristic sigmoidal shape of most such isotherms measured at 25 °C, e.g. curve (a) for epoxy resin shown in figure 2.7 (from [82]). The basic premises are as follows: (1) the sorption properties of a dry polymer depend on its initial structural state and thermodynamic compatibility with water [113, 122]; (2) in both amorphous and PC polymers, only the amorphous regions preferentially absorb water [92, 96, 123]; (3) while water sorption by non-polar polymers is low and certainly not site-specific [123], so that Flory–Huggins solution theory may be applicable, Flory–Huggins random mixing cannot describe the isotherms for polar polymers which show greatly increasing sorption with increasing relative humidity (% RH) [123, 124]; (4) for polar polymers, monolayer adsorption is thought to begin with site-specific H-bonding between individual water molecules and polar groups in the amorphous regions. [77, 114, 123] For sorption by solid polymers at $T_s < T_g$, sigmoid isotherms are said to result from two different sorption mechanisms, one operative in the region of low relative vapor pressure (RVP) and the other at high RVP. [100, 116, 122] (The

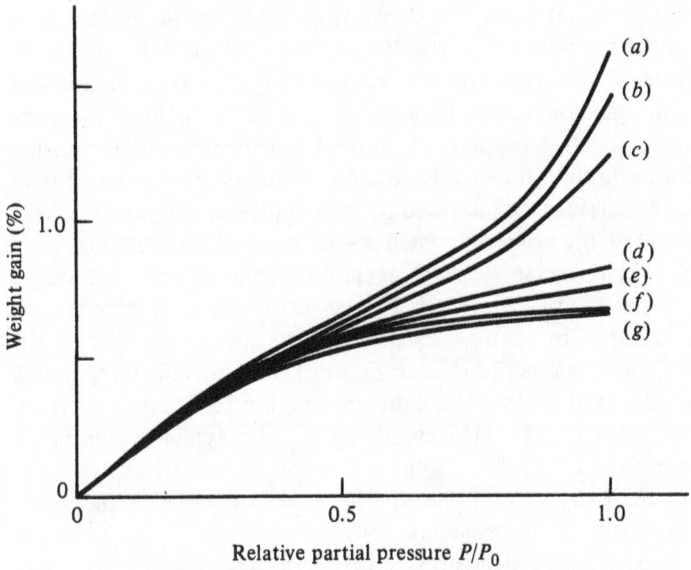

Fig. 2.7. 'Equilibrium' water vapor sorption isotherms of epoxy resin: (*a*) 25 °C; (*b*) 35 °C; (*c*) 75 °C; (*d*) 100 °C; (*e*) 125 °C; (*f*) 150 °C; (*g*) 175 °C. Reproduced, with permission, from [82].

reader will note our scrupulous avoidance of 'water activity' ('A_w'), a term derived from equilibrium thermodynamic theory and rigorously applicable only to infinitely-dilute solutions [11, 12], to describe the abscissa of a sorption isotherm for any typical water-compatible polymer at low moisture. As reviewed in detail elsewhere [5, 6, 20], we prefer RVP over a number of other popular descriptors, since it represents the parameter actually measured in a sorption experiment). At low RVP, 'dual-mode' sorption [125, 126], by a superposition of Langmuir-type monolayer site binding ('hole' filling) and Henry's law molecular solution [100, 122], is said to account for the typical concave shape [116], which alternatively has been described as classic BET Type II site-specific monolayer adsorption. [82, 123] At high RVP, large positive deviations from Henry's law [122] are said to produce the typical convex shape, which is explained on the basis of 'water clustering' [124], which is thought to follow monomolecular saturation of the polar sorption sites. [77, 87, 114] This concept of 'water clustering' is described by the Zimm–Lundberg extension of Flory–Huggins binary solution theory. [124] These clusters are imagined as chains of water molecules that extend from the H-bonding sites on the polymer and grow in size up to the compatibility limit of the polymer–water solution. [87, 124] Alternatively, this sorption behavior has been described as BET Type II multilayer formation. [113] One should be sceptical about the above explanations of isotherm shape, which were all derived originally from thermodynamic theory for equilibrium systems.

The shape of a particular isotherm depends critically on the relationship between T_s and both the initial T_g of the dry polymer and the T_g of the water-plasticized polymer during the sorption experiment. [112] We can consider the following four general cases. In case (i), for sorption at a series of increasing T_ss $\ll T_g$, sorption capacity at saturation decreases with increasing T_s, so that the isotherm shifts downward to lower values of water content at all RVPs. [108] Such behavior is illustrated by curves (*a*)–(*c*) in figure 2.7. [82] The inflection point on the isotherm, which is said to coincide with the onset of 'water clustering', occurs at a lower RVP, as the clustering tendency is thought to increase with increasing T_s. [116] This type of sorption behavior is manifested, for example, by such glassy polymers as epoxy resin [82], PVP [37], PVAc [102], and elastin. [106] Such Zimm–Lundberg clustering is also said to be indicated by the Flory–Huggins interaction parameter, χ, which decreases with increasing RVP [116], signifying that a polymer–water pair is not highly compatible. [122] The Langmuir sorption capacity factor, C'_H, which is said to be a measure of the unrelaxed free volume in the glass, decreases with increasing T_s, approaching zero at T_g. [116] C'_H is a function of ΔT [116], so that C'_H also decreases with decreasing T_g, due to water plasticization during sorption. [112, 116, 127]

In case (ii), for a series of increasing T_ss, some below and some above T_g, the vapor sorption by glassy polymers is reported to be anomalously high, due to 'water clustering', relative to rubbery polymers under the same sorption conditions. [126] The sorption capacity decreases abruptly once T_s is above T_g (as C'_H goes to zero), leading to Henry's law solution sorption in the rubbery state. [117, 127] This type of behavior has been reported for several water-sensitive polymers [97, 122], including crosslinked epoxy resin. [82] As illustrated in figure 2.7 [82], epoxy resin manifests isotherms approaching linearity at high RVP for T_s above the water-plasticized T_g, which apparently lies between 75 and 100 °C for this sample. It is interesting to note how the onset of the glass transition is indicated by a conspicuous change in a polymer's sorption behavior, which coincides with a dramatic change in the polymer's specific volume. Thus, the point (T/% moisture) at which the plot of specific volume vs. temperature (in figure 2.3) changes slope is reflected in sorption isotherms (such as those in figure 2.7) as the coincident T/% moisture condition above which the sorption curve is a flat line characteristic of solution sorption by a rubbery polymer, but below which the sigmoid sorption curve is characteristic of 'water clustering' in a glassy polymer. In another documented example of this behavior, i.e. poly(acrylonitrile) (PAN), application of the 'dual-mode' sorption model for penetrants with high solubility in glassy polymers led to the following results. [122, 125] At $C'_H = 0$, the water monolayer-saturated T_g was reported to be 42 °C, vs. a dry T_g of 97 °C, so that for $T_s = 50$ °C, the shape of the isotherm flattened suddenly, becoming much less sigmoidal than the isotherms measured at 20, 30, and 40 °C, thus indicating some but much less 'water clustering' in the rubbery polymer. [122, 125]

Case (iii) (which constitutes a variation of case (ii)) represents a single T_s, initially below T_g of the dry polymer, but subsequently above the 'wet' T_g, depressed due to water sorption. In this case, the sorptive capacity, at a specified T_s and RVP, is taken to be a measure of the kinetic free volume changes in a glassy polymer being plasticized and swollen by water. [110] During sorption, ΔT (in terms of $T_g - T_s$) decreases from positive through zero to negative, C'_H goes to zero, and free volume increases abruptly at $\Delta T = 0$ (as illustrated in figure 2.3). This can result in a sudden slope change in the isotherm, due to the structural transition at T_g. [112, 128, 129] (Similar behavior has been observed in cases where amorphous but crystallizable substrates (e.g. monomeric sugars such as sucrose [130], amylopectin in gelatinized starches [11, 131], or amorphous gelatin [24]), initially in the form of low-moisture glasses, recrystallize from the rubbery state, after plasticization due to sorption of sufficient moisture allows the glass transition to occur, either deliberately during short-term sorption experiments or inadvertently during long-term product storage. [4, 79]) Generally, a glass transition is said to cause a reduced sorptive capacity of the rubbery substrate for water. [108] However, some very mobile rubbery polymers, such as PVAc [102] and elastin [87], have been reported to show the opposite effects of increasing sorptive capacity, 'water clustering', and cluster growth once T_g falls below T_s.

The last general case (case (iv)) is that of sorption, at $T_s > T_g$, up to the thermodynamic solubility limit for water in rubbery liquid polymers. Rubbers are said to manifest Henry's law linear solution sorption, with or without additional clustering at high RVP (depending on the relative hydrophilicity of the polymer). [82, 91, 116] 'Water clustering' in a rubber is reported to be exothermic [116], and clustering is presumed to be a general characteristic of poor solvents such as water in relatively hydrophobic polymers such as elastin. [87] On the other hand, in more hydrophilic polymers such as epoxy resin [82], there is no apparent clustering, at $T_s > T_g$. Hydroxypropyl cellulose (HPC) is particularly noteworthy as the most hydrophobic of the water-compatible cellulose ethers. [114] HPC has been reported to show a great clustering tendency, starting at low RVP, and increasing with increasing RVP. [114] For sorption at $T_g < T_s <$ LCST, the sorptive capacity of HPC decreases with increasing T_s. [114] In general, it has been reported that the clustering tendency in rubbers increases, and clustering begins at a lower RVP, with increasing percentage crystallinity for PC polymers such as PAN [116] and nylons. [77]

Apart from the temperature dependence of sorption, the overall extent of 'equilibrium' vapor sorption (e.g. at 25 °C) and resulting extent of plasticization by water have been reported to vary with the following characteristics of glassy and PC polymers. Generally, sorption increases with increasing hydrophilicity [85, 98, 111] (i.e. increasing content of polar groups [76, 111]). Specifically, sorption is assumed to increase with increasing potential for water–polymer H-bonding and with decreasing interpolymer H-bonding. [123, 128] Sorption is also reported to increase with decreasing percentage crystallinity and percentage tacticity (i.e. with increasing content of amorphous regions for sorption) in PC polymers such as nylons. [92] Other polymer properties which have also been credited with producing increased water sorption include (1) increasing extent of crosslinking, e.g. in epoxy resins [83]; (2) decreasing T_g [73] and percentage neutralization [98, 103] in polymer salts; and (3) decreasing degree of substitution in cellulose derivatives. [114]

A popular subject in the recent sorption literature concerns the general molecular-level structure/property relationships of water sorbed by and dissolved in polymers at room temperature [2], specifically whether this water is strongly plasticizing, freezable, clustered, and 'bound'. The consensus view, with minor polymer-specific variations, is as follows. It is claimed that two different fractions of water molecules, which correspond roughly to the two sorption mechanisms (at low and high RVP) described above, generally exist in polymers saturated to their thermodynamic compatibility limit. [124] The first and earlier-sorbed fraction is thought to plasticize most strongly, to be always unfreezable and unclustered [125], and is often (but not always [86, 121, 129, 132]) referred to as 'bound'. These properties are said to be consequences of site-specific monolayer H-bonding to polar groups, as reported for many different synthetic and natural polymers, including PVP [121, 133], PVAc [40, 102], cellulose ethers [114, 115], nylons [73, 77], epoxy resin [83], elastin [73, 86], collagen [109, 115, 120], lysozyme [84], and keratin. [108] The second fraction has been reported to be later-sorbed water that is freezable, clustered, often referred to as 'free' [134], 'mobile' [115], or 'loosely bound' [84, 106], and is either weakly- or non-plasticizing, depending on the degree of water-compatibility of the specific polymer. Alternatively, some workers [79, 135] still employ a classical BET monolayer approach to explain this same phenomenon, as illustrated in figure 2.8 for elastin. [88] It has been reported [136] that the T_g of elastin is depressed more strongly by the first 8 % water (i.e. up to the BET monolayer value, which is determined from the point where the slope changes in the plot of log T_g vs. percentage moisture, shown in the inset of figure 2.8 [88]), and then less strongly by the BET-multilayered or Zimm–Lundberg clustered water. In fact, no such sophisticated sorption concepts and mechanisms, originally derived from equilibrium thermodynamic theory for surface adsorption by inert substrates, need be invoked (some, for example, the concept of 'water clustering', may even be misleading) to understand the sorption behavior (case (iii) above) of glassy polymers such as elastin. We see in the main plot of figure 2.8 [136] a smooth T_g curve for elastin that exemplifies the dramatic plasticizing effect of sorbed water (initially, at $T_s \ll T_g$) on many water-compatible glassy polymers at low moisture. (Note the similarity to the T_g curves for PVP in figure 2.6 and to the T_g vs. percentage moisture behavior for collagen and nylon 6 [73], alluded to earlier.) The first 8 w% moisture depresses the T_g of bone dry

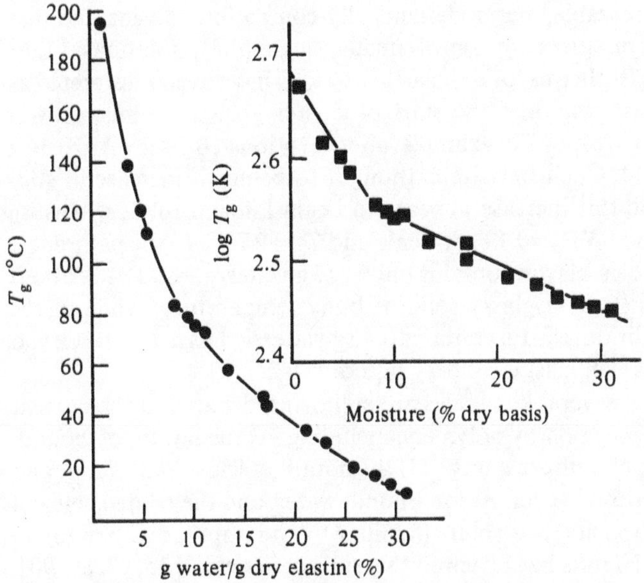

Fig. 2.8. T_g as a function of water content for elastin (from [136]). (Inset: the same data plotted as log T_g vs. percentage moisture (from [88]), to illustrate the slope change at the BET monolayer value.) Reproduced with permission.

elastin (about 200 °C) by about 110 °C, while the next 23 w% moisture depresses T_g by an additional 3 °C/w% water. By the time dry elastin has sorbed about 30 w% water, its T_g has been depressed to below room temperature, so that $T_s > T_g$, and the kinetics that govern the diffusion-controlled sorption by rubbery elastin have switched from Arrhenius- to WLF-type (i.e. increases in diffusion rates have switched from about a doubling to about a factor of 10 for each 1% increase in moisture). This fact, and recognition and quantitation of how ΔT changes during water sorption by elastin, can account for the shape of the T_g curve and the observation [87] of increased sorptive capacity of this very mobile rubbery polymer. This explanation, based on the structure–property relations for amorphous polymers at low moisture, is much more palatable to us (and others such as Franks [35, 37, 89, 90]) than one which invokes the misleading concept of water molecules in several different 'states', including 'bound', 'loosely bound', and 'free' populations.

(It is interesting to note the relationship between the reported sorption behavior of elastin and its physiological activity as a major elastic (due to plasticization by water) protein in skin and arteries. In the native state, amorphous elastin exists as a highly water-swollen, crosslinked network, which manifests rubber-like elasticity, and a dramatic increase in extensibility, at $T > T_g$. [136] For elastin at up to 0.35 g water/g, it has been reported that this water is an unfreezable, thermodynamically-compatible solvent, and that such elastin–water mixtures are homogeneous, single-phase solutions. [136] However, sorption isotherms for dry elastin at 25 °C have been interpreted as showing 'water clustering' and the start of cluster growth at much lower moisture contents. [87, 136] For example, after sorption at 0.70 RVP to 0.16 g water/g and $T_g = 40$ °C, cluster size is thought to begin to increase in still-glassy elastin, while this increase appears to begin later in rubbery elastin, after sorption at 0.80 RVP to 0.19 g water/g and $T_g = 25$ °C. In the pathologic state of arteriosclerosis, elastin contains only 0.17 g water/g, has a T_g of about 40 °C, and thus exists as a glassy solid at body temperature; while in the healthy *in vivo* condition, elastin contains 0.40 g water/g, has a $T_g < 0$ °C, and exists in the body as an elastic rubbery liquid. [96])

As implied above, several key issues are still being debated in the current literature on water sorption by polymers, including (1) the nature of 'bound' water, (2) the cause of 'unfreezability', (3) the nature of 'clustered' water, and (4) the mobility of sorbed water. As for 'bound' water, and the related concept of 'water binding capacity' of a solute (familiar to and popular with many in the food industry), Franks has reviewed this subject in detail [35, 37, 89, 90], and taken great pains to point out that this so-called 'bound' water is not truly bound in any energetic sense. While being unquestionably capable of plasticizing polymers strongly, this water is subject to rapid exchange [51], has thermally-labile H-bonds [129, 132], shows cooperative molecular mobility [115], has a heat capacity (C_p) approximately equal to that of liquid water rather than ice [86, 115, 129], and has some capability to dissolve salts. [137] 'Unfreezability' is another spurious concept which Franks has discouraged. [89] It has been conclusively demonstrated, for water-compatible polymers and monomers alike, that such 'unfreezability' is not due to tight equilibrium binding by solute [77], but to purely kinetic retardation of the diffusion of water and solute molecules (i.e. the mobility of polymer chains [115, 129]) at the low temperatures (where η increases dramatically and free volume decreases) approaching the vitrification T_g of the solute–unfrozen water mixture. [129, 132] This unfrozen water is only unfreezable during the short time frames of its measurement, be they minutes to hours in a DSC experiment or months to years in product frozen-storage. Franks [89] has reported calculations (crystal growth rate, in terms of the rate of translational diffusion of a water molecule through a glassy solid) which demonstrate that if one were to wait long enough (i.e. decades to centuries), such 'unfreezable' water would eventually form ice crystals. These facts also relate to the issue of 'clustered' water and the temperature dependence of clustering. It has been reported that even the 'unfreezable' sorbed-water fraction which is not clustered at 25 °C becomes clustered, and represents the mobile entity, as the temperature decreases to T_g of the polymer–water solution. [36, 115] In the low-temperature region around −100 °C, cooperatively-diffusing water molecules are said to exist as clusters, which represent the kinetic unit [134], near their T_g. [86, 115] As for the mobility of sorbed water, it has been reported that both clustered and Langmuir site-sorbed molecules at 25 °C are less mobile than Henry's law solution molecules associated with the same polymer. [126]

We believe that such research gaps and unanswered questions remain in the water sorption literature because polymers are almost always treated generically, without regard for the well-known dependence of T_g on MW (illustrated in figure 2.5). For most isotherms measured at 25 °C, the initial ΔT between T_g and T_s may not only be different for each type of glassy polymer, but also for each specific MW sample of a particular polymer. Sorption by different polymers or by different MW fractions of a single polymer should only be compared rigorously on the basis of characterized values of ΔT and C'_H. We could find no reported studies of sorption as a function of MW for any homologous series of glassy polymers of different initial dry T_gs and corresponding ΔTs. Nor could we find any studies (except for an attempt by Starkweather on nylon 66 [77]), of measured wet T_g vs. RVP (for samples taken at RVP increments during sorption), to augment the usual measurements of T_g vs. % moisture, in order to be able to directly correlate RVP, percentage moisture, and ΔT over the duration of a sorption experiment. The earlier discussion of these sorption parameters for elastin had to be compiled from several different references. [73, 86–8, 136]

Sorption–desorption hysteresis A major topic in the sorption literature is sorption–desorption hysteresis. It has been stated that 'hysteresis is the outstanding unexplained problem in sorption studies.' [109] Many amorphous and PC polymers, both synthetic and natural, that swell slowly and irreversibly during water sorption (with accompanying clustering [100, 122]) show marked hysteresis. [108, 109] In PC polymers (e.g. native starch), hysteresis is said to increase with increasing percentage crystallinity. [11] Water-soluble polymers such as PVP and hydroxyethyl starch (HES) show hysteresis at moisture contents $\lesssim 0.5$ g/g polymer [121], which is often referred to as their 'water binding capacity'. This moisture content corresponds to the unfreezable fraction (i.e. dynamically-constrained) of sorbed water which vitrifies at $T < T_g$ of the polymer–water solution, as illustrated in the state diagram for PVP in figure 2.6.

Characteristically, hysteresis is manifested by a desorption isotherm which occurs at higher moisture content than the corresponding sorption isotherm [108, 113, 138], such that desorption 'appears' to take place at a lower 'effective' temperature, despite constant T_s. During desorption, if a swollen polymer maintains a more open but still rigid and glassy structure (i.e. if T_g remains greater than T_s), the desorption rate may be higher than the rate of sorption, and this can produce marked hysteresis. [109] However, if sorption by a glass produces a porous, plasticized, elastic rubber (i.e. case (iii) sorption), this may lead to structural collapse (at $T_s > T_g$) during desorption, which may result in decreased desorption rate [139] and decreased hysteresis. [138] (Sorption by kinetically-metastable glassy sucrose represents a specific example of case (iii) behavior, as mentioned earlier. Not only has there been observed a sudden slope change in the sorption isotherm, but hysteresis has also been observed [131] as a consequence of irreversible collapse, followed immediately by sucrose recrystallization, from the rubbery state. The extent of hysteresis generally is reported to decrease with increasing sorption–desorption cycling [138], and with increasing T_s (due to increased chain mobility). [109] The extent of hysteresis is thought to depend on the extent of swelling during sorption [138] and on the rate of the swelling/shrinkage relaxation process (as determined by ΔT). [107–10, 122] Polymers which do not swell during water sorption (due to preexisting porosity [140]), or which can swell rapidly and reversibly (e.g. gelatinized starch [11]), due to high segmental mobility resulting from plasticization [109], show little [141] or no hysteresis. [108, 109, 122]

The extent of hysteresis at a particular T_s thus appears to be related to T_g and the magnitude of ΔT, in a manner consistent with WLF free volume theory. [107, 109, 112] Hysteresis has been seen to be most marked in swollen but rigid and still-glassy substrates whose T_g remains greater than T_s (case (i) sorption). [100, 113, 114] Conversely, hysteresis may be absent in flexible rubbery polymers [114], with T_g always less than T_s (case (iv) sorption). In the intermediate case, some hysteresis has been observed whenever the initial dry T_g is greater than T_s, but is then depressed by plasticization to a $T_g < T_s$ (case (iii) sorption). [11, 91, 111, 119] Unfortunately, as has been frequently noted [107–9, 112, 113, 121, 141], hysteresis can not be explained by any thermodynamic treatment which applies only to reversible equilibrium states. Hysteresis is not an intrinsic feature of a sorbing polymer, but depends on the experimental conditions. [108] The often anomalous nature of sorption–desorption that has been associated with hysteresis, including non-Fickian diffusion behavior, is due to the gradual plasticization of a glassy or PC polymer by water during the sorption experiment. [11, 91, 107, 109, 111–13, 119] It has been concluded that hysteresis characteristically results from a moisture/temperature/time-dependent, slow, non-equilibrium, swelling-related conformational change (involving a structural relaxation, and in some cases, even a subsequent phase change), which is facilitated by increasing free volume and mobility in a polymer which is being plasticized, during sorption that usually progresses through the stage of water clustering. [107, 108, 111, 113, 122, 141] In such systems, true thermodynamic equilibration may take weeks, months, or even years to be achieved. [101, 110, 114, 119]

Several different formal sorption theories, models, and mechanisms have been described explicitly or alluded to in section 2.2.6. For further details of these and other recent and popular theoretical sorption treatments, the reader is referred to the excellent reviews by van den Berg. [5, 6, 11, 12] The point we have tried to stress is that all these treatments suffer from serious shortcomings, due in large part to their basis on equilibrium thermodynamics. None of the available sorption treatments is capable of accounting simultaneously for: (1) the critical influence on sorption behavior of a time-dependent, water-plasticized structural relaxation occurring in a polymer at its T_g; (2) the temperature dependence of sorption, and the resulting relationship between T_s and a polymer's T_g; and (3) the relationship between time-dependent sorption parameters and the temperature location of a polymer's water-plasticized T_g. A complete mechanistic understanding of the water sorption process for a water-plasticizable polymer, and the resulting predictive capability that this would provide, would require definition of the dependence of sorption behavior on the independent variables of moisture content, temperature, time, polymer MW (and dry T_g), and the dependent variable of the structural state of a polymer, in terms of its T_g and ΔT, during sorption. Unfortunately, no single current sorption isotherm equation, model, or theory is capable of such a complete description.

2.3 Collapse phenomena in low-moisture and frozen polymeric food systems

The extensive recent literature on caking and other collapse phenomena in amorphous or PC food powders (reviewed in [4, 79, 135]) supports a conclusion that such phenomena are consequences of a material-specific structural relaxation process. We have described a unifying concept for interpreting collapse phenomena, which is based on the premise that these consequences represent the microscopic and macroscopic manifestations of an underlying molecular state transformation, from kinetically-metastable amorphous solid to unstable amorphous liquid, which occurs at T_g. [14, 26, 27] The critical effect of plasticization by water on T_g is a central element of the concept and the mechanism derived from it. This generalized physico-chemical mechanism for collapse in amorphous food polymers, derived from WLF theory [34], has been described [14, 26, 27] as follows. As the ambient temperature rises above T_g, or as T_g falls below the ambient temperature due to plasticization by water, polymer free volume increases, leading to increased segmental mobility of the polymer chains. Consequently, the viscosity of the dynamically-constrained solid falls below the characteristic η_g at T_g (i.e. the glass transition has occurred) and permits viscous liquid flow. In this rubbery state, translational diffusion can occur in practical time frames, and diffusion-controlled relaxations (including structural collapse) are free to proceed with rates defined by the WLF equation (2.1) (i.e. rates which increase exponentially with increasing ΔT).

To illustrate our concept of collapse phenomena in polymeric systems at low moisture, we have reported results for the structure–property relationships of two extensive series of food carbohydrates, (1) sugars, glycosides, and polyhydric alcohols, and (2) commercial SHPs. [26, 27] The physico-chemical properties of SHPs (which are amorphous, water-soluble polymers of glucose, such as dextrins, maltodextrins, corn syrup solids, corn syrups) represent an important, but sparsely-researched, subject within the food industry. [142, 143] One study [88] had reported a correlation between increasing T_c and increasing number-average degree of polymerization, $\overline{DP}_n$, for a series of SHPs of $2 \leqslant \overline{DP}_n \leqslant 16$ (calculated DE = 52.6 to 6.9). (DE is defined as $100/(\bar{M}_n/180.16)$, where $\bar{M}_n$ is number-average MW, since the reducing sugar content (in terms of the number of reducing end groups) of a known weight of sample is compared to an equal weight of glucose of DE 100 and $\bar{M}_n$ 180.16. [26]) In contrast, low-MW sugars and polyols have been studied extensively, and limited compilations of their characteristic T_cs are available. [37, 90] Our research has demonstrated that much can be learned about the functional attributes of SHPs, sugars, and polyols, many of which are common ingredients in both fabricated and natural foods, from a polymer physico-chemical approach to systematic studies of their thermomechanical properties and the effects thereon of plasticization by water. [26, 27] These studies have led to predictive capabilities.

We shall review here our DSC results for the T_g' values of 80 SHPs (DE values of 0.3–100) and some 60 polyhydroxy compounds. For experimental details concerning materials and DSC methodology, see [26, 27]. (As defined by Franks [35, 37, 90] and illustrated in figure 2.6 for PVP, T_g' is the particular T_g of the maximally freeze-concentrated solute–water glassy matrix surrounding the ice crystals in a frozen solution. As a consequence of freeze concentration, initially 'dilute' solutions (in our studies, always 20 w% solute) become 'low-moisture' systems.) For the SHPs, our analysis yielded a linear correlation between decreasing DE and increasing T_g', from which we constructed a calibration curve used to predict DE values for other SHPs of unknown DE. [26] The same DE vs. T_g' data were also used to construct a predictive map of functional attributes for SHPs, based on a demonstration of their classical T_g vs. $\bar{M}_n$ behavior as a homologous series of amorphous polymers. [26] Our study covered an extensive range of SHPs and provided a theoretical basis for interpreting previous, more limited results of To & Flink. [88] For some 60 polyhydroxy compounds, a linear correlation between increasing T_g' and decreasing value of 1/MW was demonstrated for the first time [27], and augmented the recent literature. [90] As expected, this correlation was not quite as good as the one for the SHPs, because these compounds did not represent a single homologous family of monomers and oligomers.

The demonstration that SHP functional behavior can be predicted from the correlation between DE (or $\overline{DP}_n$ or $\bar{M}_n$) and T_g, and that of sugars and polyols from their T_g vs. 1/MW relationship, has important implications for a better understanding of the collapse mechanism in polymeric systems at low moisture. [27] For the food industry, such predictive capabilities are valuable, because various non-equilibrium collapse phenomena (which are all sensitive to plasticization by water) affect the processing and storage stability of many fabricated and natural foods, including frozen products, amorphous dry powders, and candy glasses. We shall review the potential (and frequently demonstrated) utility of SHPs in preventing structural collapse, within a context of collapse processes which are often promoted by formulation with large amounts of low-MW saccharides. We shall also relate these insights to our interpretation of collapse phenomena, within the context of our concept of glass dynamics.

2.3.1 *Low-temperature thermal properties of polymeric vs. monomeric saccharides*

Figure 2.9 [26] shows two typical low-temperature DSC thermograms for 20 w% solutions: (*a*) glucose, and (*b*) Star Dri 10 10DE maltodextrin. In each, the heat flow curve begins at the top (endothermic down), and the analog derivative trace (zeroed to the T axis) at the bottom (endothermic up). For both thermograms, instrumental amplification and sensitivity settings were identical, and sample weights comparable. As illustrated by figure 2.9, the derivative feature of the DuPont 990 DSC greatly facilitates the identification of sequential thermal transitions, assignment of precise transition temperatures, and thus overall interpretation of thermal behavior, especially for such frozen aqueous solutions exemplified by figure 2.9(*a*). Surprisingly, we have found no other reported use of derivative thermograms, in the many DSC studies of such systems, to sort out the small endothermic and exothermic changes in heat flow that occur typically below 0 °C (see [37] for an extensive bibliography).

Despite the handicap of such instrumental shortcomings in the past, the theoretical basis for the thermal properties manifested by aqueous solutions at subzero temperatures has come to be well understood. [32, 33, 35, 37, 90, 144–7] As shown in figure 2.9(*a*), after rapid cooling of the glucose solution to below −80 °C, slow heating revealed a minor T_g at −61.5 °C, followed by an exothermic devitrification (a crystallization of some of the previously-unfrozen water) at −47.5 °C, followed by another (major) T_g, namely T_g', at −43 °C, and then finally the melting of ice at T_m. In figure 2.9(*b*), the maltodextrin solution thermogram shows only the obvious T_g' at −10 °C, in addition to T_m. These assignments of transitions and temperatures can be reconciled definitively with earlier state diagrams for such materials. [37, 145] In such diagrams (see our figure 2.6), the different cooling/heating paths which can be followed by solutions of monomeric and polymeric solutes are revealed. For the former (e.g. glucose), partial vitrification of the original solution could occur, apparently because the cooling rate was high relative to the rate of ice crystallization; whereas for the latter (e.g. a maltodextrin), the cooling rate appeared to be low relative to the rate of freezing. However, as demonstrated by the thermograms in figure 2.9, in both cases, rewarming forced the system through a glass transition at T_g'. (In many earlier DSC

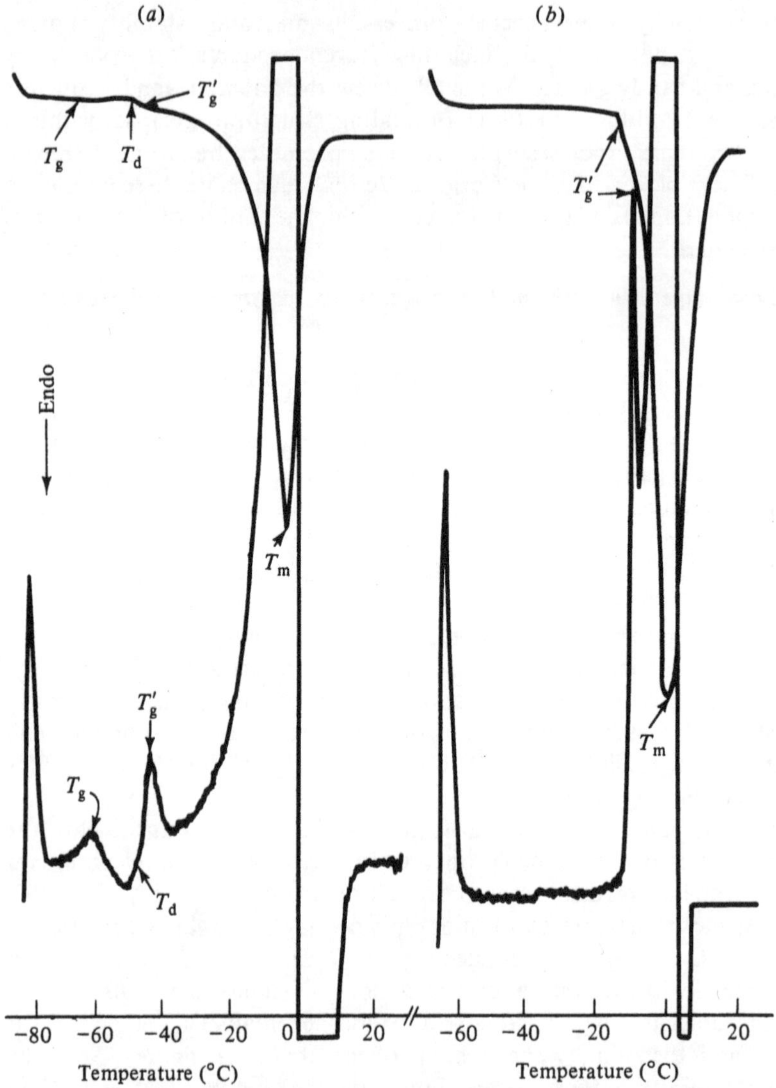

Fig. 2.9. Typical DSC thermograms for 20 w% solutions of (*a*) glucose, and (*b*) Star Dri 10 10DE maltodextrin (Staley 1984). In each, the heat flow curve begins at the top (endothermic down), and the derivative trace (zeroed to the *T* axis) at the bottom. [26]

studies [e.g. 33, 144, 148], performed without benefit of derivative thermograms, a pair of transition temperatures (each independent of initial concentration), called $T_{\text{antemelting}}$ (T_{am}) and $T_{\text{incipient melting}}$ (T_{im}) have been reported in place of a single T_g'. In fact, for the many cases that we have studied, the reported values of T_{am} [e.g. 149] and T_{im} bracket that of T_g' (as we measure it), which led us to surmise that T_{am} and T_{im} actually represent the temperatures of onset and completion of the single thermal event (a glass transition) that must occur at T_g', as defined by the state diagram. [26])

The point of greatest interest in the thermograms in figure 2.9 involves T_g'. The matrix surrounding the ice crystals in a maximally-frozen solution is a supersaturated solution (hence, in our perspective, a 'low-moisture' system) of all the solute in the fraction of water remaining unfrozen. This matrix exists as a kinetically-metastable amorphous solid (a glass of constant composition) at any $T < T_g'$, but as a viscoelastic liquid (a rubbery fluid) at $T_g' < T < T_m$ (of ice). Again with regard to a state diagram for a typical solute that does not readily undergo eutectic crystallization (see [37] or our figure 2.6), T_g' corresponds to the intersection of an extension of the thermodynamically-defined equilibrium liquidus curve and the kinetically-determined supersaturated glass curve. As such, Franks [37] described T_g' as having the appearance of a 'metastable eutectic', in that it represents a quasi-invariant point in the state diagram, invariant in both its characteristic temperature (T_g') and composition (i.e. C_g', expressed as w% solute, or W_g', expressed as grams unfrozen water/gram solute) for a particular solute. (However, 'eutectic' in this usage does not imply a phase separation. [149]) This glass, which, for example, forms on slow cooling to T_g', acts as a kinetic barrier (of high activation energy) to further ice formation (within the experimental timeframe), despite the continued presence of unfrozen water at all temperatures greater than T_g', as well as to any other diffusion-controlled process. [90] Recognizing this, one begins to appreciate why the temperature of this glass transition is important in several aspects of frozen food technology (e.g. freezer storage stability, freeze concentration, and freeze drying [37, 90]), which can be subject to various recrystallization and collapse phenomena, as we will discuss.

The measured T_g' values for 80 SHPs and some 60 polyhydroxy compounds are listed in tables 2.1 and 2.2, respectively. [27] The T_g' for glucose of -43 °C is midway between reported values for T_{am} and T_{im} [46], and within a few degrees of various values for T_c and T_r (see table 2.4). The same is true of our previously-reported T_g' for sucrose of -32 °C [25]. As shown in table 2.4, literature values for T_c and/or T_r (both always independent of initial concentration) for soluble starch of -5 or -6 °C [37, 147], and for dextrin of -9 °C [147], are also comparable to our T_g' values for similar materials.

Figure 2.10 [27] shows T_g' vs. DE for all SHPs with DE values specified by the manufacturer. There is an excellent linear correlation between increasing T_g' and decreasing DE (coefficient $r = -0.98$). Since DE is inversely proportional to $\overline{DP}_n$ and $\bar{M}_n$ for this series of SHPs [142], the results in figure 2.10 demonstrate that T_g' increases with increasing $\bar{M}_n$. Such a correlation between T_g and $\bar{M}_n$ is the general rule for any homologous family of glass-forming monomers, oligomers, and polymers. [31] The equation describing the regression line in figure 2.10 is DE $= -2.2(T_g', °\mathrm{C}) - 12.8$. We have shown [26] that figure 2.10 can be used as a calibration curve for interpolating DE

Table 2.1. *T_g' values for commercial SHPs.* [27]

SHP	Manufacturer	Starch source	DE	T_g' (°C)	Gelling
AB 7436	Anheuser Busch	waxy maize	0.5	−4	
Paselli SA-2	AVEBE (1984)	potato (Ap)	2	−4.5	yes
Stadex 9	Staley	dent corn	3.4	−4.5	yes
78NN128	Staley	potato	0.6	−5	yes
78NN122	Staley	potato	2	−5	yes
V-O Starch	National	waxy maize	?	−5.5	yes
N-Oil	National	tapioca	?	−5.5	yes
ARD 2326	Amaizo	dent corn	0.4	−5.5	yes
Paselli SA-2	AVEBE (1986)	potato (Ap)	2	−5.5	yes
ARD 2308	Amaizo	dent corn	0.3	−6	yes
AB 7435	Anheuser Busch	waxy/dent blend	0.5	−6	
Star Dri 1	Staley (1984)	dent corn	1	−6	yes
Crystal Gum	National	tapioca	5	−6	yes
Maltrin M050	GPC	dent corn	6	−6	yes
Star Dri 1	Staley (1986)	waxy maize	1	−6.5	yes
Paselli MD-6	AVEBE	potato	6	−6.5	yes
Dextrin 11	Staley	tapioca	1	−7.5	yes
MD-6-12	V-Labs		2.8	−7.5	
Stadex 27	Staley	dent corn	10	−7.5	no
MD-6-40	V-Labs		0.7	−8	
Star Dri 5	Staley (1984)	dent corn	5	−8	no
Star Dri 5	Staley (1986)	waxy maize	5.5	−8	no
Paselli MD-10	AVEBE	potato	10	−8	no
Paselli SA-6	AVEBE	potato (Ap)	6	−8.5	no
α-Cyclodextrin	Pfanstiehl			−9	
Capsul	National	waxy maize	5	−9	
Lodex Light V	Amaizo	waxy maize	7	−9	
Paselli SA-10	AVEBE	potato (Ap)	10	−9.5	no
Morrex 1910	CPC	dent corn	10	−9.5	
Star Dri 10	Staley (1984)	dent corn	10	−10	no
Maltrin M040	GPC	dent corn	5	−10.5	
Frodex 5	Amaizo	waxy maize	5	−11	
Star Dri 10	Staley (1986)	waxy maize	10.5	−11	no
Lodex 10	Amaizo (1986)	waxy maize	11	−11.5	no
Lodex Light X	Amaizo	waxy maize	12	−11.5	
Morrex 1918	CPC	waxy maize	10	−11.5	
Mira-Cap	Staley	waxy maize	?	−11.5	
Maltrin M100	GPC	dent corn	10	−11.5	no
Lodex 5	Amaizo	waxy maize	7	−12	no
Maltrin M500	GPC	dent corn	10	−12.5	
Lodex 10	Amaizo (1982)	waxy maize	12	−12.5	no
Star Dri 15	Staley (1986)	waxy maize	15.5	−12.5	no
MD-6	V-Labs		?	−12.5	
Maltrin M150	GPC	dent corn	15	−13.5	no
Maltoheptaose	Sigma		15.6	−13.5	
MD-6-1	V-Labs		20.5	−13.5	
Star Dri 20	Staley (1986)	waxy maize	21.5	−13.5	no
Maltodextrin Syrup	GPC	dent corn	17.5	−14	no
Frodex 15	Amaizo	waxy maize	18	−14	
Maltohexaose	Sigma		18.2	−14.5	
Frodex 10	Amaizo	waxy maize	10	−15.5	
Lodex 15	Amaizo	waxy maize	18	−15.5	no
Maltohexaose	V-Labs		18.2	−15.5	
Maltrin M200	GPC	dent corn	20	−15.5	

Table 2.1 (*continued*)

SHP	Manufacturer	Starch source	DE	T_g' (°C)	Gelling
Maltopentaose	Sigma		21.7	−16.5	
Maltrin M250	GPC	dent corn	25	−17.5	
N-Lok	National	blend	?	−17.5	
Staley 200	Staley	corn	26	−19.5	
Maltotetraose	Sigma		27	−19.5	
Frodex 24	Amaizo	waxy maize	28	−20.5	
Frodex 36	Amaizo	waxy maize	36	−21.5	
DriSweet 36	Hubinger	corn	36	−22	
Maltrin M365	GPC	dent corn	36	−22.5	
Staley 300	Staley	corn	35	−23.5	
Globe 1052	CPC	corn	37	−23.5	
Maltotriose	V-Labs		35.7	−23.5	
Frodex 42	Amaizo	waxy maize	42	−25.5	
Neto 7300	Staley	corn	42	−26.5	
Globe 1132	CPC	corn	43	−27.5	
Staley 1300	Staley	corn	43	−27.5	
Neto 7350	Staley	corn	50	−27.5	
Maltose	Sigma		52.6	−29.5	
Globe 1232	CPC	corn	54.5	−30.5	
Staley 2300	Staley	corn	54	−31	
Sweetose 4400	Staley	corn	64	−33.5	
Sweetose 4300	Staley	corn	64	−34	
Globe 1642	CPC	corn	63	−35	
Globe 1632	CPC	corn	64	−35	
Royal 2626	CPC	corn	95	−42	
Glucose	Sigma	corn	100	−43	

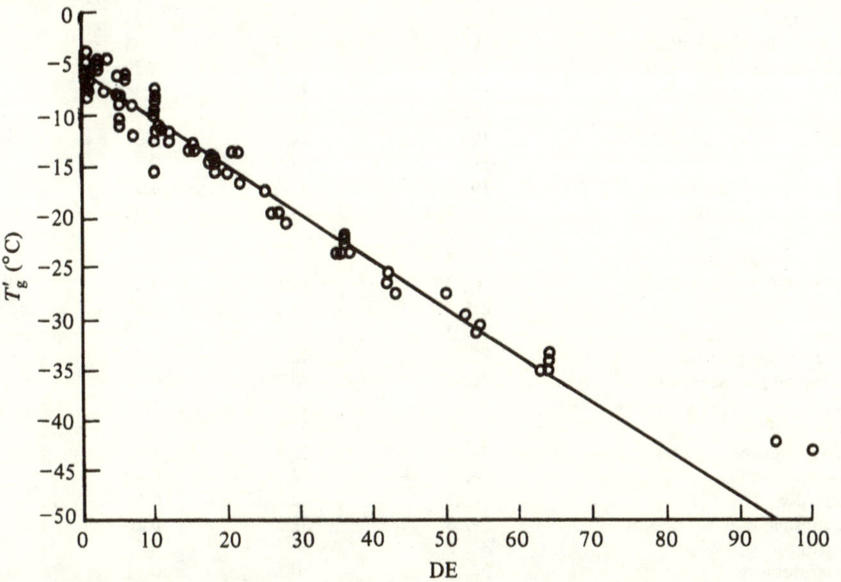

Fig. 2.10. Variation of T_g' for maximally-frozen 20 w% solutions against DE value for the commercial SHPs in table 2.1. [27]

Table 2.2. T_g' *values for sugars, glycosides, and polyhydric alcohols.* [27]

Sugar or Polyol	MW	T_g' (°C)	W_g' (g UFW/g)
ethylene glycol	62.1	−85	1.90
propylene glycol	76.1	−67.5	1.28
1,3-butanediol	90.1	−63.5	1.41
glycerol	92.1	−65	0.85
erythrose	120.1	−50	1.39
erythritol	122.1	−53.5	(eutectic)
thyminose (deoxyribose)	134.1	−52	1.32
xylose	150.1	−48	0.45
arabinose	150.1	−47.5	1.23
ribose	150.1	−47	0.49
arabitol	152.1	−47	0.89
ribitol	152.1	−47	0.82
xylitol	152.1	−46.5	
methyl riboside	164.2	−53	0.96
methyl xyloside	164.2	−49	1.01
quinovose (deoxyglucose)	164.2	−43.5	1.11
fucose (deoxygalactose)	164.2	−43	1.11
rhamnose (deoxymannose)	164.2	−43	0.90
glucose	180.2	−43	0.41
fructose	180.2	−42	0.96
galactose	180.2	−41.5	0.77
allose	180.2	−41.5	0.56
sorbose	180.2	−41	0.45
mannose	180.2	−41	0.35
tagatose	180.2	−40.5	1.33
inositol	180.2	−35.5	0.30
mannitol	182.2	−40	(eutectic)
galactitol	182.2	−39	(eutectic)
sorbitol	182.2	−43.5	0.23
2-*o*-methyl fructoside	194.2	−51.5	1.61
β-1-*o*-methyl glucoside	194.2	−47	1.29
3-*o*-methyl glucoside	194.2	−45.5	1.34
6-*o*-methyl galactoside	194.2	−45.5	0.98
α-1-*o*-methyl glucoside	194.2	−44.5	1.32
1-*o*-methyl galactoside	194.2	−44.5	0.86
1-*o*-methyl mannoside	194.2	−43.5	1.43
1-*o*-ethyl glucoside	208.2	−46.5	1.35
2-*o*-ethyl fructoside	208.2	−46.5	1.15
1-*o*-ethyl galactoside	208.2	−45	1.26
1-*o*-ethyl mannoside	208.2	−43.5	1.21
heptulose	210.2	−36.5	0.77
1-*o*-propyl glucoside	222.2	−43	1.22
1-*o*-propyl galactoside	222.2	−42	1.05
1-*o*-propyl mannoside	222.2	−40.5	0.95
2,3,4,6-*o*-methyl glucoside	236.2	−45.5	1.41
isomaltulose	342.3	−35.5	
cellobiulose	342.3	−32.5	
isomaltose	342.3	−32.5	0.70
sucrose	342.3	−32	0.56
gentiobiose	342.3	−31.5	0.26
turanose	342.3	−31	0.64
mannobiose	342.3	−30.5	0.91
lactulose	342.3	−30	0.72
maltose	342.3	−29.5	0.25

Table 2.2. T_g' *values for sugars, glycosides, and polyhydric alcohols.* [27]

Sugar or Polyol	MW	T_g' (°C)	W_g' (g UFW/g)
maltulose	342.3	−29.5	
trehalose	342.3	−29.5	0.20
cellobiose	342.3	−29	
lactose	342.3	−28	0.69
maltitol	344.3	−34.5	0.59
isomaltotriose	504.5	−30.5	0.50
panose	504.5	−28	0.59
raffinose	504.5	−26.5	0.70
maltotriose	504.5	−23.5	0.45
stachyose	666.6	−23.5	1.12
maltotetraose	666.6	−19.5	0.55
maltopentaose	828.9	−16.5	0.47
α-cyclodextrin	972.9	−9	
maltohexaose	990.9	−14.5	0.50
maltoheptaose	1153.0	−13.5	0.27

values of new or 'unknown' SHPs, in preference to the time-consuming classical methods for DE determination. [143]

For some 60 polyhydroxy compounds, the corresponding plot of T_g' vs. 1/MW is shown in the inset of figure 2.11. [27] In this case, the regression line has an r value of -0.94, which is slightly lower than the one for the homologous series of SHPs. The major contributor to the scatter in this plot is the series of chemically-different glycosides, which in no way constitutes a homologous family of glass-forming compounds. In contrast, figure 2.12 [27] shows the smooth curve that results when one plots T_g' vs. MW (from table 2.2) for the homologous malto-oligosaccharides from glucose to maltoheptaose.

Figure 2.13 [26] shows T_g' plotted vs. W_g' for 13 of the corn syrups (DE 26–95) listed in table 2.1. The composition of the glass at T_g' was calculated from the thermogram, specifically from measurement of the area (enthalpy) under the ice melting endotherm. By calibration with pure water, this measurement yields a maximum weight of ice in the frozen sample, and by difference from the known weight of total water in the initial solution, a weight of unfrozen water, per unit weight of solute, in the glass at T_g'. A few in the food industry will recognize this procedure as one of several routine methods for determining 'water binding capacity' of a solute, which was discussed earlier. As shown by the results in figure 2.13, W_g' decreases with increasing T_g' for this homologous series of corn syrup solids solutions. (The regression coefficient is -0.91.) In other words, as the average $\bar{M}_n$ of the solute(s) increases, the amount of water in the glass at T_g' generally decreases. This fact is also illustrated dramatically by the thermograms in figure 2.9. For comparable amounts of total water, the area under the ice melting peak for the glucose solution is much smaller than that for the maltodextrin solution. Once again, in the context of a typical state diagram (e.g. figure 2.6), the above results show

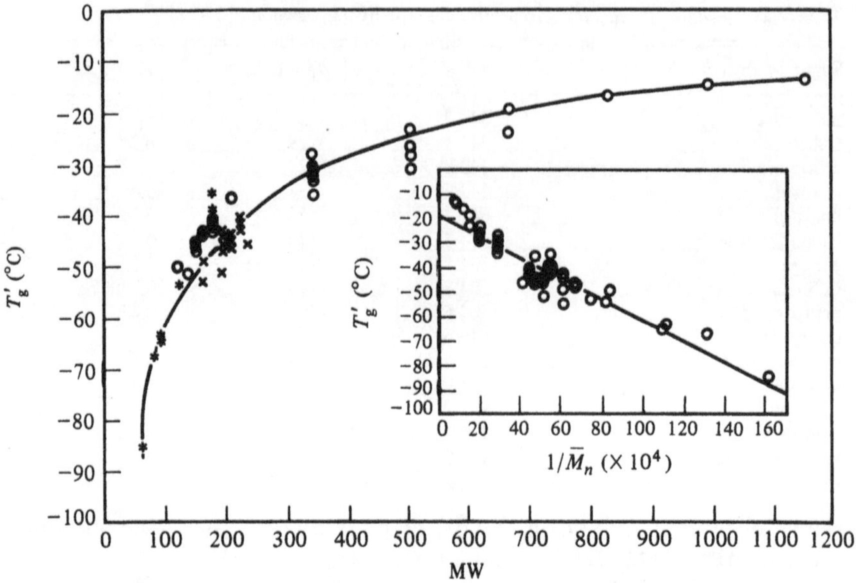

Fig. 2.11. Variation of T'_g for maximally-frozen 20 w % solutions against MW for the sugars (○), glycosides (×), and polyhydric alcohols (★) in table 2.2. (Inset: a plot of T'_g vs. $1/\bar{M}_n \times 10^4$, illustrating the theoretically-predicted linear dependence.) [27]

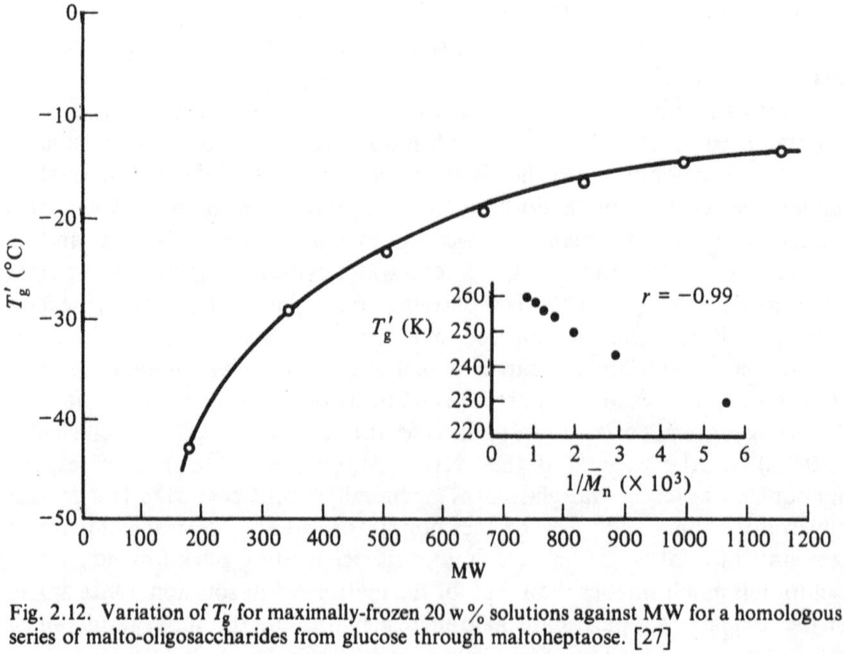

Fig. 2.12. Variation of T'_g for maximally-frozen 20 w % solutions against MW for a homologous series of malto-oligosaccharides from glucose through maltoheptaose. [27]

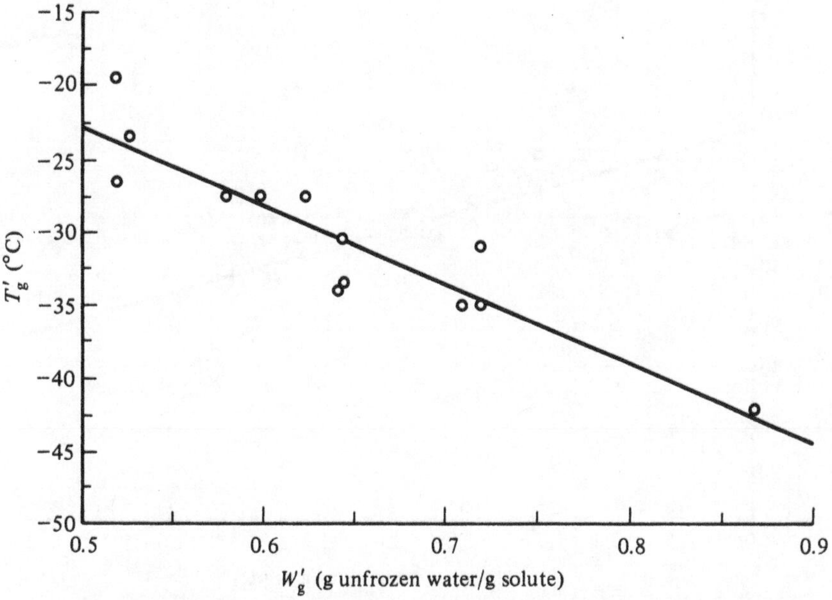

Fig. 2.13. Variation of T'_g for maximally-frozen 20 w% solutions against W'_g, the composition of the glass at T'_g, in g unfrozen water/g solute, for 13 commercial corn syrups from table 2.1. [26]

that as $\bar{M}_n$ of the solute (or mixture of homologous solutes) in an aqueous system increases, the T'_g/C'_g point generally moves up the T axis toward 0 °C and to the right along the composition axis toward 100 w% solute. The critical importance of this feature will become clear in section 2.3.4, where we describe SHP functional behavior *vis-a-vis* T'_g and dry T_g, and the possibilities of inhibiting collapse by formulating a fabricated food product with the intent of elevating T'_g and dry T_g.

In apparent contrast to the results in figure 2.13 for a homologous series of mixed glucose monomer and oligomers, the results in figure 2.14 [27], of T'_g vs. W'_g for the diverse polyhydroxy compounds listed in table 2.2, yield a regression coefficient of only −0.64. Thus, when Franks [90] notes that, among the (non-homologous) sugars and polyols most widely used as 'water binders' in fabricated foods, 'the amount of unfreezable water does not show a simple dependence on the MW of the solute', one is wise to pay heed and proceed cautiously. (In fact when the W'_g data in table 2.2 are plotted against 1/MW (not shown), $r = 0.47$.) One would conclude that the plot in figure 2.14, as shown, obviously cannot be used for predictive purposes, so the safest approach would be to rely on measured W'_g values for each potential 'water binding' candidate. However, the situation is not quite as nebulous as represented by figure 2.14. When some of the same data are plotted (actually T'_g vs. C'_g, shown in figure 2.15), but compounds are grouped by chemical classification into specific homologous series (e.g. polyols, glucose-only

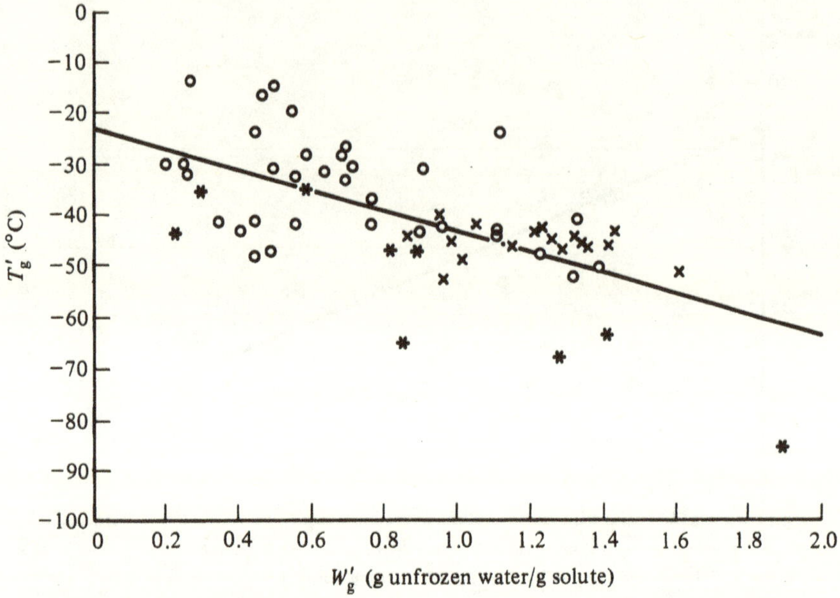

Fig. 2.14. Variation of T'_g for maximally-frozen 20 w% solutions against W'_g, the composition of the glass at T'_g, in g unfrozen water/g solute, for the sugars (○), glycosides (×), and polyhydric alcohols (★) in table 2.2. [27]

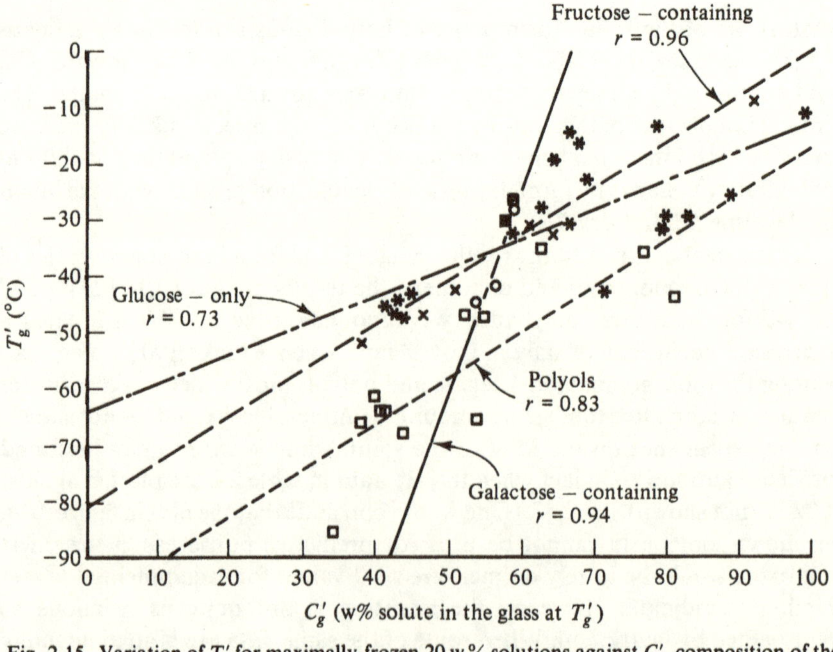

Fig. 2.15. Variation of T'_g for maximally-frozen 20 w% solutions against C'_g, composition of the glass at T'_g, in w% solute, for homologous series of polyhydric alcohols (□), glucose-only polymers (★), fructose- (×), and galactose-containing saccharides (○) in table 2.2. [18]

polymers, and fructose- or galactose-containing saccharides), better linear correlations become evident. The plots in figure 2.15 illustrate the same linear dependence of T_g' on the composition of the glass at T_g' (i.e. as the amount of unfrozen water in the glass decreases, T_g' increases) as the plot in figure 2.13. Still, Franks' suggestion that investigations of T_m and η as functions of solute concentration, and the liquidus curve as a function of solute structure, would be particularly worthwhile, is a good one.

It is interesting to note the W_g' results for the series of monomeric glycosides [27], in terms of a possible relationship between glycoside structure (e.g. size of the hydrophobic aglycone, absent in the parent sugar) and the function reflected by W_g'. Clearly, the W_g' values for all the methyl, ethyl, and propyl derivatives are much greater than those for the corresponding parent sugars. However, W_g' values appear consistently to be maximized for the methyl or ethyl derivatives, but somewhat decreased for the propyl derivatives. These results could indicate that increasing hydrophobicity (of the aglycone) leads to both decreasing W_g' and the demonstrated tendency toward increasing insolubility of propyl and larger glycosides in water.

2.3.2 *Structure–property relationships for SHPs and polyhydroxy compounds*

The straightforward presentation of the DE vs. T_g' data in figure 2.10 is not the most rigorous theoretical treatment. Yet, the linear correlation of DE with T_g' and the convenience for practical application in the estimation of DE to characterize SHP samples recommend its use. The rigorous theoretical dependence of DE on T_g' stems from the respective dependence of each of these parameters on linear *DP* and MW within a series of monodisperse (i.e. MW = $\bar{M}_n = \bar{M}_w$) homopolymers. High polymers can be distinguished from oligomers because of their capacity for molecular chain 'entanglement coupling', resulting in the formation of rubber-like viscoelastic random networks (often called gels, in accord with Flory's [150] nomenclature for disordered three-dimensional networks formed by physical aggregation) above a critical polymer concentration. [34] As summarized by Mitchell [151], 'entanglement coupling is seen in most high-MW polymer systems. Entanglements (in gels) behave as crosslinks with short lifetimes. They are believed to be topological in origin rather than involving chemical bonds.' For linear homopolymers (either amorphous or PC, and not necessarily monodisperse) with $\bar{M}_n$ values below the entanglement limit. T_g decreases linearly with increasing $1/M_n$ [31]. The onset of entanglement corresponds to a plateau region in which further increases in MW have little or no effect on T_g. [31] (There may, however, be a dramatic effect on the viscoelastic properties of the network, resulting, for example, in increased gel strength at constant temperature. [34]) The conventional presentation of such experimental data is simply T_g vs. $\bar{M}_w$ [31], which conveniently displays the plateau region. Two typical examples were shown in figure 2.5 [26], and described earlier.

To a first approximation, DE has the simple inverse dependence on $\bar{M}_n$ defined earlier. Expressing that equation in the form $\bar{M}_n = 18016/\text{DE}$, and using the conventional presentation to explore the behavior of T_g' with MW, we have shown [26, 27], in the main plot of figure 2.16, the T_g' results for the SHPs in table 2.1. After a steeply-rising portion, a plateau region is reached for SHPs with DE $\leqslant 6$ and $T_g' \geqslant -8$ °C. (The inset of figure 2.16 shows the linear relationship between decreasing T_g' and increasing $1/\bar{M}_n$, for SHPs with $\bar{M}_n$ values below the entanglement limit. To & Flink's [88] T_c/MW data showed the same correlation. In fact, the theoretical treatment of the data in the inset is simply a modified version of the straightforward presentation in figure 2.10, with the same correlation coefficient of -0.98.) We suggested [26] that the most likely explanation for this plateau behaviour is that such SHPs experience molecular entanglement in the freeze-concentrated glass that exists at T_g' and C_g'. Consequently, SHPs with DE $\leqslant 6$ and $T_g' \geqslant -8$ °C would be capable of forming gel networks (via entanglement), above a critical polymer concentration (which would be related to C'_g). Braudo *et al.* [152], in their reports on the viscoelastic properties of thermoreversible maltodextrin gels (at $T > 0$ °C), also implicated entanglement coupling above a critical polymer

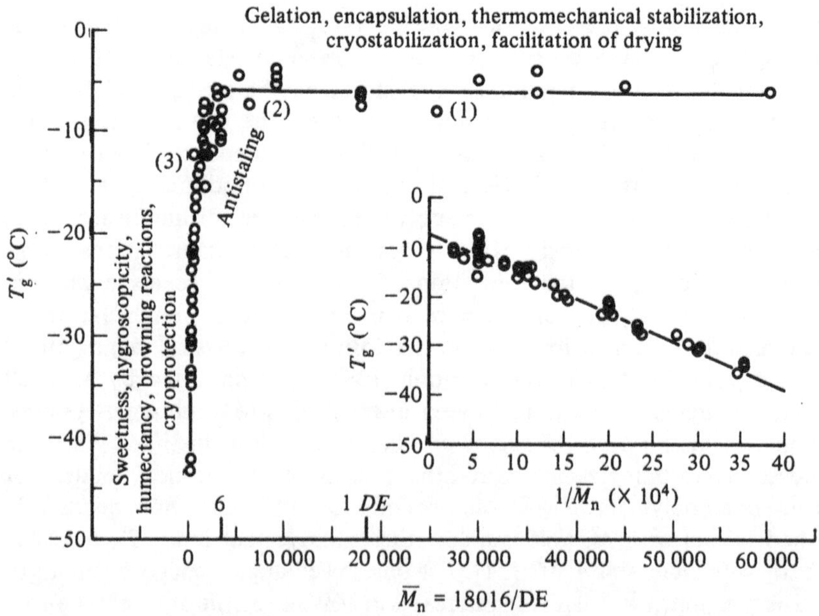

Fig. 2.16. Variation of T_g' against $\bar{M}_n$ (expressed as a function of DE) for commercial SHPs in table 2.1. DE values are indicated by numbers marked above the x axis. Data points for maltodextrin MW standards are numbered (1), (2), and (3) to provide MW markers. Areas of specific functional attributes, corresponding to three regions of the diagram, are labeled. (Inset: plot of T_g' vs. $1/\bar{M}_n$ ($\times 10^4$) for SHPs with $\bar{M}_n$ values below entanglement limit, illustrating the theoretically-predicted linear dependence. [26, 27])

concentration. They concluded that the non-cooperative gelation behavior shown by maltodextrins is characteristic of semi-rigid chain polymers. This is consistent with Ferry's [34] observation that 'molecules which are relatively stiff and extended (in concentrated solution) exhibit the effects of entanglement coupling even more prominently than do highly flexible polymers.' Additional information about thermoreversible maltodextrin (5–8 DE) gels came from Bulpin *et al.* [153], who reported that such SHP gels are apparently composed of a network of high-MW (> 10 000) branched molecules derived from amylopectin. These branched molecules represent the structural elements, which are aggregated with, and further stabilized by interactions with, short linear chains (MW < 10 000) derived from amylose.

The implications of our finding and the conclusions we have drawn from it have allowed us to explain previously-observed but poorly-understood aspects of SHP functional behavior in various food-related applications. [26, 27] The SHPs which fall on the plateau region in figure 2.16 have DEs from 6 to 0.3. These DEs correspond to DP_ns in the range 18–370, respectively, and $\bar{M}_n$s between 3000 and 60 000. (The data points for the three maltodextrin MW standards are numbered in figure 2.16 to provide MW markers. The points for (1) MD-6-40 ($\bar{M}_n = 27\,200$; $\bar{M}_w = 39\,300$) and (2) MD-6-12 ($\bar{M}_n = 6\,500$; $\bar{M}_w = 13\,000$) fall on the plateau, while (3) MD-6-1 ($\bar{M}_n = 880$; $\bar{M}_w = 1030$) is below the entanglement limit.) Within this series of SHPs, the minimum linear chain length apparently required for intermolecular entanglement corresponds to $\overline{DP}_n \approx 18$ and $\bar{M}_n \approx 3000$. This fact explains why there is no plateau region in To & Flink's [88] plot of T_c vs. $\overline{DP}_n$ for SHPs of $\overline{DP}_n \leqslant 16$ and DE $\geqslant 6.9$. Their figure 3 and the portion of our figure 2.16 for DE $\geqslant 7$ are similar in appearance; both show a steeply-rising portion for DE $\geqslant 20$, followed by a less steeply-rising portion for $20 \geqslant$ DE $\geqslant 7$. Importantly, the entanglement capability evidenced by just such SHPs of DE $\leqslant 6$ (materials analyzed by our polymer characterization method [26], but not previously analyzed by To & Flink [88]) underlies various aspects of their functional behavior. For example, as described by Slade [16], sufficiently long linear chain lengths ($\overline{DP}_n \gtrsim 18$) of SHPs have been correlated with intermolecular network formation and thermoreversible gelation, and with SHP and starch (re)crystallization by a chain-folding mechanism in dilute solution. We have suggested [26] that, in a PC SHP gel network, the existence of random interchain entanglements in the amorphous regions and 'fringed micelle' or chain-folded microcrystalline junction zones [154] each represents a manifestation of sufficiently long chain length. This suggestion is supported by recent work [155, 156] which showed that amylose gels, which are found to be partially crystalline, are formed by cooling solutions of entangled chains. Miles *et al.* [156] stated that amylose gelation requires network formation, and this network formation requires entanglement, and they concluded that 'polymer entanglement is important in understanding the gelation of amylose.'

The excellent fit of the experimental data in figure 2.16 to the conventional presentation of the behavior expected for such a family of oligomers and high polymers was gratifying, especially considering the numerous caveats that one must always mention about commercial SHPs. For example, in figure 2.16, we used $\bar{M}_n$ (and implicitly $\overline{DP}_n$) values, calculated from DE, while in the conventional form (figure 2.5), $\bar{M}_w$ is used as a basis for specifying a typical MW range for the entanglement limit. Furthermore, for highly-polydisperse solutes such as commercial SHPs (for which MW distribution, MWD $= \bar{M}_w/\bar{M}_n$, is frequently a variable [142]), the observed T_g' is actually a weight-average T_g' of the mixture of solutes. [90] Despite these facts, the entanglement limit of $\bar{M}_n \approx 3000$ for the SHPs in figure 2.16 is within the characteristic range of 1250–19 000 for the minimum entanglement MWs of many typical synthetic linear high polymers. [157] This result for the SHPs in figure 2.16 is corroborated by the behavior manifested by the polyhydroxy compounds listed in table 2.2, as illustrated by the T_g' vs. MW data in the main plot in figure 2.11. It is clear, from the shape of the curve in figure 2.11 (and the linearity of the T_g' vs. 1/MW plot in the inset), that the monidisperse sugars, glycosides, and polyols represented do not show evidence of entanglement coupling. For these saccharide oligomers, none larger than a heptamer of MW 1153, the entanglement plateau has not been reached, a result in agreement with the MW range of entanglement limits cited above. [157]

The variable polydispersity of commercial SHPs was mentioned above. Other largely-uncontrollable potential variables within the series of SHPs include (1) significant lot-to-lot variability of solids composition (i.e. saccharides distribution) for a single SHP, which would affect the reproducibility of T_g'; and (2) 'as is' moisture contents (generally in the range of 5–10 w%) for different solid SHPs, which would not affect measured T_g' (since, for example, 15, 20, and 25 w% solutions would all freeze-concentrate to the same invariant T_g' point on the state diagram), but would affect the calculated W_g'. Hence, the W_g' data in figure 2.12 were only those for some corn syrups, whose moisture contents are generally more tightly specified. Another origin of variability among different SHPs of nominally-comparable DE concerns the method of production, i.e. hydrolysis by acid, enzyme, or acid plus enzyme. [142, 158] Especially with regard to enzymatic hydrolysates, each particular enzyme produces a different set of characteristic breakdown products with a unique MWD. [159]

Still another major variable among SHPs (and even for a single SHP) from different vegetable starch sources involves the original amylose/amylopectin ratio for a starch, and the consequent ratio of linear to branched polymeric chains in an SHP. [142, 158] The influence of this variable can be particularly pronounced among a set of low-DE maltodextrins (which would contain higher-*DP* fractions), some from amylose-containing dent corn and some from all-amylopectin waxy maize. The consequent range of T_g' values can be quite broad, since, as a general rule, linear chains give rise to higher T_g' than branched chains (with multiple chain ends [7]) of the same $\bar{M}_w$. [26] This observation was illustrated by several pairs of SHPs in table 2.1. For each pair, of the same DE and manufacturer (e.g. Star Dri 1 and Star Dri 10, (1984) vs. (1986); Paselli MD-6 vs. SA-6 and MD-10 vs. SA-10: Morrex 1910 vs. 1918), the hydrolysate from amylose-free starch has a lower T_g' than the corresponding one from a starch containing amylose. [27] This type of behavior was also exemplified by the T_g' data for the thirteen 10 DE maltodextrins listed in table 2.1, where T_g' ranged from −7.5 °C for Stadex 27 from dent corn to −15.5 °C for Frodex 10 from waxy maize, a ΔT of 8 °C. [27] (Further illustration was provided by the T_g' results for some of the glucose oligomers in table 2.2. [27] Those results demonstrated that, within such a homologous series, T_g' appears to depend most rigorously on the linear $\overline{DP}_w$ of the solute. From comparisons of the significant T_g' differences among maltose (1 → 4-linked glucose dimer), gentiobiose (1 → 6-linked), and isomaltose (1 → 6-linked); and among maltotriose (1 → 4-linked trimer), panose (1 → 4-, 1 → 6-linked), and isomaltotriose (1 → 6-, 1 → 6-linked); we suggested that 1 → 4-linked (linear amylose-like) glucose oligomers manifest greater 'effective' linear chain lengths in aqueous solution (and, consequently, greater hydrodynamic volumes) than oligomers of the same MW which contain 1 → 6 (branched amylopectin-like) links. Another intereresting comparison was between the T_g' values for the linear and cyclic α-(1 → 4)-linked glucose hexamers, maltohexaose (−14.5 °C) and α-cyclodextrin (−9 °C). In this case, the higher T_g' of the cyclic oligomer led us to suggest that the ring of α-cyclodextrin apparently has a much larger hydrodynamic volume in solution (due to its relative rigidity [160]) than does the linear chain of maltohexaose, which is relatively flexible and can assume a more compact conformation in aqueous solution. [27]) We reached the obvious conclusion [26], regarding a suitable maltodextrin for a specific application, that one SHP is not necessarily interchangeable with another of the same nominal DE, but from a different commercial source. [142, 158] We advised that basic characterization of structure–property relations, for example in terms of T_g' (rather than DE, which can be a less significant [142], and even misleading quantity), be carried out before one selects such food ingredients. [26]

2.3.3 *Predicted functional attributes of SHPs and polyhydroxy compounds*

We have demonstrated how further insights into structure–function relationships may be gleaned by treating figure 2.16 as a predictive map of regions of functional behavior for SHP samples. [26, 27] For example, polymeric SHPs which fall on the entanglement plateau demonstrate certain functional attributes, some of which have been reported in the past, but not quantitatively explained from the theoretical basis of the entanglement capability revealed by our studies. The plateau region defines the useful range of gelation, encapsulation, cryostabilization, thermomechanical stabilization, and facilitation of drying processes. The lower end of the $\bar{M}_n$ range corresponds to the region of sweetness, hygroscopicity, humectancy, browning reactions, and cryoprotection. The intermediate region at the upper end of the steeply-rising portion represents the area of antistaling ingredients. [159] The map (labeled as in figure 2.16) can be used to choose individual SHPs or mixtures of SHPs and other carbohydrates (targeted to a particular T_g' value) to achieve desired complex functional behavior for specific product applications. Especially for applications involving such mixtures, use can also be made of the data for the polyhydroxy compounds represented in figure 2.11, in combination with figure 2.16. One will recognize that the area represented by the left-hand third of figure 2.11 (and the low-MW sugars and polyols included therein) corresponds to the sweetness/hygroscopicity/humectancy/browning/cryoprotection region of figure 2.16. Likewise, the tri- through hepta-saccharides occupying the right-hand portion of figure 2.11 would be predicted to function similarly to the SHPs in the antistaling region of figure 2.16. [27]

As a specific example, the production of SHPs capable of gelation from solution should be designed to yield materials of DE $\leqslant 6$ and $T_g' \geqslant -8$ °C. [26] This prediction agreed with results reported [161–3] for 25 w% solutions of potato starch maltodextrins of 5–8 DE, which produce thermoreversible, fat-mimetic gels, and for tapioca SHPs of DE < 5, which also form fat-mimetic gels from solution. [164] We also tested the accuracy of this prediction, as illustrated by the experimental results shown in table 2.1 for the gelling ability of 20 w% solutions of many of the commercial SHPs. [27] These results demonstrated a clear line of demarcation between gelling ($T_g' \geqslant 7.5$ °C and DE < 6) and non-gelling SHPs. Thus, thermoreversible gels produced by refrigerating 20 w% solutions of SHPs of DE $\leqslant 6$ and $T_g' \geqslant -7.5$ °C appear to form by a mechanism involving gelation via crystallization plus entanglement in concentrated solutions undercooled to $T < T_m$. [44, 57] As suggested previously, for SHPs of sufficiently long linear chain length ($\overline{DP}_n \gtrsim 18$), a PC gel network is formed which apparently contains random interchain entanglements in the amorphous regions and microcrystalline intermolecular junction zones. [16] The latter accounts for the thermoreversibility of such SHP gels, via a crystallization (on undercooling)–melting (on heating) process. [15]

Other experimental evidence, which supports these conclusions about the gelation mechanism for PC polymeric SHP gels, includes the following: (1) DSC analysis of 20 w% SHP gels, after setting by overnight refrigeration, revealed a small crystalline melting endotherm with $T_m \approx 60$ °C, similar to the characteristic melting transition observed in retrograded B-type starch gels [16]; and (2) the relatively small extent of crystallinity in these SHP gels was increased significantly by a two-step temperature-cycling gelation protocol (12 h at 0 °C, followed by 12 h at 40 °C), adapted from one originally developed by Ferry [69] for gelatin gels, and subsequently applied by Slade [21] to retrograded starch gels. In some fundamental respects, the thermoreversible gelation of concentrated aqueous solutions of polymeric SHPs appears to be analogous to the thermoreversible gelation and crystallization of synthetic homopolymer and copolymer–organic diluent systems. [60, 61] For PC gels of the latter type, the possibly-simultaneous presence of random interchain entanglements in the amorphous regions [60] and microcrystalline junction zones [61] has been reported. However, controversy still exists, even in these recent publications, over which of the two conditions (if either alone) might be the necessary and sufficient one primarily responsible for the structure–property relations of such polymeric PC gelling systems. The controversy could be resolved by a simple dilution test; entanglement gels can be dispersed by dilution, microcrystalline gels cannot be.

Maltodextrins to be used for encapsulation of volatile flavors/aromas and lipids should likewise be capable of entanglement and network formation ($T_g' \geqslant -8$ °C). [26] It has been reported [79, 88, 135] that effectiveness of encapsulation increases with increasing T_c, which in turn increases with increasing $\overline{DP}_n$ within a series of SHPs, although 'a quantitative relationship between T_c and MW has not been established'. [88] This finding has been confirmed recently by Reineccius [165], who tested a series of SHPs, including Maltrins M040–M365 and ARD 2326 (see table 2.1), as amorphous spray-dried substrates for encapsulation (to retard oxidation) of orange oil. His results showed that shelf life (measured in terms of extent of oxygen uptake and subsequent oil oxidation) increased with decreasing DE of the encapsulating SHP. Moreover, by far the longest shelf life (i.e. essentially no oxygen uptake) resulted for the substrate made from ARD 2326, an SHP which was predicted by our analysis to be an effective encapsulator and has been shown by our results to be capable of entanglement and gelation. [27] In other product applications, maltodextrins of DE $\leqslant$ 10 have been used as amorphous coatings for the encapsulation of crystalline salt-substitute particles [166], and maltodextrins or dextrins have also been used as coating agents for roasted nuts candy-coated with honey. [167]

With regard to the freezer-storage (in other words, 'cryo' [26]) stabilization of fabricated frozen foods (e.g. desserts such as ice cream, with smooth/creamy texture) against ice crystal growth over time, inclusion of low-DE maltodextrins elevates the composite T_g' of the mix of soluble solids, which is typically dominated by low-MW sugars. [26] In practice, a retarded rate of ice recrystallization ('grain growth') at the characteristic freezer temperature (T_f) results, along with an increase in the observed T_r. Such behavior has been documented in several soft-serve ice cream patents. [168–70] In such products, ice recrystallization is known to involve a diffusion-controlled maturation process with a mechanism analogous to 'Ostwald ripening', whereby larger crystals grow with time at the expense of smaller ones which eventually disappear. [148, 171, 172] The rate of such a process (at T_f), and thus also ΔT ($T_f - T_g'$), is reduced by formulating with low-DE maltodextrins of high T_g'. In practice, the technological utility of the T_g' and W_g' results for sugars, polyols, and SHPs (in tables 2.1, 2.2, and 2.4), in combination with corresponding relative sweetness data, has been demonstrated by the successful formulation of fabricated products [e.g. 168, 169] with an optimum combination of stability and softness at $-18\,°C$ freezer storage. Low-DE maltodextrins are also used to stabilize frozen dairy products against lactose crystallization (another example of a diffusion-controlled collapse phenomenon) during storage. [173]

Low-DE maltodextrins and other high-MW polymeric solutes (e.g. see table 2.4 [27]) are well known as drying aids for processes such as freeze, spray, and drum drying. [79, 135, 144, 174, 175] Through their simultaneous effects of increasing the composite T_g' and reducing the unfrozen water fraction (W_g') of a system of low-MW solids (with regard to freeze drying) or increasing the RVP (for spray or drum drying), maltodextrins raise the observed T_c (at any particular moisture content) relative to the drying temperature, thus stabilizing the glassy state and facilitating drying without collapse or 'melt-back'. [26] By reducing the inherent hygroscopicity of the mixture of solids being dried, maltodextrins decrease the system's propensity to collapse (from the rubbery state) due to plasticization at low moisture. These attributes are illustrated [175] by recent findings on the freeze-drying behavior of beef extract with added dextrin.

Thermomechanical stabilization refers to the stabilization of low-moisture glasses such as boiled candies (which are typically very hygroscopic and notoriously sensitive to plasticization by water [70, 176]) against such collapse phenomena as recrystallization of sugars ('graining'), mechanical deformation, and stickiness. [26] Incorporation of low-DE maltodextrins in low-MW sugar glasses (to increase average $\bar{M}_w$ of the solutes) has been shown [70, 176–8] to increase T_g, and thus storage stability at $T < T_g$. Even when such a candy 'melt' is in the unstable rubbery state at $T_g < T_{storage}$, maltodextrins are known to function as inhibitors of the diffusion-controlled propagation step in the sugar recrystallization process. [70, 142, 176] Low-DE maltodextrins are also used frequently to stabilize other amorphous solids (e.g. food powders) [142] against various collapse phenomena that are exacerbated by plasticization at low moisture. For example, 'dextrins' (SHPs $\geqslant$ tetrasaccharides of DE $\leqslant 25$) are used as anticaking/antibrowning agents in low-moisture powders [179], and 10 DE maltodextrin is used as an additive in the production of shelf-stable amorphous juice solids compositions of unusually low hygroscopicity. [180]

For the lower-T_g' SHPs in figure 2.16 (as for many of the low-T_g' reducing sugars in figure 2.11), sweetness, hygroscopicity, humectancy, and browning reactions are salient functional properties. [27, 142] A less familiar one involves the potential for cryoprotection of biological materials, for which the utility of various other low-MW sugars and polyols is well known. [35, 37] The map of figure 2.16 predicts, and DSC experiments have confirmed, that such SHPs and other low-MW carbohydrates, in sufficiently concentrated solution, can be quench-cooled to a completely-vitrified state, so that all the water is immobilized in the solute/unfrozen water glass. [26] The essence of this cryoprotective activity, indefinite avoidance of ice formation and solute crystallization in concentrated solutions of low-MW, non-crystallizing solutes which have high W_g' values, also has a readily-apparent relationship to food applications involving soft, spoonable, or pourable-from-the-freezer products. One example is Rich's patented 'Freeze-Flo' beverage concentrate formulated with high fructose corn syrup. [181] We have conducted model system experiments with 60 w% solutions of, for example, fructose and mannose (3.3 M), methyl fructoside (3.1 M), ethyl fructoside, ethyl mannoside, and ethyl glucoside (2.9 M), in which these samples have remained pourable fluids (completely ice- and solute-crystal-free) under −18 °C freezer storage (i.e. at a temperature below the theoretical freezing point due to colligative depression) for about four years to date.

The literature on SHPs as antistaling ingredients for starch-based foods (reviewed by Slade [16]), including the recent work of Krusi & Neukom [159], reports that (non-entangling) SHP oligomers of DP_n 3–8 are effective in inhibiting, and not participating in, starch recrystallization.

We have also postulated from the map of figure 2.16 that addition of a low-MW sugar to a gelling maltodextrin could produce a sweet and softer gel. [26] Addition of a glass-forming sugar to an encapsulating maltodextrin should promote collapse of the entangled network around the absorbed species (if collapse were desirable) but decrease the ease of spray drying. Furthermore, the map led us to two other intriguing postulates. [26] The freeze-concentrated glass at T_g' of an SHP cryostabilizer (of DE $\leqslant$ 6) would contain entangled polymer molecules, while in the glass at T_g' of a low-MW SHP cryoprotectant, the solute molecules could not be entangled. By analogy, various high-MW polysaccharide gums are claimed to be capable of improving freezer-storage stability of ice-containing fabricated foods, in some way which is poorly understood. The effect has been attributed to increased viscosity. [172, 182] Such gums may owe their limited success not only to their viscosity-increasing ability, which would be common to all glass-formers, but to their possible capability to undergo entanglement in the freeze-concentrated, non-ice matrix of a frozen food. Entanglement might enhance a mouthfeel which masks their limited ability to inhibit diffusion-controlled processes. In a related vein, the effects of entanglement coupling on the viscoelastic and rheological properties of random-coil polysaccharide concentrated solutions [183] and gels [151, 152], at $T > 0$ °C, have been reported recently.

2.3.4 *The role of SHPs in collapse processes and their mechanism of action*

In the remaining parts of this section, we shall explore the critical role of polymeric SHPs in preventing structural collapse, within a context of the

Table 2.3. *'Collapse'-related phenomena which are governed by T_g and involve plasticization by water.* [27]

Processing and/or storage	References
$T > 0\,°C$	
(1) Cohesiveness, sticking, agglomeration, sintering, lumping, caking, and flow of amorphous powders $\geqslant T_c$	[4, 72, 79, 88, 135, 176, 179, 180, 184, 187, 192, 195, 196, 199, 200]
(2) Plating of, for example, coloring agents or other fine particles on the amorphous surfaces of granular particles $\geqslant T_g$	[185, 201]
(3) (Re)crystallization in amorphous powders $\geqslant T_c$	[4, 79, 88, 118, 135, 176, 192]
(4) Structural collapse in freeze-dried products (after sublimation stage) $\geqslant T_c$	[4, 79, 88, 135, 184]
(5) Loss of encapsulated volatiles in freeze-dried products (after sublimation stage) $\geqslant T_c$	[4, 72, 79, 88, 135, 174]
(6) Oxidation of encapsulated lipids in freeze-dried products (after sublimation stage) $\geqslant T_c$	[4, 79, 88, 118]
(7) Enzymatic activity in amorphous solids $\geqslant T_g$	[84, 193, 197]
(8) Maillard browning reactions in amorphous powders $\geqslant T_g$	[179]
(9) Stickiness in spray drying and drum drying $\geqslant T_{sp}$	[4, 72, 79, 88, 135, 184]
(10) Graining in boiled sweets $\geqslant T_g$	[36, 70, 79, 176–8, 188, 189, 191]
(11) Sugar bloom in chocolate $\geqslant T_g$	[186, 194]
$T < 0\,°C$	
(1) Ice recrystallization ('grain growth') $\geqslant T_r$	[32, 33, 37, 90, 148]
(2) Lactose crystallization ('sandiness') in dairy products $\geqslant T_r$	[37, 79, 173, 176]
(3) Enzymatic activity $\geqslant T_g'$	[14, 193]
(4) Structural collapse, shrinkage, or puffing (of amorphous matrix surrounding ice crystals) during freeze drying (sublimation stage) = 'melt back' $\geqslant T_c$	[33, 37, 79, 88, 135, 149, 175, 176, 184, 190]
(5) Loss of encapsulated volatiles during freeze drying (sublimation stage) $\geqslant T_c$	[79, 88, 135, 174]
(6) Reduced survival of cryopreserved embryos, due to cellular damage caused by diffusion of ionic components $\geqslant T_g'$	[198]

various collapse processes listed in table 2.3. [26, 27] These phenomena include ones pertaining to processing and/or storage at $T > 0\,°C$ as well as ones involving the frozen state, all of which are governed by the particular T_g relevant to the system and its content of plasticizing water. While for frozen systems, T_g' of the freeze-concentrated, 'low-moisture' glass is the relevant T_g for describing the T_g–MW relationship (as illustrated by figure 2.16), for amorphous dried powders and candy glasses, the relevant T_g pertains to a higher temperature/very-low-moisture state. We have assumed, as explained in section 2.3.1, that a plot of T_g vs. $\bar{M}_n$ for dry SHPs would reflect the same fundamental behavior as that shown in figure 2.16. Since T_c for low-moisture samples is known to represent a good quantitative approximation of dry T_g, To & Flink's [88] results substantiate this assumption.

All the collapse phenomena mentioned in table 2.3 are translational diffusion-controlled (many are also nucleation-limited) processes, with a mechanism involving viscous flow in the rubbery liquid state. [72, 79] These kinetic processes are controlled by the variables of time, temperature, and moisture content. [184] At the relaxation temperature, moisture content is the critical determinant of collapse and its concomitant changes [135], through water's plasticizing effect on T_g of the amorphous regions in both glassy and PC polymeric systems.

Whenever the glass transition and the resultant structural collapse occur on the same time scale [37], T_g equals the minimum onset temperature for the collapse processes listed in table 2.3. [36] Thus, a system is stable against collapse, within the period of the experimental measurements of T_g and T_c, at $T < T_g$. Increasing moisture content is known to lead to decreased stability and shelf life, at a particular storage temperature. [135] The various phenomenological threshold temperatures (e.g. $T_c \approx T_r \approx T$ sticky point (T_{sp})) are all equal to the particular T_g (or T_g') which corresponds to the solute(s) concentration for the situation in question. [26] Thus, in figure 2.6, for PVP-44, $T_g' = T_r = T_c \approx -21.5\,°C$ and $C_g' \approx 65$ w% PVP ($W_g' \approx 0.54$ g unfrozen water/g PVP) [14, 33, 37, 145]; while for PVP-700, $T_g = T_c = T_{sp} \approx 120\,°C$ at $\approx 5\%$ residual moisture. [14, 76] The equivalence of T_r for ice or solute recrystallization, T_c for collapse, and the concentration-invariant T_g' for an ice-containing system explains why T_r and T_c have always been observed in the past to be concentration-independent for all initial solute concentrations lower than C_g' [37], as illustrated in figure 2.6. [26]

Our conclusion regarding the fundamental equivalence of T_g, T_c and T_r [26] represented a departure from the previous literature. For example, while To & Flink [88] acknowledged that 'the relationship between T_c and MW is identical to the equation for T_g of mixed polymers' and that 'collapse and glass transition are (clearly) phenomenologically similar events', they differentiated between T_g and T_c by pointing out that 'while glass transitions in polymeric materials are generally reversible, the collapse of freeze-dried matrices is irreversible.' We pointed out [26] that, while the latter facts may be true, the argument is misleading. At the molecular level, the glass-to-rubber transition for an amorphous thermoplastic material is reversible. That is, the glass at T_g'/C_g' can be repeatedly warmed and recooled (slowly) over a reversible T/C path between its solid and liquid states. The same is true for an amorphous (and non-crystallizable) freeze-dried material. The reason collapse is said to be irreversible for a porous matrix has nothing to do with reversibility between molecular states. Irreversible loss of porosity is simply a macroscopic, morphological consequence of viscous flow in the rubbery state at $T > T_g$, whereby the porous glass relaxes to a fluid (incapable of supporting its own weight against flow) which then becomes non-porous and more dense. Subsequent recooling to $T < T_g$ yields a non-porous glass of the original composition, which can thereafter be temperature-cycled reversibly. The only irreversible aspect of T_g-governed collapse is loss of porosity. [26]

Recently, our conclusion about the fundamental identity of T_g' with T_r and T_c was corroborated by Reid. [198] He reported a study in which T_g', measured by DSC, corresponded well with the temperature at which a frozen aqueous solution, viewed under a cryomicroscope, became physically mobile. Reid remarked that 'T_g', the temperature at which a system would be expected to become mobile due to the appearance of the solution phase, has also been related to the T_c in freeze drying, again relating to the onset of system mobility, which presumably allows for the diffusion of solution components.' Reid's study revealed another phenomenon related to collapse, governed by T_g' of a frozen system, that was added to table 2.3: slow warming of cryopreserved embryos to $T > T_g'$ facilitates the detrimental diffusion of ionic components (salts), resulting in cellular damage due to high ionic strength and much reduced embryo survival.

Table 2.4 [27] shows a comparison of DSC-measured T_g' values, for a variety of water-compatible monomers and polymers, and literature values [37, 147] for other collapse transition temperatures. These results for observed T_c and T_r, which are usually measured (on an experimental time scale similar to that of our DSC method) by cryomicroscopy of frozen or vitrified aqueous solutions, are generally very close to, but almost always at a slightly higher temperature than, our values for T_g'. We have taken this fact as further support of our contention that T_g' represents the minimum onset temperature for these subzero collapse phenomena. [27]

The physico-chemical mechanism of collapse, and its prevention A universal, quantitative mechanism for collapse, derived from WLF theory and based on the WLF equation, was described earlier. It was also noted that the WLF equation depends critically on the appropriate reference T_g for a specific glass-forming system, be it T_g' for a frozen system, or T_g for a low-moisture one.

The controlled agglomeration of amorphous powders (e.g. low-MW carbohydrates) represents a specific example of a WLF-governed kinetic process related to caking, in low-moisture systems sensitive to plasticization by water and/or heating. [26] It has been demonstrated [72, 200] that spontaneous agglomeration of solid powder particles occurs when the viscosity of the liquid phase at the surface of the particle drops to $\approx 10^7$ Pa s. This viscosity is $\approx 10^5$ lower than η_g. From the WLF equation, this $\Delta\eta$ of 10^5 Pa s corresponds to a ΔT of ≈ 21 °C between T_g and the T_{sp} for spontaneous agglomeration. Thus, on a state diagram of temperature vs. percentage moisture, the T_g and T_{sp} curves would represent parallel isoviscosity lines. The T_{sp} curve for fast agglomeration during processing [72] would lie above the T_g curve for slow caking during storage, and the ΔT of 21 °C would reflect the different time scales for these two surface phenomena. [26]

In practice, collapse (and all its different manifestations) can be prevented, and product quality and stability maintained, by the following three

Table 2.4. *Comparison of T_g' values and literature values for other 'collapse' transition temperatures.* [27]

Substance	T_r (°C)[a]	T_c (°C)[b]	T_g' (°C)
ethylene glycol	−70[c]		−85
glycerol	−58, −65[c]		−65
ribose	−43		−47
glucose	−41, −38	−40	−43
fructose	−48	−48	−42
sucrose	−32, −30.5	−32, −34[d]	−32
maltose		−32	−29.5
lactose		−32	−28
raffinose	−27, −25.4	−26	−26.5
inositol		−27	−35.5
sorbitol		−45	−43.5
glutamic acid, sodium salt		−50	−46
gelatin	−11	−8	
gelatin (300 Bloom)			−9.5
gelatin (250 Bloom)			−10.5
gelatin (175 Bloom)			−11.5
gelatin (50 Bloom)			−12.5
bovine serum albumin	−5.3		−13
dextran		−9	
dextran (MW 9400)			−13.5
soluble starch	−5, −6[c]		
soluble potato starch			−3.5
hydroxyethyl starch	−21	−17[d]	−6.5
PVP	−22, −21, −14.5	−23, −21.6[d]	
PVP-10			−26
PVP-40			−20.5
PVP-44			−21.5
PEG	−65, −43	−13	
PEG (MW 200)			−65.5
PEG (MW 300)			−63.5
PEG (MW 400)			−61

[a] Recrystallization temperatures, from Franks [37], p. 297.
[b] Collapse temperatures during freeze-drying, from Franks [37], p. 313.
[c] From Luyet [147], p. 564.
[d] Antemelting temperatures, from Virtis Co. [149], p. 4.

fundamental countermeasures [26]: (1) storage at $T < T_g$ [176]; (2) deliberate formulation to increase T_g to a temperature greater than the processing or storage temperature, by increasing the overall $\bar{M}_w$ of the water-soluble solids in a product mixture. As described earlier, this is often accomplished by adding polymeric stabilizers such as low-DE SHPs (or other polymeric carbohydrate, protein, or cellulose and polysaccharide gum stabilizers, some of which are included in table 2.4) to a formulation dominated by low-MW sugars and/or polyols. [70, 72, 88, 135, 142, 173, 176–80, 184, 190] The effect of increased MW on the T_g of PVPs is also illustrated in figure 2.6; (3) in low-moisture amorphous food powders and other hygroscopic glassy solids especially prone to the detrimental effects of plasticization by water (the latter including 'candy' glasses such as boiled sweets [70, 142, 176, 177, 186], extruded melts [188], candy coatings [178], sugar in chocolate [194], and supersaturated sugar syrups [36, 72, 191]), (*a*) reduction of residual moisture content to $\lesssim 3\%$ during processing, (*b*) packaging in superior moisture-barrier film or foil to prevent moisture pickup during storage, and (*c*) avoidance of high temperature/high relative humidity ($\gtrsim 20\%$ relative humidity) conditions during storage. [70, 79, 176, 178, 188] (In a related vein, a 'state of the art' computer model was recently described [202], which can be used to predict the shelf life of particular moisture-sensitive products, based on the moisture-barrier properties of a packaging material and the temperature/humidity conditions of a specific storage environment.)

On the other hand, Karel [4] has pointed out that water plasticization (to depress $T_g < T$ of the phenomenon) is not always detrimental to product quality. Examples of applications involving deliberate moisturization to produce desirable consequences include (1) controlled agglomeration or sintering (by limited heat/moisture/time treatment) of amorphous powders (as described above and in [79, 184, 199]), and (2) compression (without brittle fracture) of freeze-dried products after limited replasticization. [135]

Prevention of enzymatic activity and other chemical reactions at $T < T_g$ One collapse-related phenomenon listed in table 2.3 but not yet discussed involves enzymatic activity, in amorphous substrate-containing media, which occurs only at $T > T_g$. [26] Enzymatic activity represents a pleasing case study with which to end this section, because it is potentially important in many food applications which cover the entire spectrum of processing/storage temperatures and moisture contents, and because examples exist [4] which elegantly illustrate the fact that activity is inhibited in low-moisture amorphous solids at $T < T_g$, and in frozen systems at $T < T'_g$. Bone & Pethig [84] studied the hydration of dry lysozyme powder at 20 °C, and found that, at 20 w% water, lysozyme becomes sufficiently plasticized so that measurable enzymatic activity commences. We interpreted their results to indicate the following [14, 26]: a diffusion-controlled enzyme–substrate interaction is essentially prohibited in a glassy solid at $T < T_g$, but sufficient water plasticization depresses T_g of lysozyme to <20 °C, allowing the onset of enzymatic activity in the rubbery lysozyme solution at $T > T_g$, the threshold temperature for activity. Our interpretation was supported by the results of a related study of 'solid glassy lysozyme samples' by Poole & Finney [197], who noted conformational changes in the protein as a consequence of hydration to the same 20 w% level, and were 'tempted to suggest that this solvent-related effect is required before (enzymatic) activity is possible'. Recently, Morozov & Gevorkian [193] also noted the critical requirement of low-temperature, water-plasticized glass transitions for the physiological activity of lysozyme and other globular proteins.

Within the context of our concept of cryostabilization, and the critical role of low-DE SHPs as cryostabilizers, we have verified the above conclusion in maximally-frozen biological systems. [26] By analogy, in such cases, the threshold temperature for onset of enzymatic activity is T_g'. Cryostabilization [26], as a practical technology derived from our concept of glass dynamics, is a means of protecting freezer-stored and freeze-dried foods from the deleterious changes in texture (e.g. grain growth of ice, solute crystallization), structure (e.g. shrinkage, collapse), and chemical composition (e.g. flavor degradation, fat rancidity, as well as enzymatic reactions) typically encountered. The key to this protection lies in controlling the physico-chemical properties of the freeze-concentrated matrix surrounding the ice crystals. If this matrix is maintained as an amorphous mechanical solid (at $T_f < T_g'$), then the diffusion-controlled changes that typically result in reduced storage stability can be prevented or at least greatly retarded. If, on the other hand, a natural food is improperly stored at too high a T_f, or a fabricated product is improperly formulated, and thus the matrix is allowed to exist in the freezer as a rubbery fluid (at $T_f > T_g'$, within the domain of water dynamics), then freezer-storage stability would be reduced. Moreover, the rates of the various deleterious changes would increase exponentially with the ΔT between T_f and T_g' [26], as dictated by WLF theory.

The prevention of enzymatic activity at $T < T_g'$ was demonstrated experimentally *in vitro* in a model system consisting of glucose oxidase, glucose, methyl red, and bulk solutions of sucrose, Morrex 1910 (10DE maltodextrin), and their mixtures, which provided a range of samples with known values of T_g'. [26] The enzymatic oxidation of glucose produces an acid which turns the reaction mixture from yellow to pink. Samples with a range of T_g' values from -9.5 to -32 °C were stored at various temperatures: 25, 3, -15, and -23 °C. All the samples were fluid at the two higher temperatures, while all looked like colored blocks of ice at -15 and -23 °C. However, only the samples for which the storage temperature was above T_g' turned pink. Even after two months storage at -23 °C, the samples containing the maltodextrin, with $T_g' > -23$ °C, were still yellow. The frozen samples which turned pink, even at -23 °C, contained a concentrated enzyme-rich fluid surrounding the ice crystals, while in those which remained yellow, the non-ice matrix was a glassy solid. Significantly, enzymatic activity was prevented by storage below T_g', but the enzyme itself was not inactivated. When the yellow samples were thawed, they quickly turned pink. Thus, cryostabilization with a low-DE SHP preserved the enzyme during storage, but prevented its activity below T_g'.

2.4 Structural stability of intermediate moisture foods

Dilute aqueous systems can exist in equilibrium, governed by energetics and appropriately described by thermodynamic treatments. For such systems, the measurement of water activity (A_w), in terms of an equilibrium vapor pressure relative to the partial pressure of pure water, may have some relevance. This legitimate use of A_w as a true thermodynamic property for equilibrium systems which are composed of a given solute or ratio of solutes with varying water content may be predictive of technological performance and physiological viability. [5, 6, 11, 12, 20, 203] However, the specific physico-chemical and biochemical contributions from different solutes may prevent intersystem predictions. [20]

In contrast, intermediate-moisture (IM) systems are not equilibrium systems, are not governed by energetics, and cannot be described by thermodynamic treatments. [18, 20, 22] In particular, the measured vapor pressure for these non-equilibrium IM systems is not a function of A_w, and A_w as a thermodynamic property is not relevant to the description of their behavior. This is true even for *in vivo* biological systems under extreme stress of drought, freezing, or salinity. At best, the measured value of RVP of these systems represents a stationary state vapor pressure. More typically, the measured RVP is an instantaneous value for a dynamic property of a system that is under kinetic control. The use of RVPs as a measurement of 'water availability' to predict product quality and biological viability in IM systems would require a detailed physico-chemical description of their behavior. The spectacular lack of predictive utility [5] of RVP (so-called A_w), in the absence of such a description, is demonstrated by, for example, the hysteresis of sorption–desorption isotherms. [6] Recent discussions [5, 18, 20] about the utility of A_w as a credible measure of technological performance and physiological viability have resulted in new conclusions and guidelines for more credible criteria of IMF product quality and biological viability to replace the current popular usage of 'A_w'. [20]

IM systems should be treated as 'low-moisture' systems, according to the context of this review defined previously. Hence, we have described a set of fundamental principles for the structural stabilization and moisture management of IMFs, from the perspective of our concept of water dynamics. [18, 22] These principles have been derived from structure–property relationships for such non-equilibrium IM systems, viewed generically as amorphous or PC polymer systems, consisting of homologous series from monomers to oligomers to high polymers with their solvents and plasticizers. As described earlier, the behavior of these systems is controlled by kinetics, rather than energetics, and this kinetic control (i.e. the rate of approach to equilibrium) varies according to distinct temperature/percentage moisture domains. On a 'dynamics map' (e.g. the state diagram in figure 2.6), the domain of water dynamics corresponds to the region of $T > T_g$ and $W > W_g$. [18, 20, 22]

The first question to ask in order to determine the contribution of water 'availability' to quality, stability, and viability of IM systems should be 'is water required *per se* or is the requirement simply for the mobility resulting from the presence of a low-MW species (plasticizer)?' Other molecules (e.g. sugars and sugar alcohols, which act as cryoprotectants in nature) can replace water in maintaining biopolymers (e.g. proteins) in a viable but inactive native state, even under conditions of extreme dehydration. [35] If there is no chemical requirement for water molecules, then it is appropriate to consider the factors that govern mobility. [18]

Mobility of an IM system is governed by temperature and composition, the latter expressed in terms of $\bar{M}_w$, of the total system, i.e. polymer(s) and plasticizer(s). [18] At constant concentration, which defines $\bar{M}_w$ of a system, a change in temperature changes mobility. At constant temperature, a change in concentration changes mobility. The extent of mobility can be measured in terms of viscosity, diffusion, and rates of relaxation or reaction. The temperature domains which influence mobility and kinetic control are determined by the absolute temperature with respect to the two characteristic transition temperatures of PC systems, T_g and T_m. As previously illustrated in figure 2.4 [37], T_g is typically 0.5–0.8 (K) of T_m [57] and only at $T < T_g$ and $T > T_m$ are Arrhenius kinetics observed. [18] It bears repeating here that, in contrast, WLF kinetics are observed at $T_g < T < T_m$, which represents the most significant temperature region (within the domain of water dynamics) for the study of water availability. [18] While Arrhenius kinetics typically involve a doubling of rate for an increase in temperature of 10 K (without identifying any particular reference temperature), WLF kinetics typically involve an order of magnitude increase in rate (equivalent to an order of magnitude increase in diffusion or decrease in viscosity) for each 3–5 K increase in temperature above T_g as a reference temperature. [34]

T_g depends on MW (i.e. linear *DP*) for a single substance within a homologous series and on $\bar{M}_w$ for a mixture. [31] Thus, a 'glass curve' with constant viscosity can be defined in terms of temperature and composition; the high-temperature tie-point is determined by T_g of the pure high-MW component and the low-temperature tie-point by that of the low-MW component, which serves as a plasticizer (as shown earlier in figure 2.6 for the water–PVP-44 system). [18] Other typical solute–water glass curves are highlighted in figure 2.17, where schematic state diagrams, which illustrate the relationship between T_g and T_m, are represented for the common sugars sucrose, glucose, and fructose. [18] Another interesting illustration of what can be gleaned from an analysis of glass curves for complex aqueous mixtures is shown in figure 2.18. [18] In this artist's rendering of a three-dimensional state diagram for a hypothetical three-component system, both solutes (e.g. a polymer, 2, and its monomer, 1) are non-crystallizing, interacting (i.e. compatible), and plasticized by water, which is the crystallizing solvent, 3. The diagram reveals the postulated origin of a sigmoidal curve of T_g' vs. w% solute composition, the T_g' glass curve *ABCDEF*. [18] In fact, similar sigmoidal curves of T_c vs. w% concentration, for collapse during freeze drying of analogous three-component aqueous systems, have been reported [205,

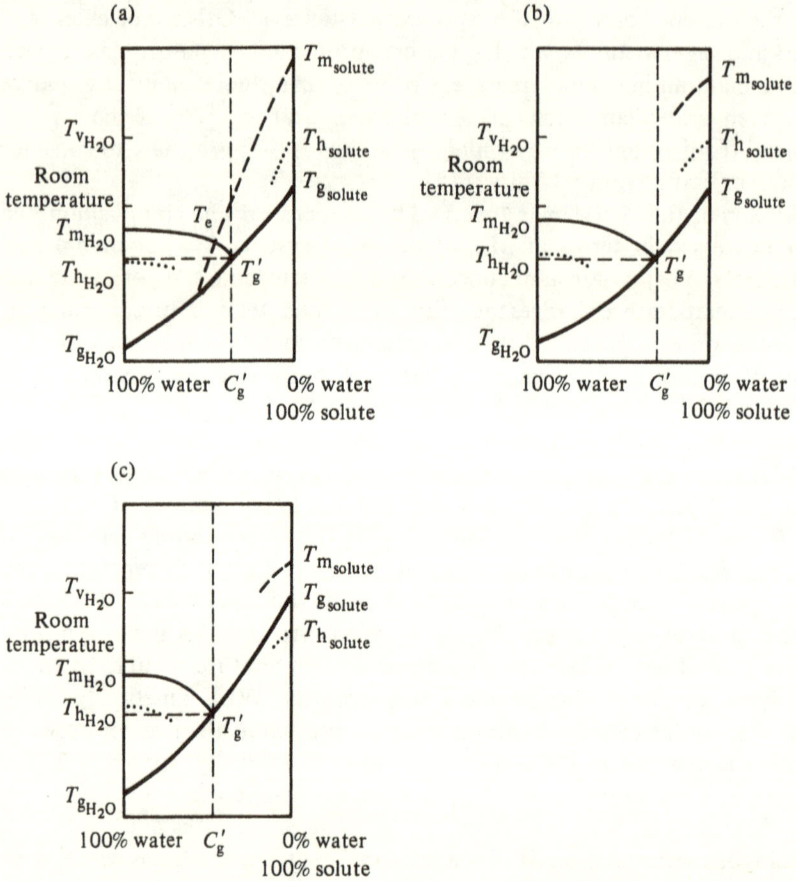

Fig. 2.17. Schematic state diagrams of temperature vs. composition for (*a*) sucrose, (*b*) glucose, and (*c*) fructose, which emphasize the solute–water glass curve, and the relationship between T_m, T_g, and T_h, the estimated homogeneous nucleation temperature, for each pure solute. T_e is the eutectic melting temperature and T_{vH_2O} is the boiling point. [18]

p. 288], but 'the basis for (their non-linearity) has not yet been determined'. [205]

As was demonstrated in the previous section on collapse phenomena, T_g is a useful diagnostic tool. It is a reference temperature which defines a viscosity of $\approx 10^{14}$ Pa s, for systems with the typical ratio of $T_m/T_g \approx 1.25$–2. [36, 57, 74] It is manifested by a dramatic change in viscosity, diffusion, and specific heat over a small temperature interval near T_g. [34] Isoviscosity curves differing by an order of magnitude can be drawn at about 3–5 K intervals in the region between T_g and T_m. While this rubbery domain is typically about 100 K for 'well-behaved' systems (e.g. pure high polymers) with the typical ratio of T_m/T_g, there are some notable exceptions and qualifications to this situation. For example, in frozen aqueous solutions, such as discussed in the previous

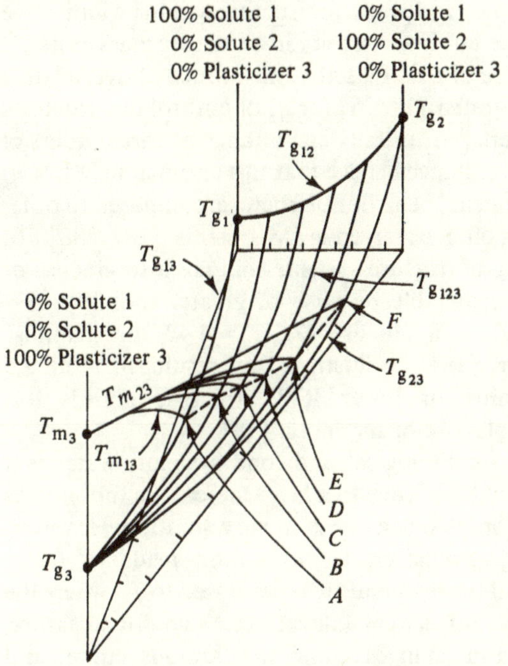

Fig. 2.18. A schematic three-dimensional state diagram for a hypothetical three-component aqueous system. The two solutes (e.g. polymer + monomer) are both non-crystallizing, interacting, and plasticized by water, which is the crystallizing solvent. The diagram illustrates the postulated origin of a sigmoidal curve of T_g' vs. w% solute composition. [18]

section, the magnitude of the WLF region, in terms of ΔT between T_g' and the T_m of ice, can be as small as 4 °C for a 0.5DE dextrin. [18, 26] A synthetic polymer with a very low value of 1.18 for T_m/T_g, reportedly due to anomalously-large free volume at T_g, is bisphenol polycarbonate. [74] Fructose is another important exception. [18]

In an amorphous, quench-cooled melt of pure crystalline β-D-fructose, we have observed (by DSC) two separate glass transitions. The first T_g occurs around 30 °C, which is approximately the same temperature reported for the T_g of other monosaccharides such as glucose and mannose. [18, 36] However, the second T_g for fructose occurs at 100 °C, only 26 °C below its T_m. [18, 36] The observation of two T_gs may indicate the presence of two different major anomeric forms of fructose immobilized in the quenched melt. Similar findings of multiple anomeric forms (which is a common situation in aqueous sugar solutions) in quenched melts have been reported from DSC analyses of other low-MW crystalline carbohydrates. [211] The higher T_g of fructose would translate to a T_m/T_g ratio of 1.06, much lower even than that for bisphenol polycarbonate. [18] Since the mobility of a glass or high-viscosity rubber depends on the T_m/T_g ratio, which is a predictor of relative free volume for pure substances [34], the very low T_m/T_g ratio for fructose would be indicative of high mobility and free volume but low viscosity in a fructose glass at its T_g. Thus, instead of the typical value of 10^{14} Pa s at η_g mentioned above, a value of only 10^{11} Pa s has been suggested [18, 36] for η_g of amorphous fructose. This explains why we and others [36] usually cite a range of three orders of magnitude for η_g at T_g. Importantly, we believe that this anomalously-low η_g may account for much of the unusual behavior of fructose, compared to other more typical monosaccharides, often observed in IM systems. [207, 208] For example, the relative instability of fructose systems compared to glucose or mannose systems would be predictable because of greater mobility in a fructose glass or rubber than in a glucose ($T_m/T_g = 1.42$) or mannose ($T_m/T_g = 1.36$) glass or rubber. [18] As illustrated by results in table 2.5 (described below), at the same or lower RVP, fructose affords less microbiological stability than glucose or mannose. [18]

As the universal plasticizer for biological and food systems, water is a 'mobility enhancer'. [20] Its low MW leads to a large increase in mobility as moisture content is increased from that of a dry solute to a solution. However, since water is a crystallizing plasticizer, ice formation leads to freeze concentration of a solution, until the temperature is depressed to T_g, where the limited rate of diffusion prevents further crystallization. As described earlier, upon slow freezing, composition changes along the liquidus curve, and maximum freeze concentration is observed at T_g'. [37] For each solute or mixture of solutes, W_g' is the characteristic maximum water content of the aqueous glass at T_g'. [18, 26] Even though this water is kinetically metastable and does not freeze within a practical time frame, it is not bound at subzero temperature and is not energetically stable. [20] The aqueous glass identified by T_g' and W_g' is as homogeneous as the same solution at room temperature. These parameters serve as a third point on the diagnostic glass curve, so that the shape as well as the end-points can be described for each system. The state diagrams in figure 2.17 illustrate the glass curves for sucrose, glucose, and fructose. [18] Comparison of these diagrams reveals differences in the kinetically-metastable domains between T_m and T_g, with respect to the estimated homogeneous nucleation temperatures (T_h) of these sugars. In each case, T_h was estimated from the ratio of T_h/T_m (K), which, for many PC polymers, is typically 0.8, with a reported range of 0.78–0.85. [44, 204] The relationship between T_h and T_g allows prediction about the stability towards recrystallization of concentrated and supersaturated aqueous solutions. [18] For sucrose and glucose, which are known to crystallize readily by undercooling such solutions, $T_h > T_g$, so homogeneous nucleation can occur before vitrification on cooling from $T > T_m$. In contrast, fructose cannot be crystallized (in a realistic time) by the same mechanism, because $T_g > T_h$. On cooling, vitrification occurs first, thus immobilizing the system and preventing the possibility of crystallization.

The following experimental approach, based on WLF kinetics of PC polymer systems, has been suggested for investigation of water dynamics as a predictive parameter for quality and stability of IM systems. [20]

Measure T_g of anhydrous solute.

Measure T_g' and W_g' of freeze-concentrated glass.

Use literature estimate [37] for T_g of water.

Construct the T_g as a function of concentration ($T_g(c)$) curve to define the reference state for kinetic metastability ($\eta \approx 10^{14}$ Pa s).

Measure T_m/T_g to estimate departure from reference of $\eta \approx 10^{14}$ Pa s for typical glass behavior.

Then shelf life is determined by a combination of both ΔT and ΔW, where

$\Delta T = T - T_g$	for constant water content	at $W < W_g'$;
$\Delta W = W - W_g$	for constant temperature	at $W < W_g'$ and any T;
$\Delta W = W - W_g$	for constant temperature	at $W > W_g'$ and $T > T_g'$;
$\Delta T = T - T_g'$	for	$W > W_g'$;
$\Delta W = W - W_g'$	for constant temperature	at $W > W_g'$ and $T < T_g'$.

For a convenient estimation of relative shelf life, a vector can be constructed from $\Delta T = T - T_g'$ and $\Delta W = W - W_g'$. [18]

Two examples of the use of this approach to describe water availability and product quality are shown in figure 2.19 for a low-temperature system and table 2.5 for a system near room temperature. At low temperature,

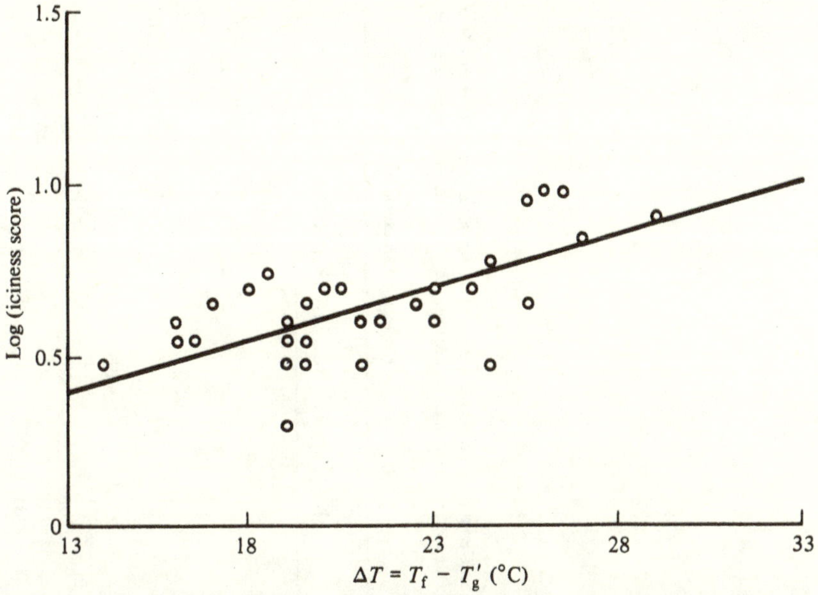

Fig. 2.19. Log (iciness score) (determined organoleptically, on a 0–10 point scale) as a function of ΔT ($= T_f - T_g'$), for experimental ice cream products, after 2 weeks of deliberately abusive (temperature-cycled) frozen storage in a so-called 'Brazilian Ice Box'. [18, 168, 169]

Table 2.5. *Germination of mold spores of Aspergillus parasiticus in IM-moisture systems.* [18]

Design parameters						IM sample		
RVP[a] (30 °C)	T_g' (K)	W_g'[b] (w% H_2O)	T_g (K)	T_m (K)	T_m/T_g	Conc. (w% H_2O)	Solute type	Days required to germinate at 30 °C
Controls								
1.0						100	none	1
~1						99	glucose (α-D)	1
~1						99	fructose (β-D)	1
~1						99	PVP-40	1
~1						99	glycerol	2
0.92	251.5	35	373			50	PVP-40	21
0.92	227.5	49.5	302	444.5	1.47	60	α-methyl glucoside[c]	1
0.83	231	49	373	397	1.06	50	fructose	2
0.83	208	46	180	291	1.62	60	glycerol	11
0.99	243.5	20	316	402	1.27	60	maltose	2
0.97	241	36	325	465	1.43	60	sucrose	4

0.95	250	31	349	406.5	1.16	50	maltotriose	8
0.93	232	26	303	412.5	1.36	50	mannose	4
0.95	250	31	349	406.5	1.16	50	maltotriose	8
0.92	251.5	35	373			50	PVP-40	21
0.93	232	26	303	412.5	1.36	50	mannose	4
0.87	231	49	373	397	1.06	54	fructose	2
0.92	227.5	49.5	302	444.5	1.47	60	α-methyl glucoside	1
0.87	231	49	373	397	1.06	54	fructose	2
0.92	227.5	49.5	302	444.5	1.47	60	α-methyl glucoside	1
0.70	231	49	373	397	1.06	30	fructose	2
0.85	230	29	304	431	1.42	50	glucose	6
0.83	231	49	373	397	1.06	50	fructose	2
0.82	230.5	48	293			40	1/1 fructose/glucose	5
0.98	247	36	339			50	PVP-10	11
0.98	231	49	373	397	1.06	60	fructose	2
0.93	247	36	339			40	PVP-10	11
0.95	251.5	35	373			60	PVP-40	9
0.99	247	36	339			60	PVP-10	11
0.99	243.5	20	316	402	1.27	60	maltose	2

[a] RVP measured after 7 days ‘equilibration’ at 30 °C.

[b] W_g' expressed here in terms of w% water, for ease of comparison with IM sample concentration.

[c] Commercial material from Staley – Technical grade, used as received. Different from in-house synthesized and purified sample listed in Table 2.2.

development of iciness (due to grain growth of ice) during freezer storage of ice cream is detrimental to product quality. The rate of development of iciness, which limits useful shelf life, depends on the temperature of the freezer relative to T_g' [26], since this recrystallization process is an example of the WLF-governed collapse phenomena described earlier. Because typical ice cream products have T_g' values in the range −30 to −43 °C, they would exist as rubbery fluids (with embedded ice and fat crystals) under typical freezer storage at −18 °C [168, 169], and WLF kinetics would describe the rate of ice crystal growth. Figure 2.19 contains a WLF plot of log(rate of iciness development) vs. ΔT, which shows that iciness increases with increasing ΔT, with a linear regression coefficient of 0.68. [18] Considering that iciness scores were obtained by sensory evaluation, we believe this unprecedented experimental demonstration of WLF behavior in a frozen system to be remarkable. However, we were amazed to find that this non-Arrhenius behavior of freezer-stored ice cream is apparently recognized, at least empirically, by the UK frozen foods industry. We have seen on ice cream packages in England the following shelf life code: in a one-star home freezer (15 °F), storage life = 1 day; two-star freezer (0 °F), 1 week; and three-star freezer (−10 °F), 1 month.

Near room temperature, germination of mold spores of an *Aspergillus* depends only on water availability, not on the presence of nutrients. [206] The observed rates of germination at 30 °C could not be predicted by measured RVPs, but an increased understanding was gained from the suggested approach based on state diagrams to describe the kinetics of these PC polymer systems. The illuminating results shown in table 2.5 [18] represent a dramatic experimental demonstration of the failure of 'A_w' to predict the relative usefulness of additives for antimicrobial stabilization. The experimental protocol, adapted from a microbiological assay used by Lang [206], compared the inhibitory effects on conidia germination for a series of IM solutions of selected glass-formers. The germination is essentially an all-or-nothing process, with a massive appearance of short hyphae surrounding the previously-bare spores occurring within 24 h at 30 °C in pure water or dilute solution (RVP $\approx$ 1.0). Various glass-formers were assayed in pairs, deliberately matched as to the individual parameters of approximately equal RVP (at 30 °C), solute concentration, $\bar{M}_w$, T_g', and/or W_g'.

For fructose vs. glucose (at equal solute concentration, $\bar{M}_w$, and T_g'), fructose produced a less stable system (i.e. faster germination), even at slightly lower RVP. Likewise for the fructose vs. glycerol, maltose vs. sucrose, and mannose vs. fructose pairs, the solute system with the lower ratio of T_m/T_g was faster to show germination, regardless of RVP values. Thus, apparently due to higher free volume, mobility, and lower viscosity in the rubber (as governed by T_m/T_g), water availability was greater for fructose ($T_m/T_g = 1.06$) > mannose ($T_m/T_g = 1.36$) > glucose ($T_m/T_g = 1.42$) > glycerol ($T_m/T_g = 1.62$) and maltose ($T_m/T_g = 1.27$) > sucrose ($T_m/T_g = 1.43$), so that greater antimicrobial stabilization was observed for glycerol > glucose > mannose > fructose and sucrose > maltose. The extraordinary mobility and water availability of IM fructose rubbers was manifested by the same fast germination time observed for solutions of 40–70 w% fructose and corresponding RVPs of 0.98–0.70. Other noteworthy results in table 2.5 involved the PVP-40 vs. methyl glucoside, maltotriose vs. mannose, and PVP-10 vs. fructose pairs, for which solute MW, as reflected by T_g', appeared to be the critical variable and determinant of effectiveness. In each case, the solute of higher MW and T_g' produced lower water availability in its IM rubber (regardless of RVP values), and thus greater stabilization against germination, in direct analogy with the structure/property principles described earlier for cryo- and thermomechanical stabilization in the glass dynamics domain. [26, 27]

The definitiveness of these results contrasts sharply with the controversial, contradictory, and confusing state of the IMF literature on the use of fructose vs. glucose to lower 'A_w' in moisture management applications. [207, 208] Since this subject is one of great practical and topical interest and importance to the food industry, it represents a good example of the potential utility of our concept of water dynamics, in place of 'A_w', for predicting functional behavior from structural properties. [18] The current IMF literature [e.g. 207, 208] can be summarized by the following paradoxical 'truths'. On an equal weight basis, replacement of glucose by fructose results, for many IMF products, in a lower RVP and increased shelf stability. However, if a product has been formulated with glucose to achieve a certain RVP, which has been found to result in satisfactory microbiological stability and organoleptic quality, then reformulation with fructose to the same RVP often has not produced as stable a product. This is so because less fructose than glucose is required to achieve the same RVP, while, as has been shown by our model studies, at the same RVP, fructose solutions are less stable than glucose solutions. (Unfortunately, it has not even been possible for agreement to be reached in the IMF literature on correct RVP values for 50 and 60 w% fructose solutions. [207, 208]) We feel that the proposed concept of water dynamics can explain such apparently contradictory results and will allow the IMF field to advance beyond its current state of affairs. [20]

Another aspect of our experimental approach to understanding water dynamics in IM systems has involved investigating the possible correlation between RVP and W_g'. RVP was measured, after 9 days 'equilibration' at 30 °C, for a series of 67.2 w% solids solutions, and plotted against W_g' for maximally-frozen 20 w% solutions of the same solids, as shown in figure 2.20 [18]. The samples represented a quasi-homologous family of sugar syrups, including high fructose corn, ordinary corn, sucrose, and invert syrups, all of which are ingredients commonly used in IMF products. Figure 2.20 illustrates the fair linear correlation ($r = -0.71$) between decreasing content of unfrozen water in the glass at T_g' and increasing RVP, which is typically assumed to be

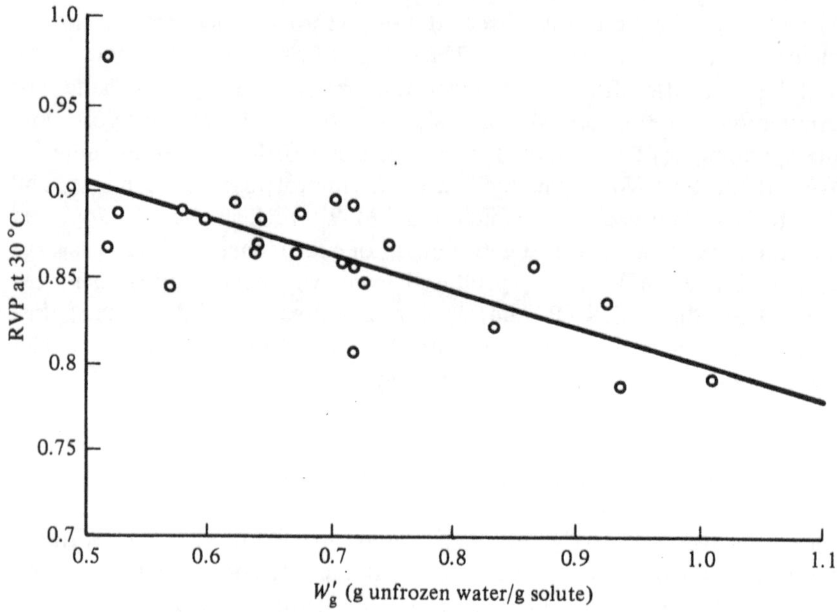

Fig. 2.20. Variation of RVP (measured for 67.2 w% solutions of various corn, sucrose, and invert syrup solids, after 9 days at 30 °C) against W_g', the composition of the glass at T_g', in g unfrozen water/g solute, for maximally-frozen 20 w% solutions of the same syrup solids. [18]

an indicator of free water content in IM systems at room temperature. The obvious scatter of these data prohibited any fundamental insights into the question of water availability in such systems. This was not unexpected, since many of these samples represented heterogeneous, polydisperse mixtures of polymeric solutes of unknown $\bar{M}_W$, molecular weight distribution, and T_m/T_g.

A general set of structure–property relationships for IM systems has begun to emerge from experimental results such as those in table 2.5, considered from the perspective of water dynamics in low-moisture polymeric systems. [20] Other results, such as those in table 2.6 [18], have also been integrated in the development of these new guidelines for more credible criteria (with predictive capability) of quality, stability, and viability in IM systems to replace 'A_w' or RVP. Table 2.6 contains literature data [209, 210] for relative germination times of plant seeds in IM systems at equivalent osmotic pressures (analogous to equivalent RVPs) and temperatures near room temperature. Such results suggest, at best, a rough correlation between increasing solute MW and increasing germination time, but have left unanswered questions about the observed differences between glucose, sucrose, and mannitol. The emerging guidelines are as follows: (1) for solutes of equal solids content at T_g, greater stability is achieved for one with a higher ratio of T_m/T_g; (2) for solutes with equal ratios of T_m/T_g, greater stability at T_g is achieved by higher solids content; (3) at a given ambient temperature,

Table 2.6. *Relative germination times of seeds in intermediate moisture systems at equivalent osmotic pressures.* [18]

IM system	Relative germination times
Winter wheat seeds [209]	
Water control	1[a]
Mannitol	2.75[a]
Poly(ethylene glycol) (MW 20 000)	5.5[a]
Scarlet globe radish seeds [210]	
Water control	1[b]
Mannitol	1.3[b]
Sucrose	1.5[b]
Glucose	1.8[b]
PVP (MW not specified)	9[b]

[a] At 20 °C.
[b] At 24 °C.

greater stability results for a smaller ΔT above T_g; and (4) for reasons of efficacy and cost-effectiveness of a stabilizing agent for IM systems, choose a material for which vitrification is achieved with low solids content (which implies high W_g'), ΔT is small (which implies high T_g), and the T_m/T_g ratio is high (which implies low free volume in the glass). [18, 20, 22]

2.5 Starch as a PC polymer plasticized by water: thermal analysis of non-equilibrium melting, annealing, and recrystallization behavior

Starch and gelatin are water-compatible polymers, which, in terms of abundance and utility, represent important industrial biopolymers, particularly as food ingredients. They also serve as outstanding models to illustrate our conceptual approach to the study of structure–property relationships of food molecules, which are treated as homologous systems of polymers, oligomers, and monomers with their plasticizers and solvents. Thermal analysis by DSC is particularly well suited to demonstrate that both starch and gelatin exhibit non-equilibrium melting, annealing, and recrystallization behavior characteristic of kinetically-metastable, water-plasticized, PC polymers with relatively small extents of crystallinity. [13, 15] As detailed in section 2.2.1, their aqueous gels have been represented by the 'fringed micelle' model for a three-dimensional, metastable network composed of amorphous regions containing plasticizing water and hydrated microcrystalline regions which serve as junction zones. This model also describes the morphology of native granular starch, for which melting is kinetically controlled due to the existence of contiguous microcrystalline regions (the crystallizable short branches) and amorphous regions (backbone segments plus branch points) in the amylopectin molecules. [16, 21] (Refer to section 2.2.2 for earlier remarks on starch gelatinization as a non-equilibrium melting process, retrogradation as a recrystallization process, mechanism of gelation/recrystallization, and kinetics of recrystallization and its acceleration.)

2.5.1 *Thermal properties of starch–water model systems*

Figure 2.21 [13, 15] shows DSC (Perkin–Elmer DSC-2c) thermal profiles of isolated wheat starch, with water added to a final moisture content of 55 w%. The native starch with no pretreatment (curve (*a*)) exhibited a major glass transition and subsequent superimposed crystalline transitions in the temperature range 50–85 °C, which comprise the events of gelatinization (initial swelling) and pasting (second-stage swelling) of a starch granule. Slade was the first to recognize that the glass transition of the amorphous regions of amylopectin was observed at the leading edge of the first melting peak,

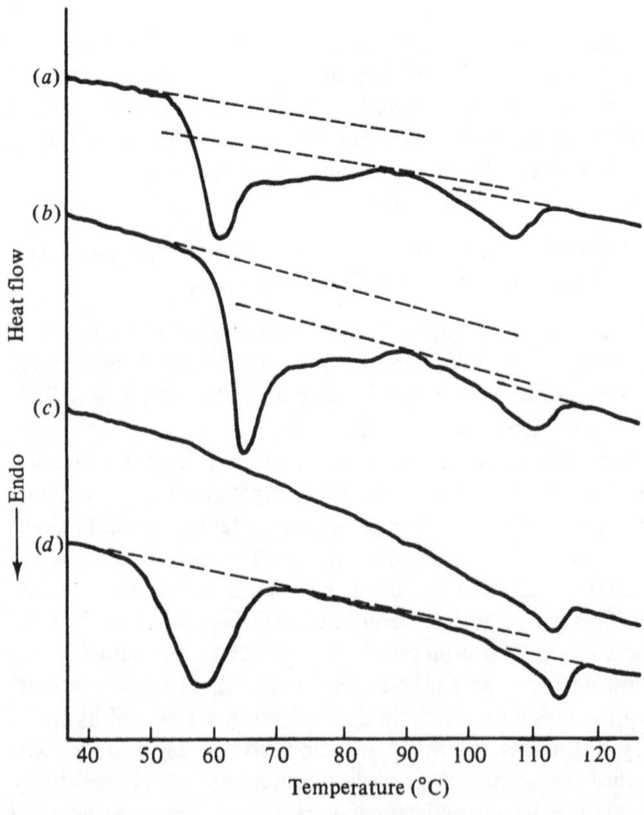

Fig. 2.21. DSC heat flow curves of wheat starch–water mixtures (45:55 by weight): (*a*) native; (*b*) native, after 55 days at 25 °C; (*c*) immediate rescan after gelatinization of sample in (*a*); (*d*) sample in (*c*), after 55 days at 25 °C. Dashed lines represent extrapolated baselines. [13, 15]

between 50 and 60 °C. This near superposition of the glass transition on the crystalline melt, due to imbalance of moisture content inside (10 w %) and outside (100 w %) the granules and of heating rate, was revealed by the expected characteristic change in C_p shown by extrapolated baselines and by derivative thermal curves. [13, 15] This important insight into basic starch thermal properties has recently been corroborated by DSC results for granular rice starches. [9] A minor glass transition and subsequent melting of crystalline amylose–lipid complex was seen in the temperature range 85–115 °C. [10, 19] When native starch was allowed to anneal at 55 w % sample moisture (initially 10 w % inside the granules and 100 w % outside) for 55 days at 25 °C (curve (*b*)), transition temperatures and the extent of crystallinity increased, but melting of the microcrystallites was still governed by the requirement for previous softening of the glassy regions of amylopectin. Annealing is used here to describe a crystal growth/perfection process, in a metastable PC polymer [44], which is carried out optimally at a temperature, T_a, above T_g and typically 0.75–0.88 (K) of T_m. [74] In contrast, recrystallization is a process which occurs in a crystallizable but amorphous metastable polymer at $T_g < T_r < T_m$. [54] After gelatinization by heating to 130 °C, and quench-cooling to 25 °C, the sample in curve (*c*) showed no transitions in the range 50–85 °C. However, when this amorphous (i.e. no remaining amylopectin microcrystals) sample was allowed to recrystallize (now at uniformly distributed 55 w % moisture content) for 55 days at 25 °C (curve (*d*)), it showed a major T_m at about 60 °C, which was not immediately preceded by a T_g. This T_m characterizes the symmetrical melting transition observed in retrograded wheat starch gels with excess moisture (which are partially crystalline and contain hydrated B-type starch crystals [16]) and in stale bread and other high-moisture wheat starch-based baked goods. [67, 213–16, 220]

Low-temperature DSC analysis [13, 15] revealed (curves not shown) why no T_g was observed to occur immediately before the starch T_m in curve (*d*). Native wheat starch, at 55 w % total moisture, showed only a T_m of ice in a low-temperature thermogram. In contrast, a gelatinized sample, at the same water content, showed a major glass transition of fully-plasticized amorphous starch at about −5 °C, preceding and superimposed on the ice melt. (This thermogram looked quite similar to figure 2.9(*b*), since the T_g at −5 °C is actually T_g' for gelatinized wheat starch in excess moisture, which is defined by $W_g' \approx 27$ w % water. [16, 21]) The critical conclusion was that knowledge of total sample moisture alone cannot reveal the state of plasticization of the amorphous regions of a starch granule. The amorphous regions of a native granule are only partially plasticized by excess water in a sample at room temperature, so that softening of the glassy regions must occur at about 50 °C (during heating at 10 °C/min in the DSC), before the microcrystallites can melt. (This situation of partial, and dynamically-changing, plasticization at room temperature also explains why slow annealing was possible for the native starch sample in curve (*b*) of figure 2.21.) After gelatinization, the homogeneous, amorphous starch is fully plasticized at 55 w% moisture, and the metastable amorphous matrix exists at room temperature as a mobile rubber (with ΔT about 30 °C above T_g), in which recrystallization can proceed. Another insight revealed by these DSC results concerns the dynamic effects on starch caused by the DSC measurement itself. During a DSC heating run, plasticizer (water) content increases from 6–10 w% in a native sample before heating to 55 w% at the end of melting, and this kinetically-controlled moisture uptake leads to swelling of starch granules above T_g, which is not reversible on cooling. The same behavior is manifested in volume expansion measurements performed by thermomechanical analysis (TMA), results of which have appeared to show several stages of swelling below T_m. This volume expansion is due primarily to swelling which depends linearly on the amount of water taken up, and only secondarily on thermal expansion of amorphous starch. Thus, the predominant mechanism of swelling is indirectly related to the role of water as a plasticizer of starch, while the minor mechanism is directly related. The use of TMA results for native starch, which

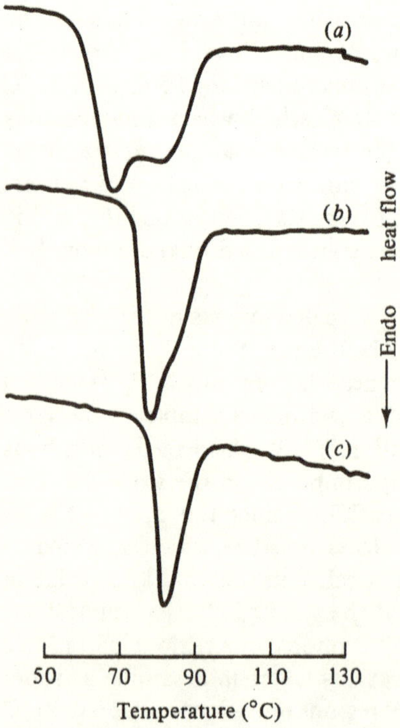

Fig. 2.22. DSC heat flow curves of waxy maize starch–water mixtures (45:55 by weight): (*a*) native; (*b*) native, after 15 min at 70 °C; (*c*) native, after 30 min at 70 °C. Reproduced, with permission, from [17].

is a polydisperse system of branched and linear polymers, in the presence of excess water [9] to support a proposal for a 'three-microphase' model for native granular starch, based on a theoretical model for thermal expansion of undiluted amorphous and PC pure polymer systems, is of questionable validity. [21]

After Slade [16] had established that the thermal behavior of native starch at 55 w% total moisture in the temperature range 50–100 °C represents the superposition of a second-order glass transition followed by a first-order crystalline melting transition, it was shown (figure 2.22) that it is possible to accelerate plasticization of the amorphous regions by water without melting the crystalline regions. [17] In figure 2.22, waxy maize starch was used as a model to study amylopectin in the absence of amylose. Like wheat starch, native waxy maize starch (curve (*a*)) exhibited the non-equilibrium melting of a PC glassy polymer, with a requisite glass transition preceding multiple crystalline transitions in the temperature range 50–100 °C, when total sample moisture was 55 w%. When this sample was annealed for 15 min at 70 °C (curve (*b*)), total excess heat uptake below the baseline was reduced by about 25% and the temperature range of the multiple transitions was shifted upward and became narrower, but the glass transition immediately before the crystalline melt was still evident. A similar result had been seen, in figure 2.21 curve (*b*), for wheat starch annealed at 25 °C for 55 days. (Note the apparently exponential dependence on ΔT, dictated by WLF kinetics, of rates for the two different annealing conditions.) In contrast, when native waxy maize starch was annealed for 30 min at 70 °C (curve (*c*)), total excess heat capacity below the baseline was reduced by 50% and represented only the enthalpy of the first-order crystalline melting transition, because the preceding glass transition had been depressed to less than 0 °C, due to full plasticization of the amorphous regions by water. [16, 21]

The amorphous regions of the starch granule represent a continuous phase, and the covalently-attached microcrystalline branches of amylopectin plus discrete amylose–lipid crystallites represent a discontinuous phase. For each polymer class (distinguished by an arbitrarily small range of linear *DP*), water added outside the granule acts to depress T_g of the continuous amorphous regions, thus permitting sufficient mobility for the metastable crystallites to melt on heating to T_m above T_g. [13, 15] The effect of changing the amount of added water on the thermal behavior of native rice starch can be seen in figure 2.23. [17] At the 'as is' 10 w% moisture content, with no added water (curve (*a*)), the glass transition of the amylopectin occurred above 100 °C and multiple crystalline melting transitions occurred above 150 °C. Similar profiles would have been seen at increasing total sample moisture contents up to about 30 w%, with the initial glass transition occurring at decreasing temperatures. At moisture contents higher than about 30 w% (curves (*b*)–(*f*)), the initial glass transition occurred at about the same temperature, and the subsequent cooperative events occurred at lower and narrower

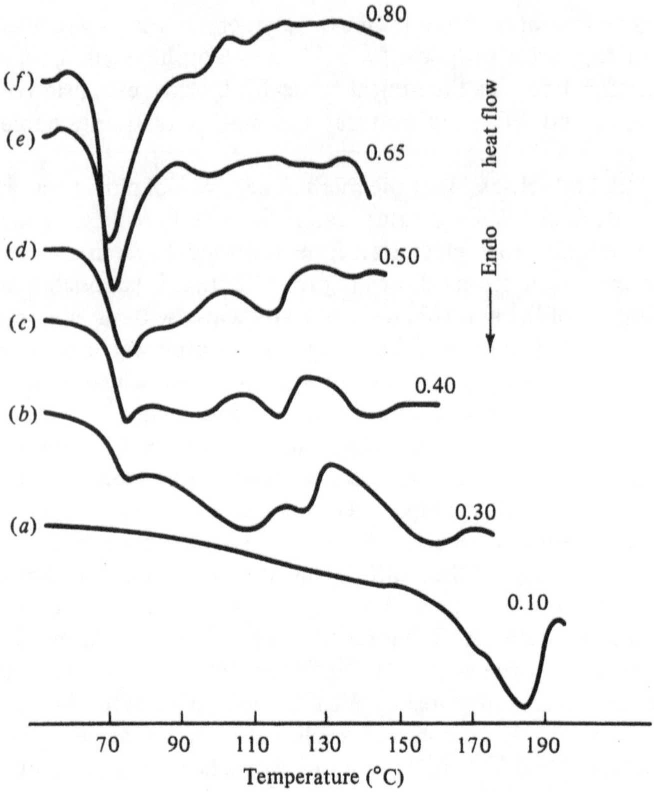

Fig. 2.23. DSC heat flow curves of native rice starch at various water contents: (*a*) starch with 'as is' moisture of 10 w%; (*b*)–(*f*) starch with moisture added to water weight fractions indicated. Reproduced, with permission, from [17].

temperature ranges as water content was increased. The moisture content which is sufficient to plasticize starch completely after gelatinization is about 27 w% (i.e. W_g'). However, unlike gelatin gels (to be described in section 2.6), which can be dried to different moistures so that water is uniformly distributed throughout the amorphous regions [13, 15], native starch starts out typically at 6–10 w% moisture, and the added water is outside the granule (and so initially non-plasticizing), prior to moisture uptake and swelling during DSC heating.

These results, and our conclusions from them [13, 15], have been confirmed by recent DSC results of Biliaderis *et al.* [9] on several other varietal rice starches. In addition to several composite thermograms similar in appearance to figure 2.23, they presented a graph of T_g vs. starch concentration (w%), for native starch–water mixtures, which starts at about 240 °C for the dry sample, and decreases with increasing moisture to 68 °C at about 30 w% total moisture ('as is' plus added). Beyond this moisture content (i.e. for further

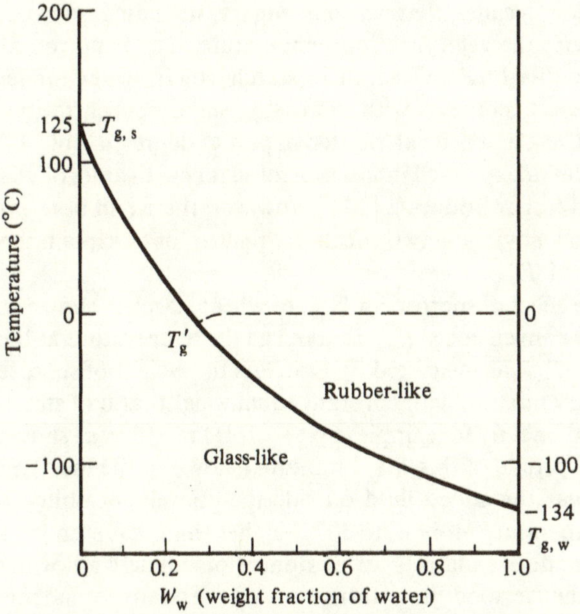

Fig. 2.24. State diagram, showing 'the approximate T_gs as a function of mass fraction, for the starch–water system.' The subscript w denotes water and s denotes starch. [6] Reproduced with permission.

additions of (initially non-plasticizing) water outside the granules), and prior to gelatinization during DSC heating, initial T_g appears to remain constant. [9] This graph of the effect of water on the dynamically-measured value of T_g for PC native starch should not be confused with the T_g curve in a state diagram for a homogeneous starch–water system, i.e. amorphous gelatinized starch–water. Such state diagrams have recently been presented by van den Berg [6] and ourselves [21], and the former's is shown here in figure 2.24. It illustrates the smooth glass curve which connects T_g of dry starch with T_g of amorphous water, and passes through T'_g (−5 °C); W'_g (27 w% water) for gelatinized starch.

2.5.2 *Thermal properties of three-component model systems: the antiplasticizing effect of added sugars on the gelatinization of starch*

Slade extended the description of the effect of water as a plasticizer on native starch, from the perspective of starch as a PC glassy polymer system, to the next level of complexity, i.e. three-component model systems of native starch, water, plus added sugars. [13, 15] When the plasticizer of the amorphous regions is not water alone, but an aqueous sugar solution, this sugar–water cosolvent actually exerts an antiplasticizing effect, relative to water alone, on the gelatinization of starch. [16]

It has been known for decades that various sugars, including sucrose, fructose, and glucose, raise the gelatinization temperature (T_{gelat}) and retard the increase in viscosity ('pasting') of starch in starch–sugar–water model systems, and that this effect increases with increasing sugar concentration. [64, 217] Recently, this effect had been attributed in part to depression of 'A_w' by sugars and in part to an unexplained interaction of sugars with amorphous areas of starch (so-called 'sugar bridges'). [212] However, there had been no successful attempt to show how these two effects are related, or to explain the mechanism of elevation of T_{gelat}.

As an illustration, the effect of sucrose on T_{gelat} of wheat starch is shown in figure 2.25. [13, 15] For convenience, T_{gelat} is taken as the temperature at the peak of heat uptake (i.e. T_{min}) as measured by DSC. As the weight of sucrose was increased in a ternary mixture with constant equal weight ratio of starch and water, T_{min} increased linearly for samples up to a 1:1:1 mixture, as shown in figure 2.25. The DSC profile of this 1:1:1 mixture (shown in the inset), in which 50 w% sucrose was the added fluid outside the granule, exhibited a glass transition at a temperature more than 30 °C higher than that seen for a 1:0:1 mixture, when the added fluid was water alone. Immediately following and superimposed on the elevated glass transition was a relatively narrow crystalline melting transition, as seen when native starch was annealed either for 55 days at 25 °C (figure 2.21(*b*)) or for less than 30 min at 70 °C (figure 2.22(*b*)).

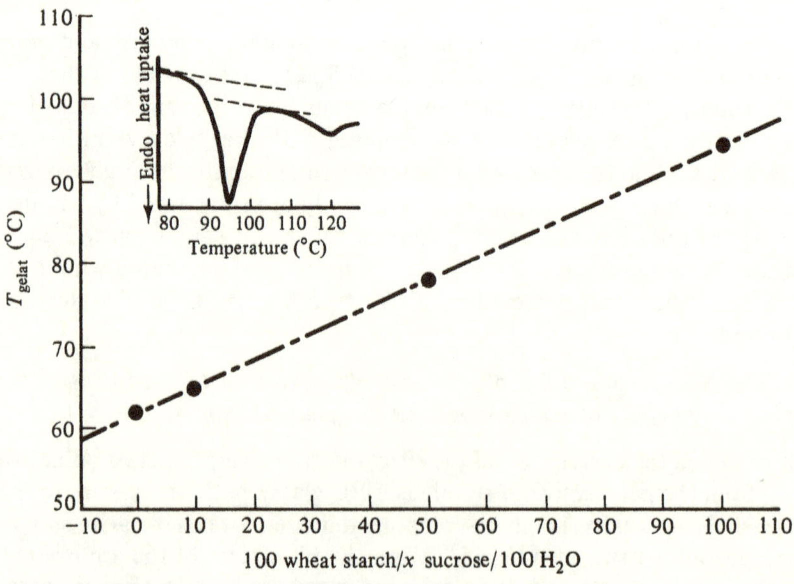

Fig. 2.25. Gelatinization temperature, T_{gelat}, as a function of added sucrose content for three-component mixtures of native wheat starch–sucrose–water (100:*x*:100 parts by weight). Inset: DSC heat flow curve of 100:100:100 mixture. [13, 15]

Slade reported [13, 15, 16] how the effect of sucrose on T_{gelat} can be explained, within the predictions of our conceptual framework of starch as a PC glassy polymer, by applying WLF theory. If a sugar–water solution is viewed as a plasticizing cosolvent, it is evident that such a coplasticizer, of greater average MW and molecular volume than water alone, would be less effective in mobilizing and increasing the free volume of the amorphous fringes in the 'fringed micelle' structure of a starch granule. Less effective plasticization would result directly in less depression of T_g, and thus indirectly, in less depression of non-equilibrium crystalline melting transition temperatures. In this sense, in comparing efficiencies of aqueous solvents as plasticizers of glassy regions of native starch, water alone is the best plasticizer, and sugar–water cosolvents are actually antiplasticizers relative to water itself. By most effectively depressing the requisite T_g which initiates gelatinization, added water results in the lowest T_{gelat}. Increasing concentrations of a given sugar result in increasing average molecular volumes of the cosolvent and thus increasing antiplasticization compared to water alone.

Of course, increasing the concentration of a given sugar in an aqueous cosolvent also decreases the 'A_w' (actually, RVP), but it has already been demonstrated that the temperature of the initial glass transition of native starch in the presence of added water is independent of total sample moisture above about 30 w% total water content. [13–17] Moreover, according to WLF theory, it would be expected that antiplasticization would increase with increasing $\bar{M}_w$ of cosolvent, within a homologous series of cosolvent components, from monomer to dimer to oligomer to polymer. Yet, in such a case, the RVP of cosolvents at equal weight concentrations would generally increase with increasing cosolvent $\bar{M}_w$. For the homologous series glucose, maltose, maltotriose, 10 DE maltodextrin, RVPs of 50 w% solutions increase from 0.85 to 0.95 (for both maltose and maltotriose) to 0.99. [13, 15]

Experimental results for the effect of this homologous series of cosolvents on T_{gelat} of starch confirmed the prediction based on WLF theory, as shown in figure 2.26. [13, 15] As the $\bar{M}_w$ of the antiplasticizing cosolvent increases, free volume decreases, and the resultant T_g of the cosolvent increases. As a consequence of this, the cosolvent becomes less efficient in depressing the T_g of native starch, which initiates the gelatinization transitions, so T_{gelat} increases with increasing cosolvent $\bar{M}_w$.

More recently, Slade reported [21] a comparison of degree of elevation of the T_{gelat} of native wheat starch, for 1:1:1 starch–sugar–water mixtures of a larger and non-homologous series of sugars. Her DSC results showed that T_{gelat} increases in the following order: water alone < galactose < xylose < fructose < mannose < glucose < maltose < lactose < maltotriose < 10DE maltodextrin < sucrose. She suggested that the same structure–property principles which act as determinants of 'water availability' in the water dynamics domain (section 2.4) appear to influence the elevation of T_{gelat}. She

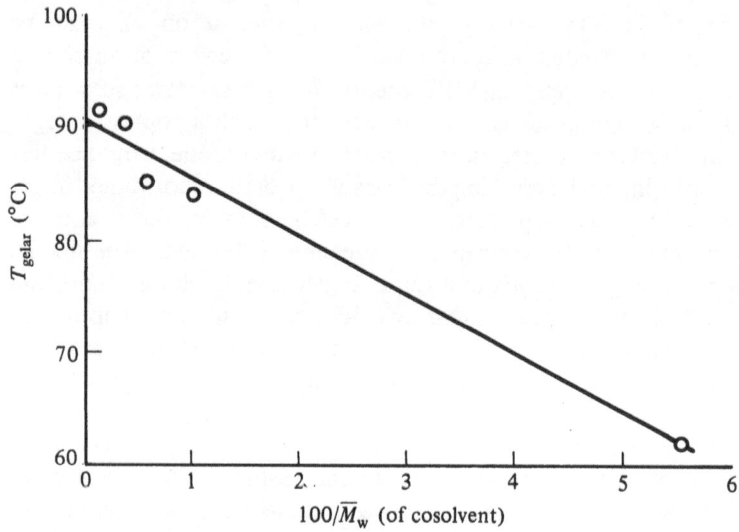

Fig. 2.26. Gelatinization temperature, T_{gelat}, as a function of $100/\bar{M}_w$ (of cosolvent, water + glucose polymer), for three-component mixtures of native wheat starch–glucose polymer–water (1:1:1 parts by weight) and a two-component mixture of native wheat starch–water (1:2 parts by weight). [13, 15]

concluded that no single parameter, e.g. T_g, W_g', or RVP, can explain completely the mechanism of elevation of T_{gelat} by sugars, but a combined parameter which incorporates a free volume contribution may prove useful to explain why the elevating effect on T_{gelat} is greater for sucrose than glucose than fructose.

As pointed out earlier with regard to effects of water content in binary starch–water systems [13, 15], attempts to treat effects of sugar content in ternary starch–sugar–water systems on observed T_m of starch by Flory–Huggins theory for melting-point depression by diluents [30] would be inappropriate. [16, 21] Aside from theoretical problems introduced by the facts that the interaction parameter is concentration-dependent and amylopectin microcrystallites are not a monodisperse system, Flory–Huggins theory only describes an equilibrium melting process. Regardless of the amount of water or sugar solution added to native starch, initial melting of a native granule is a non-equilibrium process, in which melting of microcrystallites is controlled by previous plasticization of glassy regions via heat/moisture treatment to $T > T_g$.

2.5.3 *Retrogradation/staling as a starch recrystallization process*

The rate and extent of staling in high-moisture (>27 w% water), low sugar/fat, starch-based baked products (e.g. breads, rolls, and English muffins) have been correlated with rate and extent of starch retrogradation. [67] Starch retrogradation has been demonstrated to typify a non-equilibrium recrystallization process in an amorphous but crystallizable polymer system, which exists in a kinetically-metastable rubbery state, and is sensitive to the plasticizing effects of water and temperature. [16, 21] In their retrogradation behavior, baked crumb of wheat starch-based breads and experimental starch model systems are known to be analogous. [67] If adequate packaging prevents simple loss of total moisture, the predominant mechanism of staling in bread crumb or wheat starch gels is recrystallization of amylopectin from the amorphous state of a freshly-heated product to the PC state of a stale product, with concomitant redistribution of moisture and increased firmness. [16] This recrystallization depends strongly on the previous history of a sample, since both initial heating and subsequent aging are non-equilibrium processes. [16] (Starch gelatinization during baking constitutes a heat/moisture/time treatment. [217, 218]) Moisture content in the amorphous regions of a native starch granule determines the temperature of the glass transition that must precede melting of the crystalline regions (A-type in wheat) during gelatinization; complete melting during typical baking eliminates residual seed nuclei available for subsequent recrystallization. [16] The extent of swelling and release of protruding and extragranular polymer available for subsequent three-dimensional network formation by recrystallization depend on total moisture content during gelatinization. [64]

Immediately after baking and cooling to room temperature, the amorphous gelatinized starch network in high-moisture bread begins to recrystallize to a PC structure containing disperse B-type crystalline regions. [16] B-type starch is a higher-moisture (vs. A-type) crystalline hydrate polymorph. [55, 56] Its recrystallization requires incorporation of water molecules into the crystal lattice, and this incorporation must occur while starch chains are becoming aligned. Thus, this recrystallization necessitates internal moisture migration within the crumb structure, whereby some of the (previously homogeneously-distributed) water must diffuse from the surrounding amorphous matrix and be incorporated into crystalline regions. [16] Since crystalline hydrate water can neither plasticize the starch network nor be perceived organoleptically, the overall consequence of this phenomenon is a drier and firmer texture characteristic of stale bread. [67] An implicit requirement of starch recrystallization is the availability of sufficient moisture, at least locally within the matrix, both for mobilizing long polymer chains (by plasticization) and being incorporated in crystal lattices. For gelatinized wheat starch, a moisture content $\gtrsim 27$ w% (i.e. W_g') represents the minimum requirement. [16] In fact, in low-moisture products, native starch granules in dough are not even gelatinized during baking. [7] Slade reported, from DSC results for model wheat starch gels, that the percentage recrystallization increases monotonically with increasing percentage total moisture in the range 27–50 w% water (due to increasingly effective plasticization), then decreases with further increases in moisture up to 90 w% (apparently due to a dilution effect). [16] Others [219] recently confirmed Slade's findings.

Staling has been studied in bread crumb and model starch gels using DSC, compression tests, and X-ray crystallography to monitor formation and aging of the recrystallized starch network. [16, 21, 67, 213–16, 220] A quantitative DSC method has been developed to measure rate and extent of recrystallization, as functions of ingredients, time, temperature, and product moisture content during baking and storage, in terms of increasing content of retrograded B-type crystalline starch (measured from the area of the characteristic melting endotherm at 60 °C). Typical DSC results are illustrated in figure 2.27 [16] for two commercial bakery products: (*a*) shows white pan bread, immediately after baking (completely amorphous = 'fresh') and after 7 days storage at 25 °C (extensively recrystallized = stale); (*b*) shows

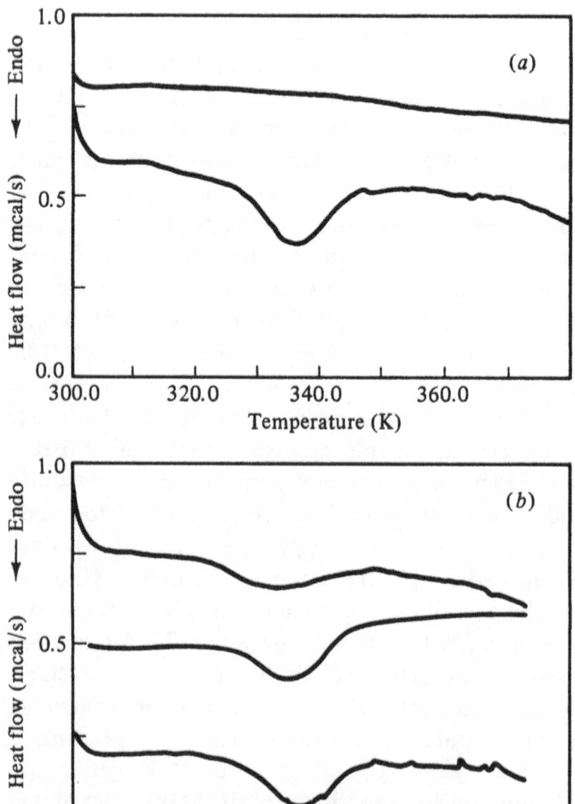

Fig. 2.27. DSC heat flow curves illustrating the rate and extent of starch recrystallization in two types of commercial baked goods: (*a*) white bread, immediately after baking (upper curve) and after 1 week at room temperature (lower curve); (*b*) English muffins, on days 1 (upper curve), 7 (middle curve), and 13 (lower curve) after baking. [16]

progressively increasing recrystallization in English muffins after 1, 7, and 13 days ambient storage after baking.

Mechanism of anti-staling by sugars As described in section 2.5.2, sugars, acting as antiplasticizers relative to water alone, raise T_{gelat} of native starch. Sugars are also known to function as antistaling ingredients in starch-based baked products. For example, as mentioned in section 2.3.3, glucose oligomers of *DP* 3–8, used as antistaling additives, are reported to be effective in inhibiting, and not participating in, starch recrystallization. [159] Slade has suggested that these two effects are related as follows. [16, 21] Analogous to the elevation of T_g and thus T_m of native PC starch, and therefore T_{gelat} (relative to the plasticizing action of water alone), sugar solutions produce an elevated network T_g of the resulting three-dimensional, amorphous, entangled starch gel matrix in a freshly-baked product. This elevated network T_g (relative to that of the corresponding network plasticized by water alone) controls subsequent recrystallization of B-type starch in the undercooled rubbery gel, by controlling the rate of propagation at ΔT above T_g, according to WLF kinetics. For storage at temperatures greater than the network T_g, there is sufficient mobility for devitrification and subsequent formation of crystalline junction zones, resulting in a PC polymer system which constitutes the retrograded starch gel. Relative to typical storage at ambient temperature, higher network T_g (due to the addition of sugars) translates to smaller ΔT and therefore a lower rate of propagation of starch recrystallization at $T_{storage}$. Thus sugars act to retard the rate and extent of starch staling during ambient storage. Furthermore, WLF theory predicts that greater MW of a sugar would translate to greater antiplasticization by the sugar solution, and so a greater antistaling effect.

Recently, Slade reported [21] the first systematic study of antistaling by a large and non-homologous series of common sugars. Her DSC results compared the degree of elevation of T_{gelat} of native wheat starch (described in section 2.5.2) with the degree of inhibition of recrystallization of gelatinized starch, for the same series of sugars. She analyzed 1:1:1 starch–sugar–water mixtures after 8 days storage at 25 °C after gelatinization, and found the extent of recrystallization increased as follows: fructose > mannose > water alone > galactose > glucose > maltose > sucrose > maltotriose > xylose > lactose > maltooligosaccharides (enzyme-hydrolyzed, *DP* > 3). For the glucose homologs within this sugar series, MW and resultant T_g are the apparent primary determinants of antistaling activity. However, for the other sugars, Slade suggested, as she had with regard to their effect on gelatinization, that water 'availability', as determined by mobility and free volume, appears to play a key role in their antistaling effect. In their inhibitory action on starch recrystallization, as for their elevating action on T_{gelat}, sucrose was more effective than glucose than fructose. The fact that fructose–water, relative to water alone, actually accelerated starch staling was a particularly surprising and salient finding.

Mechanism and kinetics of recrystallization: acceleration of nucleation Since retrogradation in gelatinized starch–water (>27 w %) systems is a nucleation-controlled, gelation-via-crystallization process, the crystallization kinetics can be described as shown in figure 2.2, within the temperature range between T_g (i.e. T_g') $= -5$ °C and $T_m = 60$ °C. [16, 21] The overall rate of crystallization (i.e. rates of nucleation and propagation), at a *single* storage temperature, is maximized about midway between T_g and T_m, and this for starch is about room temperature. Typical DSC results for the extent of starch staling in freshly-baked white bread crumb stored at 25 °C for 39 days are illustrated in figure 2.28, curve (*a*). [21]

Starch recrystallization can also be treated as a time/temperature-governed polymer process which can be manipulated. [16] For PC polymers in general [44], and starch in particular [21, 221], the rate-limiting step in the overall crystallization process is nucleation (enhanced at low temperatures) rather than propagation (enhanced at high temperatures). Curves (*b*)–(*d*) in figure 2.28 illustrate how the rate and extent of staling can be influenced by separating these mechanistic steps and maximizing nucleation rate. Compared to nucleation and propagation at 25 °C (curve (*a*)), faster

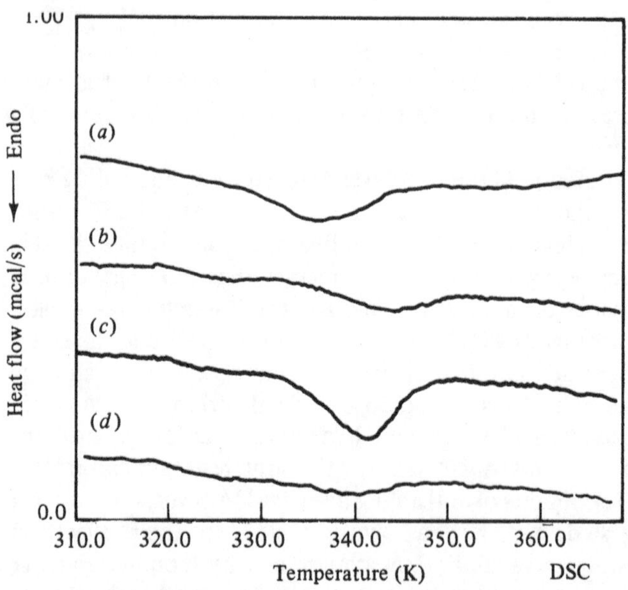

Fig. 2.28. DSC heat flow curves of freshly-baked white bread crumb staled at different storage temperatures and times: (*a*) 25 °C for 39 days; (*b*) 25 °C for 42 h, then 40 °C for 2.5 h; (*c*) 0 °C for 42 h, then 40 °C for 2.5 h; (*d*) −11 °C for 42 h, then 40 °C for 2.5 h. [21]

propagation at 40 °C produced significant staling (curve (*b*)) in much less time. Initial storage at −11 °C ($< T_g'$) inhibited nucleation (except during cooling) and produced less staling (curves (*d*) vs. (*b*)). However, the greatest rate and extent of staling were achieved (curve (*c*)) by faster nucleation at 0 °C (for long enough to allow extensive nucleation), followed by faster propagation at 40 °C.

Figure 2.29 [21] shows DSC results for model wheat starch–water (1:1) mixtures, exposed to different temperature/time storage protocols immediately following gelatinization. This study represented an exploration of optimum nucleation temperature for maximum rate of recrystallization. For single temperature storage conditions, the rate of nucleation and thus overall crystallization increased with decreasing T ($40 < 25 < 4$ °C, curves (*a*)–(*c*)), as long as the temperature was greater than T_g'. Freezer storage at $T < T_g'$ inhibited nucleation (curves (*d*), (*e*)), and so retarded recrystallization, even when followed by propagation at 40 °C. Once again, as for white bread crumb in figure 2.28(*c*), gels first held at 4 °C to promote rapid nucleation, then held at 40 °C to allow rapid propagation (curve (*g*)), showed by far the greatest overall rate and extent of starch retrogradation.

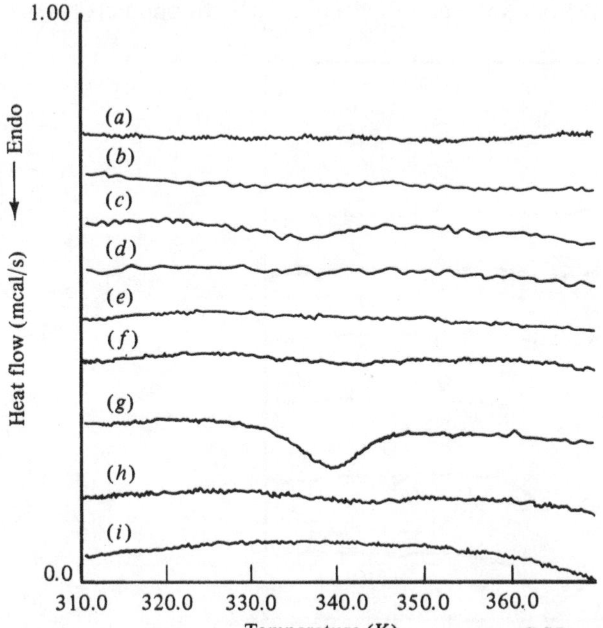

Fig. 2.29. DSC heat flow curves of wheat starch–water 1:1 mixtures, stored under different T/time conditions immediately following gelatinization: (*a*) 40 °C for 4.5 h; (*b*) 25 °C for 3.5 h; (*c*) 4 °C for 3 h; (*d*) −23 °C for 2.5 h; (*e*) −196 °C for 1 min, then −23 °C for 2 h; (*f*) 25 °C for 2 h, then 40 °C for 5 h; (*g*) 4 °C for 2 h, then 40 °C for 6.5 h; (*h*) −23 °C for 2 h, then 40 °C for 6 h; (*i*) −196 °C for 1 min, then −23 °C for 2 h, then 40 °C for 5.5 h. [21]

Figure 2.30 [21] shows DSC results for model wheat starch–water (1:1) mixtures, examined for the effect of increasing nucleation time at 0 °C, immediately after gelatinization, and prior to 30 min of propagation at 40 °C. It is apparent from the trend of the steadily-increasing endotherm areas in curves (*a*)–(*f*) that the extent of nucleation and overall crystallization increased monotonically with increasing time of nucleation. Moreover, recrystallization was already measurable after only 1 h total storage, and quite extensive after 5.5 h. (Compare the endotherm areas in figure 2.30 with others in figures 2.27–2.29, all of which are plotted with 1.0 mcal/s full scale.)

These results were used to design a patented industrial process for accelerated staling of bread (for stuffing) and other starch-based foods. [222] By a two-step temperature-cycling protocol involving, first, a several-hour holding period at 4 °C (to maximize nucleation rate), followed by a second several-hour holding period at 40 °C (to maximize propagation rate), a much greater overall extent of staling is achieved vs. the same total time spent under constant ambient storage, equivalent to staling bread for several days at room temperature.

2.5.4 *Amylopectin–lipid crystalline complex formation at low moisture*

It has long been known that amylose forms a helical complex with lipids, which crystallizes readily from water as anhydrous (and hydrophobic) crystals

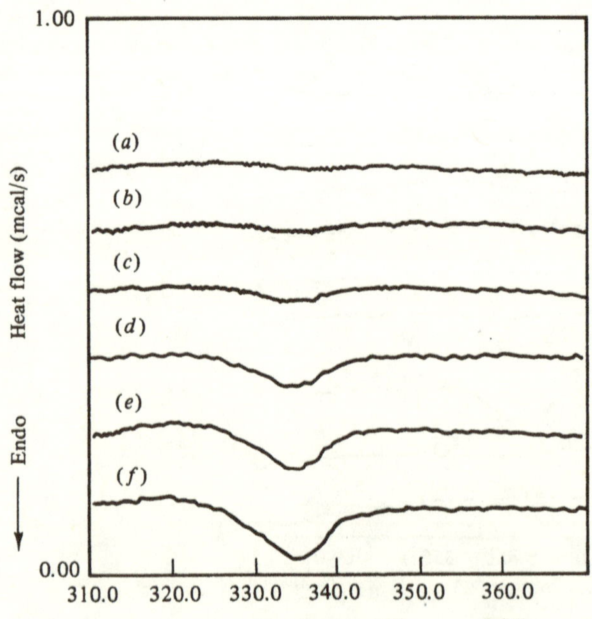

Fig. 2.30. DSC heat flow curves of wheat starch–water 1:1 mixtures, nucleated for different times at 0 °C, then propagated for 30 min at 40 °C, immediately following gelatinization: (*a*) 10 min; (*b*) 30 min; (*c*) 60 min; (*d*) 180 min; (*e*) 240 min; (*f*) 300 min. [21]

which give rise to V-type X-ray diffraction patterns. [223–7] T_m of the crystalline amylose–lipid complex is about 110 °C (see figure 2.21 for typical DSC thermograms), for endogenous lipids of cereals, such as wheat, corn, and rice. [10, 16, 19, 21, 226, 227] The amylose–lipid complex is more stable than the well-known amylose–iodine complex [224], and amylose–lipid crystals are more thermostable than amylose-hydrate polymorphs (B-type, $T_m \approx 60$ °C, in retrograded amylose gels, or A-type, $T_m \approx 85$ °C, in native cereal grain starches). [10, 16, 19, 21] For a given MW of amylose, pure anhydrous amylose crystals are most thermostable of all, with $T_m > 140$ °C. [10, 19]

Normal native starch contains about 70–80% amylopectin and 20–30% amylose in the form of a layered granule, which is partially crystalline but not a polymer spherulite. [55, 56] Depending on the endogenous lipid content of the granule, amylose may exist as a glassy hydrate, crystalline hydrate, or crystalline amylose–lipid complex. Upon heating starch for DSC analysis at total sample moisture content >27 w%, preexisting crystalline amylose–lipid is seen as a melting endotherm at ≈110 °C. Preexisting glassy amylose, in the presence of but not pre-complexed with lipid, is seen as a crystallization exotherm near 95 °C. For starches with low endogenous lipid, addition of exogenous lipid results in a crystallization exotherm on the first heating, and a melting endotherm on the second heating. [10, 19] In contrast, for waxy starches, which contain essentially no amylose, even when the endogenous lipid content is high or exogenous lipid is added, DSC analysis in the conventional temperature range from 20–130 °C had never before revealed evidence of crystallization or melting of amylopectin–lipid complexes [227], until Slade's recent report. [21]

It is known that amylose can be precipitated with butanol, but typical amylopectin cannot; this is the basis for the traditional distinction between the two polymers. [225, 227] The longest accessible linear chains in amylopectin are branches about 16–20 glucose units long. [228] These branches, which are responsible for the microcrystalline regions of amylopectin, are not long enough to form stable complexes with iodine or butanol. [227] It had been assumed until recently [21, 227] that amylopectin branches are also too short to form stable complexes with lipids, and the absence of DSC transitions [227] had supported this assumption. The most stable complexes of linear amylose with iodine are formed by chains of $DP > 40$ glucose units. [224] Despite the lack of previous evidence from DSC and other analytical methods for interactions between amylopectin and lipid, addition of stearoyl lipids (e.g. sodium stearoyl lactylate, SSL), is known to affect the texture of waxy maize starch samples. It was believed possible that amylopectin–lipid complexes do occur, but the T_m of the crystals, if determined by the low MW of amylopectin branches, might be well below the temperature range typically examined by DSC for starch–lipid complexes.

Slade used native waxy maize starch, with only 'as is' moisture (<10 w%), as the amylopectin source. She heated a starch–SSL (10:1 w/w) mixture at 120 °C for 15 min (under 15 psi pressure), to assure comelting of reactants. Starch alone, treated the same way, produced the featureless thermogram shown in figure 2.31(*a*) [21], while SSL alone melted at 45–50 °C. The rationale for this experimental approach was that, in the presence of only enough moisture (< 10 w %) to permit plasticization and melting of starch at high temperatures, but not enough to allow formation of amylopectin A- or B-type crystal hydrates, amylopectin–lipid crystalline complex formation would be possible and favored. The starch–SSL comelt was then nucleated at 4 °C for 24 h, heated to 120 °C at 10 °C/min and recooled, then analyzed by DSC, as shown in figure 2.31(*c*). The thermogram shows a small crystallization exotherm at 55 °C, followed by a large and narrow melting endotherm at $T_m \approx 70$ °C. This was presented [21] as the first DSC evidence of a crystalline amylopectin–lipid complex produced at low moisture. The low T_m, relative to that for amylose–lipid complex, was suggested to indicate a lower-MW complex, formed with the short, crystallizable branches of amylopectin.

2.6 Gelatin: polymer physico-chemical properties

The functional properties of gelatin, which make it a preferred ingredient in many different industrial applications, can be understood from an analysis of

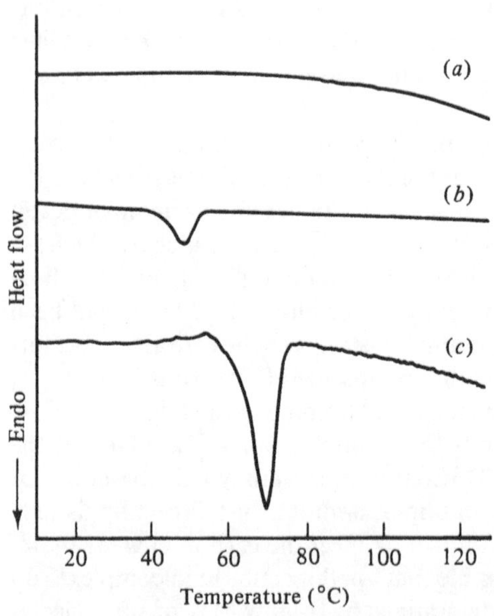

Fig. 2.31. DSC heat flow curves of: (*a*) waxy maize starch (< 10 w % water), after heating at 120 °C (15 psi pressure) for 15 min; (*b*) SSL alone, same heat treatment (middle); and (*c*) 10:1 (w/w) waxy maize starch (< 10 w % water)–SSL, same heat treatment, followed by 24 h at 4 °C, then heating to 120 °C at 10 °C/min and recooling, before rescanning. [21]

the structure–property relationships of this PC glassy polymer. Various aspects of the polymer physical chemistry of gelatin, including thermomechanical, viscoelastic, and structural properties in the solid state, and its mechanism of gelation in water, can influence the material properties, processing, and storage stability of typical commercial gelatin and its products. [13, 15, 23, 24] Many major food-related, pharmaceutical, photographic, and other industrial uses of gelatin [229, 230], which are based on the structure–property relationships of this PC glassy polymer, involve the use of water as an excellent plasticizer of the predominant amorphous regions [1] and a thermodynamically-compatible solvent (at $T > T_m$ of the crystalline regions). (Refer to section 2.2 for earlier remarks on gelatin gelation by three-step polymer crystallization mechanism, crystallization kinetics of gelation, acceleration of gelation, and non-equilibrium melting of gelatin gels.)

2.6.1 *Structure–property relationships*

There is general agreement in the literature that gelatin can be appropriately described as a PC glassy polymer [41, 46, 63], with characteristic thermoplastic [41] and viscoelastic properties. [231, 232] Typical commercial gelatin is derived from animal collagen as a denatured protein gel, which is air-dried to about 10 w% moisture. [233] The three-dimensional gel network, which represents the supramolecular level of structure, is held together by microcrystalline junction zones composed of intermolecularly associated regions of individual gelatin molecules. [41, 45–7, 57, 234] The mechanism of thermoreversible gelation of gelatin in water, as a crystallization process, involves side-by-side association ('disordered aggregation' [49]) of helical chain segments stabilized by neutral, non-polar –Gly–Pro–Hypro– sequences. [41, 46, 232, 238, 239, 241] At the intramolecular level, chain segments within microcrystalline junctions are in a left-handed poly-L-proline II helical conformation, in which intramolecular peptide bonds are in a *trans* isomer configuration. [48, 134, 231, 234–7, 240] *Trans* is the preferred, low-energy configuration of a polyproline helix in water, so water favors helical association of the polyproline II conformation, leading to formation of intermolecular crystalline junctions in an aqueous gelatin gel. [234, 237] Within the small dimensions (< 10 nm [41, 47]) of microcrystallites in a gelatin gel formed at $T \ll T_m$, the familiar triple helix of collagen is thought not to be relevant. [49, 239] Especially for gelatin solutions above a critical gelling concentration (0.5 w%), 'complete renaturing of gelatin to collagen, by three-strand helix formation, is rather unlikely'. [46, 241]

The viscoelastic behavior of kinetically-metastable gelatin gels in an undercooled liquid state has been described in the context of WLF theory. [46, 47] At $T > T_g$, gelatin gels manifest a characteristic rubber-like elasticity [232], due to existence of a network of entangled, randomly coiled chains. [231] With increasing temperature, a gelatin gel traverses five regions of viscoelastic behavior characteristic of synthetic, partially-crystalline polymers

[231]: (1) at $T < T_g$, vitrified glass; (2) at $T = T_g$, glass-to-rubber transition; (3), (4) at $T_g < T < T_m$, rubbery plateau to rubbery flow; and (5) at $T > T_m$, viscous liquid flow.

2.6.2 *Thermal properties of PC gelatin*

Thermal analysis by DSC has proven well-suited to characterize the gelatin–water system, and especially the 'solid-state' properties of gelatin at low moisture. [13, 15, 23, 24, 46, 63, 231, 241] The thermal properties of PC gelatin gels have been described in terms of the T_g of the amorphous regions and the T_m of the crystalline junction zones. [13, 15, 23, 24, 63] (T_m also represents the solid-state counterpart of the helix-to-coil transition temperature, typically observed for dilute gelatin solutions below the critical gelling concentration. [231]) For a gelatin gel with excess moisture (i.e., $W > W_g' = 35$ w% water), T_m is the familiar 'gel melting temperature' of ≈ 37 °C [49, 239], while T_g (actually T_g') is about -12 °C for typical high-MW (high 'Bloom') gelatin. [13, 15] (Reutner *et al.* [241] recently reported similar values, which we recognize as $T_g' = -10$ °C and $W_g' = 30$ w% water for an industrial gelatin.) As the gel is dried, as in a commercial air-drying process [233], and plasticizing water is removed, the transition temperatures increase until, in a typical gelatin powder at 10 w% moisture, T_g is 80–90 °C and T_m is 110–115 °C. [13, 15, 23, 24, 63] (As for other amorphous polymers, T_g of gelatin increases with increasing $\bar{M}_w$, and thus with Bloom value.)

A DSC thermal profile for such a low-moisture gelatin sample (e.g. figure 2.32 for a type B calfskin gelatin of 10.2 w% moisture) reflects the non-equilibrium melting of a water-plasticized gel. At T_g, amorphous regions of the 'fringed micelle' network soften and take on the mobility and viscoelasticity of a rubber. With further heating, microcrystallites, now surrounded by this viscoelastic fluid, are free to melt (i.e. dissociate) at $T_m > T_g$, thereby rendering the gelatin–water solution a molten liquid of randomly coiled macromolecules. [13, 15]

Based on these thermal properties, one can understand why commercial gelatin powder requires boiling water to be dissolved rapidly to a molecularly-dispersed solution. The mechanism of dissolution can be described as follows. [13, 15] Initially, the temperature of boiling water is greater than T_g, but less than T_m. As amorphous regions pick up water and become more plasticized (as W approaches W_g'), their T_g begins to fall rapidly toward T_g'. This allows the T_g-controlled T_m to fall quickly below 100 °C. At this point, the temperature of the water is greater than the T_m of the microcrystallites that hold the network together, so the junctions dissociate and the gel network dissolves rapidly. To dissolve PC gelatin completely and rapidly, the water temperature must remain above 37 °C. Solubilization in water at room temperature has been observed, but the process of hydration, swelling, and the ultimate reformation of an extended gel network took months to occur. [13, 15] (Gelatin in cold-cast and dried films, as in gels and dried powders, is also a

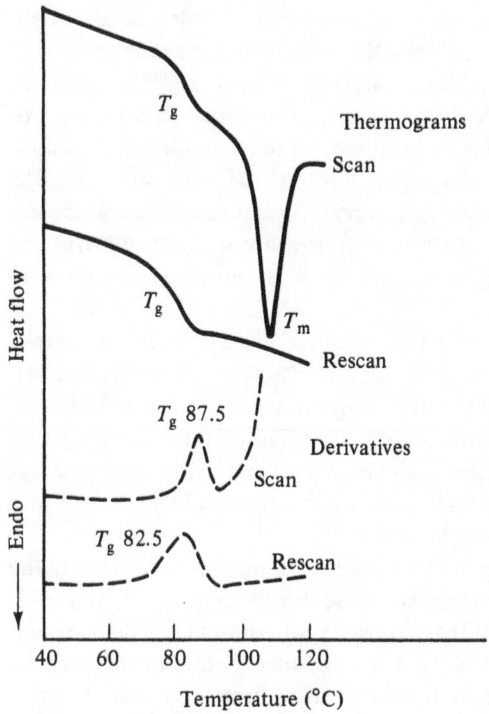

Fig. 2.32. DSC heat flow curves (top – scan and immediate rescan) and derivative thermograms (bottom – scan and immediate rescan) of type B calfskin gelatin (10.2 w% water). [13, 15, 24]

PC glassy polymer, less-perfectly crystalline than collagen [63, 78, 231], with ≈20% crystallinity. [41]

It has been known since 1932 [78] that readily cold-water-soluble gelatin can be obtained by drying from the sol rather than gel state, at $T > T_m$ [41, 242], e.g. by hot-casting/drying of films. [63, 78, 231] In this fashion, gelation (i.e. crystallization at $T < T_m$) can be prevented, and an amorphous material (a network of randomly coiled, entangled chains [231]) can be produced. [41] As reviewed in section 2.6.4, common industrial drying processes suitable for this purpose include spray, drum, freeze, microwave, vacuum oven, and extrusion drying. Such an amorphous gelatin is mechanically a solid (of $\eta > \eta_g$) at room temperature, despite the absence of crystalline crosslinks. Glassy gelatin is actually a solid solution, which is rapidly soluble in cold water by a mechanism involving dilution by plasticizing solvent. [14] Jolley [41] has reported that powdered amorphous gelatin is readily dispersible and soluble in water at $T \ll 37$ °C.

One can also produce a gelatin glass in a calorimeter, as illustrated by the typical DSC thermogram in figure 2.32. [13, 15, 24] In the initial heating scan, a T_g at 87.5 °C and T_m at 111 °C were clearly evident. Once heated through the crystalline transition, the sample was a molecularly-dispersed plastic melt. Upon quench cooling with liquid nitrogen, the sample became immobilized in an amorphous form. An immediate rescan showed only a T_g at 82.5 °C, but no T_m. The T_g in the rescan was 5 °C lower than in the initial scan because of increased plasticization by water. Crystalline regions in air-dried gelatin contain more water than amorphous regions (about twice as much, at 50% relative humidity [41]), so upon melting the crystals and releasing this crystal hydrate water for redistribution throughout the amorphous matrix, the overall moisture content of the glassy phase was increased, and its T_g decreased. [63]

The thermogram in figure 2.32 reflects the state of knowledge in the current literature on the thermal properties of PC gelatin at low moisture, as assessed by DSC. [24, 63] A key feature of this thermogram is that, in the initial scan, heating was stopped at 125 °C, immediately after completion of the melting transition. This is standard practice, because of a belief that 'gelatin samples degrade rapidly when they are heated to temperatures slightly above those over which gelatin structural order is lost'. [63]

The DSC thermal profiles in figure 2.33 illustrate a new finding made in the course of our development of a diagnostic DSC assay for amorphous, cold-water-soluble gelatin. [245] Both thermograms, of commercial high Bloom calfskin and pigskin gelatins of 8–10 w% moisture content, showed the existence of a third thermal event. It was a large first-order endothermic

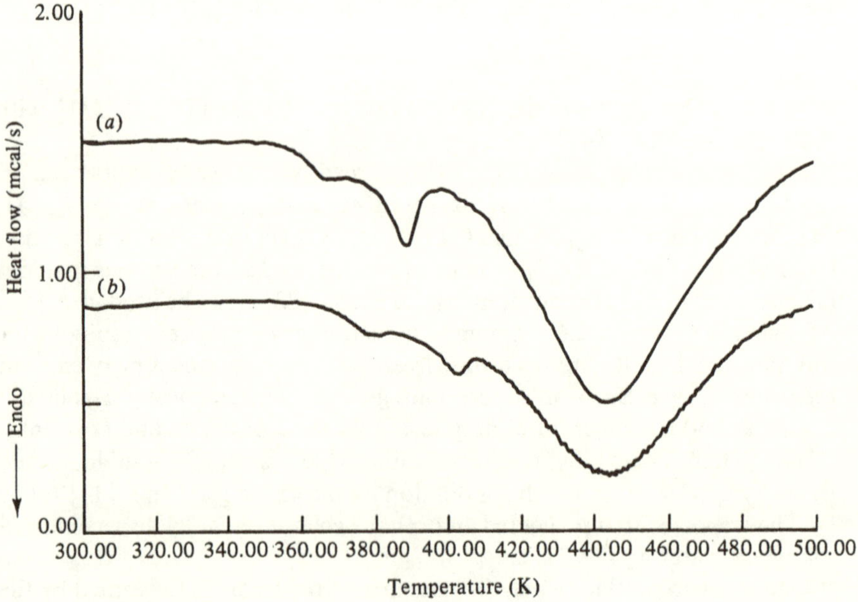

Fig. 2.33. DSC heat flow curves of (*a*) 275 Bloom calfskin gelatin (9.8 w% water) and (*b*) 295 Bloom pigskin gelatin (8.0 w% water).

transition, occurring at about 170 °C, well above T_g and T_m, with an enthalpy generally ranging from 20–50 cal/g. In contrast, the enthalpy associated with T_m of commercial gelatins of low crystallinity typically ranges from 1–3 cal/g. [63] We believe this third transition has never before been described in the DSC literature on gelatin. (However, Borchard may have been referring to the same phenomenon, when he mentioned [243] that 'a fast isomerization process has been found for gelatin in water'.)

From an analysis of literature on collagen, gelatin, and polyproline structure [48, 235–7], we have hypothesized that this thermal transition represents an intramolecular *trans–cis* isomerization of peptide backbone bonds in polyproline II helical segments of gelatin molecules. We refer to this high temperature, high energy isomerization transition as T_i. While the *trans* configuration of the polyproline II helix is the preferred low-energy form in water [237] and the conformation that favors helical association and formation of intermolecular crystalline junctions in an aqueous gelatin gel, the *cis* configuration is a higher energy form. [237] A *trans–cis* isomerization would require input of heat energy, and would be manifested as a first-order endothermic transition. Since the *cis* configuration does not favor helix formation or intermolecular gelation-via-crystallization in water, an absence of gel-forming ability associated with this configuration might explain the earlier reference to 'degradation by heating to $T > T_m$.' [63]

The high-temperature, high-energy nature of this transition, and the effects of different solvents, favor our hypothesis of its origin. As low-moisture gelatin is heated from room temperature, it passes through T_g and then T_m, at which point it exists as a molten liquid. In this form, each gelatin molecule is mobile and free to undergo a configurational transition, a *trans–cis* isomerization, if sufficient heat energy (i.e. 20–50 cal/g) is supplied to the system at $T_i > T_m$.

We have also found that non-aqueous solvents such as glacial acetic acid and dimethyl formamide (DMF) have a striking, and predictable, effect on the T_i of gelatin. DMF is a poor solvent for gelatin, in that it favors an all *cis*, rather than all *trans*, configuration in the polyproline helix. [235] When gelatin is pretreated with DMF, a DSC thermogram (not shown) shows that T_i is completely absent. In contrast, acetic acid is a better solvent for gelatin than water, in promoting the all *trans* configuration. [235] A thermogram for gelatin pretreated with acetic acid (not shown) shows a T_i with an enthalpy much greater than that characteristic of aqueous gelatins.

The graphs in figure 2.34 are state diagrams for PC gelatin–water. They show the three characteristic thermal transitions, T_g, T_m, and T_i, plotted vs. w% moisture content, for two high Bloom calfskin and pigskin gelatins obtained by a commercial air-drying process. While similar graphs of T_g and T_m vs. moisture content have been reported [13, 15, 24, 63, 241], the T_i curves are new, and key to our diagnostic DSC assay for amorphous gelatin. As indicated by the dashed portions of the curves for w% moisture >35% (i.e. W_g'), all three transition temperatures are independent of moisture content in the presence of excess moisture. ($T_g = T_g' = -12\,°C$ [241], and $T_m = T_m^0 = 37\,°C$: T_m^0 is the equilibrium T_m.) However, at $W < W_g'$, all three transition temperatures increase with decreasing moisture, converging to a temperature range >200 °C at moisture contents <10 w%. [63, 241] The

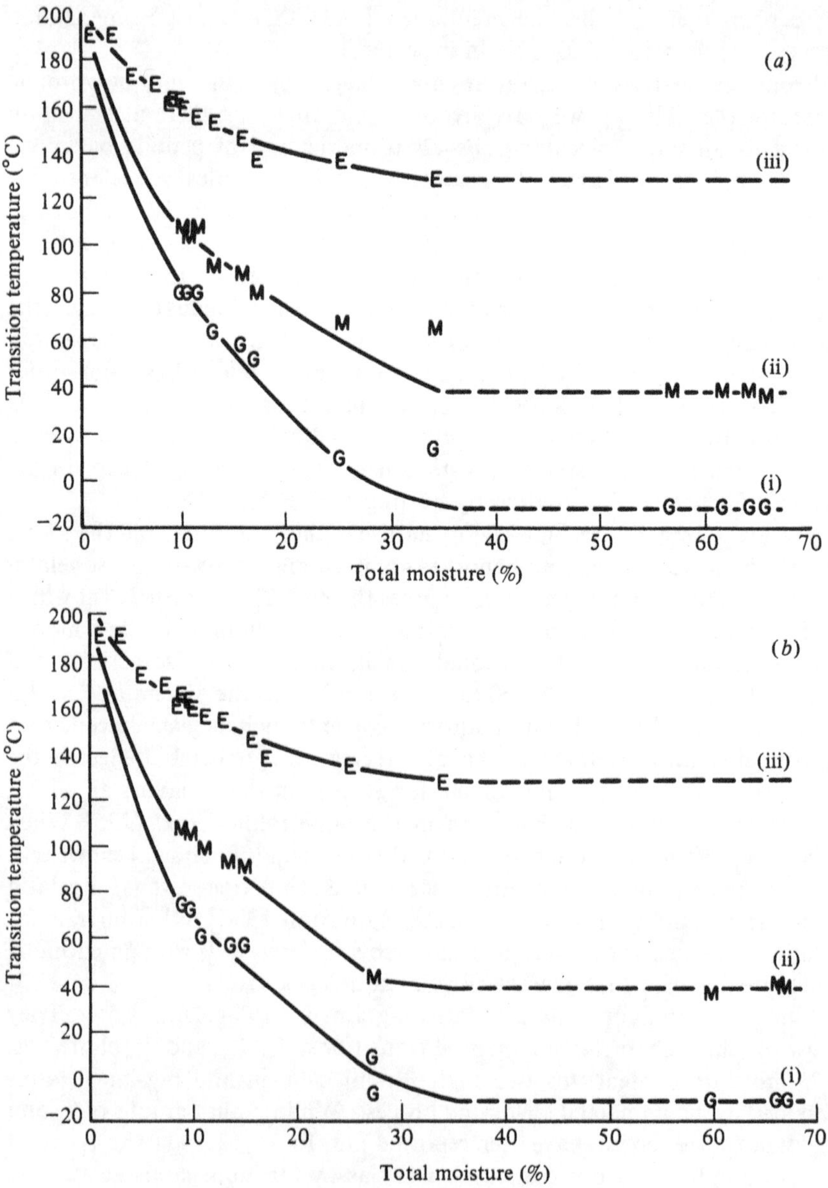

Fig. 2.34. State diagrams of (i) T_g (bottom curve), (ii) T_m, and (iii) T_i (top curve) vs. w% moisture content for (*a*) calfskin and (*b*) pigskin high Bloom gelatins. [13, 15, 24]

temperature spacing between transitions is also determined by sample moisture, such that, as moisture decreases, the temperature interval between transitions also decreases. However, in terms of a diagnostic fingerprint provided by a DSC thermogram, T_m for a PC gelatin must always be located between T_g and T_i. If this temperature space is devoid of a melting endotherm, a gelatin sample must be completely amorphous.

These results illustrate two important facts. First, water is an effective plasticizer of the amorphous regions of PC gelatin. [41, 244] Other well-known but less effective gelatin plasticizers include polyols such as glycerol, sorbitol, and propylene glycol. [231] Second, the melting of gelatin, in the DSC or in macroprocesses, represents an example of non-equilibrium melting of a kinetically-metastable solid material, which is controlled by the state and extent of plasticization of the amorphous regions. [46, 47, 49, 73] The direct effect of water plasticization on metastable glassy regions leads to increased mobility, which allows crystalline junctions to dissociate at T_m only slightly above T_g. An uncharacteristically high ratio for near-anhydrous T_g/T_m of more than 0.9, which indicates this influence of metastable supramolecular structure with non-uniform moisture distribution [73], has also been attributed, unnecessarily, to the special kind of irregularly alternating tercopolymer structure of gelatin. [241] In contrast, the 'equilibrium' T_m^0 for hydrated gelatin microcrystallites, in the absence of contiguous rigid amorphous regions, or in a fully-matured gel with excess moisture (where $W = W_g'$ and $T_g = T_g'$), is $\approx$35–45 °C.

2.6.3 *Thermomechanical behavior of amorphous gelatin*

Our DSC assay for amorphous gelatin is illustrated by the thermal profiles in figure 2.35. These thermograms also illustrate the extreme sensitivity of amorphous gelatin to moisture pickup during ambient storage, and the consequences of plasticization, in terms of increased mobility and a propensity toward reversion to partial crystallinity and concomitant loss of cold-water solubility. Curve (*a*) shows an amorphous gelatin sample (no T_m between T_g and T_i) containing <3 w% moisture, produced by spray drying a 25 w% gelatin sol. [245] This sample manifested excellent cold-water solubility. Curve (*d*), which shows a T_m between T_g and T_i, represents the PC starting material for sample (*a*). Curve (*b*) represents sample (*a*) after moisture uptake to 6.5 w% during storage at ambient temperature. This sample, while still cold-water-soluble, manifested a small exothermic relaxation transition, at $T < T_g$, attributed to nucleation during processing, which was exacerbated by the higher-moisture, greater-mobility condition of the stored sample. The intensity of this relaxation increased with increasing moisture pickup, until, at 14 w% moisture, a pronounced crystalline T_m appeared, as shown in curve (*c*), and this sample was no longer cold-water-soluble. (Thermograms similar to curve (*c*), but with even larger crystalline melting transitions, have been observed for gelatins extruded at 15–20 w% moisture. These samples were

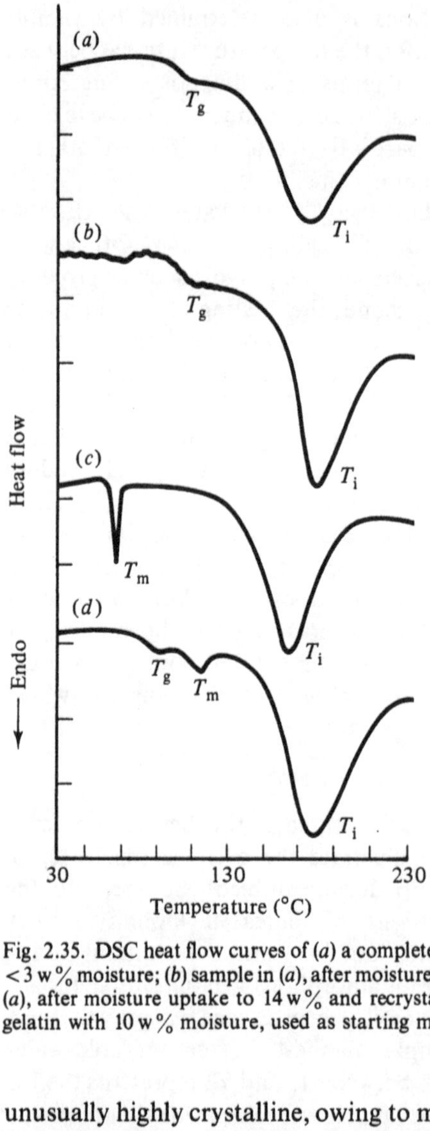

Fig. 2.35. DSC heat flow curves of (*a*) a completely amorphous spray-dried gelatin sample with <3 w% moisture; (*b*) sample in (*a*), after moisture uptake to 6.5 w% during storage; (*c*) sample in (*a*), after moisture uptake to 14 w% and recrystallization during storage; (*d*) an air-dried, PC gelatin with 10 w% moisture, used as starting material for sample in (*a*).

unusually highly crystalline, owing to molecular orientation effects caused by high-temperature extrusion of such high-moisture, mobile gelatins.)

Nucleation in amorphous gelatin can occur due to shear orientation or stress development during various drying or post-drying processes. Generally, nucleated samples of originally amorphous gelatin show an inherent tendency toward instability, such that exposure to plasticization by moisture and/or heat can result in crystal growth, leading to loss of cold-water solubility, and thus decreased storage stability. Such nucleated samples require low-temperature storage in excellent moisture-barrier packaging material to prolong their shelf life.

A common problem with amorphous gelatins involves a phenomenon called 'Bloom loss' (i.e. reduced gel strength, compared to that of corresponding PC starting materials). While Bloom loss is reversible with time, and has been shown by fluorescence analysis for free amino groups not to result from depolymerization due to hydrolysis of peptide bonds, it does necessitate higher product weights of amorphous gelatin. While this countermeasure compensates for initial loss of gel strength, it leads to intensified maturation problems only hours after product preparation. We have hypothesized that Bloom loss can be correlated with the enthalpy of the T_i transition. Typical high-Bloom, PC gelatins show enthalpies approaching 50 cal/g, indicative of highly *trans* character, which would be expected in commercial gelatins of high gel strength. In contrast, some amorphous gelatins, which have manifested the greatest extents of Bloom loss, have shown much lower enthalpies for T_i. If their processing had been temperature-abusive, leading to more extensive *trans–cis* isomerization, so that their *trans* character, and thus the enthalpy of T_i, was reduced, one would expect diminished initial gel strength. However, such Bloom loss would be reversible, because gelatin molecules in a gel with excess moisture would eventually seek a lower energy state, and would revert from *cis* to *trans*.

2.6.4 *Functional properties of gelatin in industrial applications*

Food applications of gelatin Gelatin's largest single food use is in gel desserts, because of the unique 'melt at mouth temperature' character of thermoreversible gelatin gels with excess moisture. [233, 246] In response to a modern trend toward more convenient consumer products, in this case ones that would be both cold-water-soluble and quick setting, coupled with the 50-year-old knowledge about drying from the sol state (at $T > T_m$) in order to produce amorphous gelatin, many recent US patents for cold-water-soluble, quick-setting gelatins and their processes have appeared. [242, 245, 247–56] Such gelatin materials, in powdered form, are reported to be rapidly soluble in 50–60 °F water and quick setting, the latter characteristic being due to faster heat transfer on cooling from the temperature of the cold sol to the setting temperature in a refrigerator.

Such gelatin products are of two general types. One is amorphous straight gelatin, produced by spray, drum, or hot-air tunnel drying of a gelatin sol [242, 245], or microwave drying of a thin gelatin film at 8–16 w% moisture. [248] The other type is exemplified by amorphous gelatin–sugar–acid 'comelts', produced by extruding a dry mix at low moisture [247], or extruding a viscous solution as a foam [254], or vacuum drum drying [253, 256] or spray drying a solution. [255] The use of carbohydrate and/or other diluents in the latter type of process permits manufacture of amorphous comelts which manifest a reduced tendency for gelatin to revert to a PC form, since such diluents inhibit amorphous gelatin's propensity to recrystallize. [247] Especially for straight amorphous gelatin products, there are stringent requirements for low residual moisture, good moisture-barrier packaging, and storage conditions of low ambient temperature and relative humidity, in order for gelatin to maintain its stability against water/heat-plasticized reversion from a kinetically-metastable glass to a PC material no longer cold-water-soluble. [24, 242, 245]

Another modern approach to cold-water-soluble, quick-setting gelatin products is exemplified by patented shelf-stable liquid gelatin concentrates (sols) containing urea [250] or some other edible, non-acid lyotropic agent [251, 252], or refrigerated concentrates containing edible acids. [249] Such products are technologically based on the 'lyotropic effect' of the Hofmeister series of ionic salts, acids, and denaturants such as urea. [257] T_m of an aqueous gelatin solution is depressed below 20 °C by urea, thereby rendering such a gelatin concentrate a liquid at room temperature, but settable to a gel at refrigerator temperature. However, dilution of the concentrate to eating strength with cold water dilutes urea's effect, so that T_m rises back toward 37 °C. [24] Thus, the diluted product is quick setting in a refrigerator, but a gel at room temperature. [250–2]

Still another modern approach to quick-setting gelatin materials, suitable for food as well as photographic applications, involves a technologically-sophisticated fractionation of ordinary commercial acid- or base-treated skin gelatins. As detailed in a recent US patent [258], it has long been known that high-MW gelatin fractions (optimally, a monodisperse population of alpha chains of 95 000 MW) have higher T_ms and so faster gel setting rates (due to greater extent of undercooling, $T_m - T_{set}$, for the same setting temperature) than more polydisperse gelatins of lower MW. Tomka's patent [258] teaches that observed gel T_m is 1–3 °C higher for a fractionated gelatin, containing only 1% of material lower in MW than alpha chains, than for commercial gelatins, which typically contain about 30% of material smaller than alpha chains. Correspondingly, gel setting time for such a fractionated gelatin is reported to be 10–100 times faster (as governed by WLF kinetics for diffusion-controlled crystallization from the rubbery state) than comparable setting times for commercial gelatins.

Another major historical use of gelatin in foods is as a stabilizer for frozen dairy products and other frozen desserts and novelties. [182] Since before World War II, high-Bloom gelatin has been used as a protective colloid in ice cream, frozen yogurt, and cream pies. This usage continues today, as exemplified by several recent US patents. [259–61] In such products, gelatin functions as an inhibitor of ice crystal growth and lactose recrystallization during frozen storage. [182, 233, 246]

Postulated inhibition mechanisms have been based on several consequences of gelatin's physico-chemical properties as a water-compatible polymeric stabilizer. One popular postulate concerns gelatin's functionality as a simple viscosity enhancer of macroscopic viscosity (even when used at very low concentration) at typical freezer-storage temperature (−18 °C). [182]

Increased viscosity is supposed to result in decreased mobility of the freeze-concentrated product matrix, and so decreased rates of all diffusion-controlled processes, including migration and recrystallization of water and lactose molecules (within the product's unfrozen fluid phase), via Ostwald ripening.

Another postulated mechanism [24], which has been derived [23] from pioneering research results of Franks [37] and MacKenzie [33] on the properties of aqueous solutions at subzero temperatures, involves the behavior of high-Bloom gelatin as a high-polymeric solute in water, as reviewed in section 2.3. Gelatin's characteristic T_g' value of about -12 °C (which corresponds to its T_r [23]) has a significant elevating (and thereby stabilizing) effect on the composite T_g' (and thus T_r) of the freeze-concentrated aqueous matrix in a typical freezer-stored dessert product (a matrix ordinarily dominated by low-MW, low T_g' sugars). This thermomechanical property actually represents the underlying molecular physico-chemical basis for gelatin's microscopic viscosity-enhancing behavior.

A third stabilization mechanism suggested [262] for gelatin involves its gelling behavior at freezer temperatures. Even for high-Bloom gelatin concentrations <0.5 w% (i.e. non-gelling) in a product mix, freezing of a product produces a higher gelatin concentration, which may exceed 0.5 w% in the freeze-concentrated aqueous matrix, thus resulting in formation of a gelatin gel at freezer temperatures. Presence of this gel (as a physical barrier, in the form of a network with crystalline crosslinks) has essentially no effect on the diffusion of water and lactose molecules through the unfrozen fluid phase of a product, and so does not inhibit migratory recrystallization that may occur. However, it has been suggested that the porosity of a gelatin gel network may serve to limit physically the extent of ice crystal growth. [262]

A fourth possible stabilization mechanism for gelatin involves its well-known surface adsorption behavior. Via surface adsorption of individual gelatin macromolecules onto active growth sites on an ice crystal lattice, gelatin may inhibit the crystal growth and maturation steps of ice crystallization [24], just as fish and insect 'antifreeze' proteins and glycoproteins are thought to do in nature. [269] This postulated functionality of gelatin may be similar to its familiar behavior of crystal growth and maturation control in photographic applications [23], where gelatin is thought to adsorb to surfaces of silver halide crystals (which are very similar to ice in their crystal structures) in a photographic emulsion. [263]

Other industrial uses of gelatin Comparison of the functional properties of gelatin in food applications and other non-food industrial uses points up common aspects among diverse technologies and apparently-universal correlations between gelatin's structure–property relationships and functional behavior. [24]

In the pharmaceutical industry, gelatin is used to manufacture both hard- and soft-type drug capsules. [264] The high-Bloom gelatin powder used to make hard two-piece capsules is a high-MW, PC glassy polymer plasticized only by water. [265] The performance of such gelatin during capsule production and finished-product storage depends critically on its so-called 'equilibrium' moisture content and mechanical conditions of preparation. Optimal performance requires a PC glassy polymer with maximum modulus for maximum strength and stability. [78] In contrast, soft, elastic, one-piece shells are produced from a flexible film of low-Bloom, water-plasticized, PC gelatin, which is plasticized further by added polyols such as glycerol, sorbitol, or propylene glycol to enhance workability during capsule forming. [266]

One of the oldest, most classical uses of gelatin is in the manufacture of adhesives (e.g. 'animal glue' from horses' hooves) from low-MW, extensively-hydrolyzed gelatins. [264] In applications requiring functional properties of adhesiveness and rapid tackiness, such as gummed paper tapes [267], this industry capitalizes on the structure–property relationships manifested by hot-cast/dried, amorphous gelatin films described previously. [24] When an application requires a permanent, tough, flexible glue film (e.g. bookbinding), a polyol plasticizer such as sorbitol is added to a gelatin–water adhesive. [264] When an application requires that a gelatin glue should not be heated, but should remain a liquid sol at room temperature and produce its functional properties at subambient temperatures, a lyotropic T_m-depressant such as urea is added to a gelatin–water adhesive. [264]

Industrial use of gelatin binders for light-sensitive 'emulsions' in the manufacture of photographic films represents another historical (over 100-years-old) application of gelatin technology [263], but one that is still widely practised and of intense current interest and effort, as exemplified by more than 50 new US patents issued since 1982. [e.g. 258] Gelatin's functionality as a binder for silver halide layers in photographic film is clearly based on the polymer physico-chemical properties previously described for aqueous gelatin solutions. [23] In a typical four-step film manufacturing process [263], gelatin

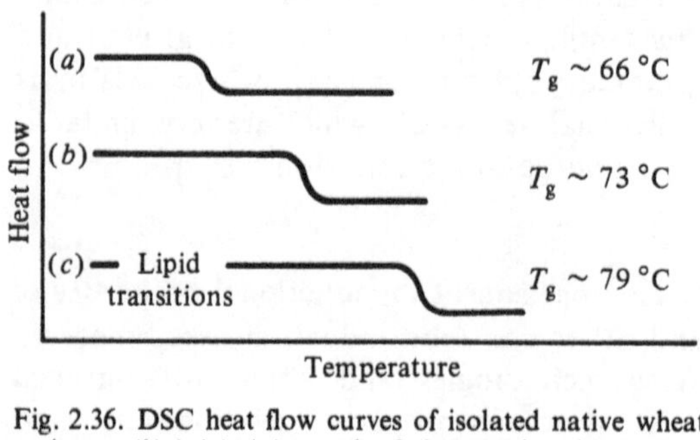

Fig. 2.36. DSC heat flow curves of isolated native wheat gluten: (a) as received, with 6 w% moisture; (b) 1:1 (w/w) comelt of gluten–triacetin; (c) 1:1 (w/w) comelt of gluten–B12K (lipids blend). [16]

first serves as a protective colloid in the preparation of a silver halide emulsion, by stabilizing the crystallization process and so insuring formation of a suspension of uniform crystals. In subsequent manufacturing steps, gelatin is required to gel rapidly and thermoreversibly at $T < T_m$, then to become a highly swollen and permeable, but insoluble, gel in water at $T < T_m$, and finally to form thin, uniform, crosslinked films (i.e. layers) of high gel strength. Obviously, gelatin's unique combination of physico-chemical properties has made it the best available material for this application, as witnessed by the fact that no superior substitute has been identified in over 100 years. [263] The same has been said about gelatin's use in gel desserts and other industrial applications. [23] These facts have led us to ask the moot question – is gelatin simply well suited to these applications, or were these applications actually designed around gelatin as a PC, water-compatible/plasticizable polymer? [24]

2.7 Gluten: amorphous polymeric behavior

In 1984, Slade [16] presented DSC results of T_g measurements from the first study of native wheat gluten (a plant storage protein), approached as an amorphous, water-compatible polymer. As shown in figure 2.36, isolated native wheat gluten at low moisture (i.e. 6 w% 'as is' moisture content) manifested a T_g at 66 °C (curve (*a*)). Once heated through this glass transition, and due to the mobility produced by plasticization by water, gluten becomes a thermoset material, via covalent disulfide crosslinking. [16] This thermosetting reaction is thought to be analogous to chemical 'curing' of epoxy resin and 'vulcanization' of rubber, which are also possible only at $T > T_g$. [270] Slade also reported that, in contrast to its response to plasticization by water, gluten remains thermoplastic (i.e. its glass-to-rubber transition is reversible) in 1:1 comelts of non-aqueous plasticizers such as triacetin (curve (*b*)) and B12K (curve (*c*)), a blend of lipids commonly used as a bread ingredient. However, the T_g values of 73 °C in curve (*b*) and 79 °C in curve (*c*) demonstrated the dramatic plasticizing effect on gluten of only 6 w% water, compared to 50 w% triacetin or B12K. [16] In the context of effects on bread-baking performance, Slade deduced that the interaction of gluten with lipids represents another example of antiplasticization of a polymer by a higher-MW plasticizer (analogous to the effect of sugars on starch), relative to plasticization by water alone. We have also analyzed 1:1 (w/w) gluten–water mixtures by low-temperature DSC, and found values of $T_g' \approx -5$ °C and $W_g' \approx 0.35$ g unfrozen water/g, which are similar to those for many other high-polymeric food proteins and carbohydrates. [26]

Slade's conclusion, that the thermomechanical properties of wheat gluten can be explained by treating gluten protein as a water-plasticizable, amorphous polymer, was verified by T_g measurements from a subsequent DSC study by Hoseney *et al.* [268] They reported a graph of T_g vs. w% water, which is a smooth curve from $T > 160$ °C for the glassy polymer at < 1 w% moisture to $T \sim 20$°C for the rubbery polymer at 16 w% moisture. These results exemplified the typical extent of plasticization (about 10 °C/w%) of gluten by water manifested by many other water-compatible, amorphous polymers and monomers, several of which have been discussed earlier in this review. However, they failed to mention the thermosetting behavior of aqueous gluten, reported by Slade.

2.8 Conclusion

As emphasized in this review, we have seen, especially since 1980, a growing awareness among a small but increasing number of food scientists of the value of a polymer science approach to the study of food materials and systems. In this respect, food science has followed the compelling lead of the synthetic polymers field. Recognition of two key elements of this research approach, (1) the critical role of water as a plasticizer of amorphous materials and (2) the importance of the glass transition as a physico-chemical parameter which can govern product properties, processing, and stability, has also increased markedly during this decade. Today, the first question we are asked most frequently is 'so what is the T_g of sucrose (or starch, gelatin, gluten, etc.)?' The necessary and appropriate response to this question is another question 'at what moisture content (W_g)?'. With the answer to the second question plus information on the value of W_g' in relation to W_g for the specific material and situation, one can begin to answer the original question intelligently. In this review, we have tried to illustrate, with examples from our experiences with food materials, how such questions and answers can allow one to understand and explain complex behavior, to design processes, and to predict product quality and storage stability, all based on fundamental structure–property relationships defined by studies which employ a polymer science approach to low-moisture polymeric systems plasticized by water. In future years, we can expect to see much progress reported from this emerging, crossdisciplinary research area.

Acknowledgements

We thank General Foods Corporation for permission to publish; our colleagues at General Foods, Timothy Schenz, Allen Bradbury, and Terry Maurice, for their contributions to our research program; Cornelis van den Berg and John Blanshard for encouragement of our work; and especially our consultant, Prof. Felix Franks of the University of Cambridge, for invaluable suggestions, discussions, and much encouragement and support over the years.

References

1. J. K. Sears & J. R. Darby. *The Technology of Plasticizers*. Wiley-Interscience: New York, 1982.
2. S. P. Rowland (ed.). *Water in Polymers, ACS Symp. Ser. 127*. American Chemical Society: Washington, D.C., 1980.
3. A. Eisenberg. In *Physical Properties of Polymers* (eds. J. E. Mark, A. Eisenberg, W. W. Graessley, L. Mandelkern & J. L. Koenig). American Chemical Society: Washington, D.C., 1984, pp. 55–95.
4. M. Karel. In *Properties of Water in Foods* (eds. D. Simatos & J. L. Multon). Martinus Nijhoff: Dordrecht, 1985, pp. 153–69.
5. C. van den Berg. At *Faraday Div., Royal Soc. Chem., Ind. Phys. Chem. Group, Disc. Conf. on Concept of Water Activity, July 1–3*, 1985, Girton College, Cambridge.
6. C. van den Berg. In *Concentration and Drying of Foods* (ed. D. MacCarthy). Elsevier Applied Science: London, 1986, pp. 11–36.
7. J. M. V. Blanshard. In *Chemistry and Physics of Baking* (eds. J. M. V. Blanshard, P. J. Frazier & T. Galliard). Royal Society of Chemistry: London, 1986, pp. 1–13.
8. S. Ablett, G. E. Attenburrow & P. J. Lillford. In *Chemistry and Physics of Baking* (eds. J. M. V. Blanshard, P. J. Frazier & T. Galliard). Royal Society of Chemistry: London, 1986, pp. 30–41.
9. C. G. Biliaderis, C. M. Page, T. J. Maurice & B. O. Juliano. *J. Agric. Food Chem.* **34** (1986), 6–14.
10. C. G. Biliaderis, C. M. Page & T. J. Maurice. *Carbohydr. Polym.* **6** (1986), 269–88.
11. C. van den Berg. Doctoral Thesis, Agricultural University: Wageningen, 1981.
12. C. van den Berg & S. Bruin. In *Water Activity: Influences on Food Quality* (eds. L. B. Rockland & G. F. Stewart). Academic Press: New York, 1981, pp. 1–61.
13. L. Slade & H. Levine. *ACS NERM 14*, June 12, 1984, Fairfield, CT, abs. 152.
14. H. Levine & L. Slade. In *Royal Society of Chemistry Residential School – Water Soluble Polymers: Chemistry & Application Technology course manual*, July 16–20, 1984, Girton College, Cambridge, pp. 132–52, 259–73.
15. L. Slade & H. Levine. In *Proceedings of the 13th NATAS Conference* (ed. A. R. McGhie). Sept. 23–6, 1984, Philadelphia, PA, p. 64.
16. L. Slade. At *AACC 69th Annual Meeting, Sept.* 30–*Oct.* 4, *1984*. Minneapolis, MN, abs. 112.
17. T. J. Maurice, L. Slade, C. Page & R. Sirett. In *Properties of Water in Foods* (eds. D. Simatos & J. L. Multon). Martinus Nijhoff: Dordrecht, 1985, pp. 211–27.

18. L. Slade & H. Levine. At *Faraday Div., Royal Soc. Chem., Ind. Phys. Chem. Group Disc. Conf. on Concept of Water Activity*, July 1–3, 1985, Girton College, Cambridge.

19. C. G. Biliaderis, C. M. Page, L. Slade & R. R. Sirett. *Carbohydr. Polym.* **5** (1985), 367–89.

20. L. Slade, H. Levine & F. Franks. *CRC Crit. Revs. Food Sci. Nutr.*, in press.

21. L. Slade & H. Levine. In *Recent Developments in Industrial Polysaccharides* (eds. S. S. Stivala, V. Crescenzi & I. C. M. Dea). Gordon and Breach Science: New York, 1987, pp. 387–430.

22. L. Slade & H. Levine. In *Food Structure – Its Creation and Evaluation* (eds. J. R. Mitchell & J. M. V. Blanshard). Butterworths: London, 1987, chap. 8.

23. H. Levine & L. Slade. In *Royal Society of Chemistry Residential School – Water Soluble Polymers: Chemistry & Application Technology course manual*, July 16–20, 1984, Girton College, Cambridge, pp. 274–84.

24. L. Slade & H. Levine. In *Advances in Meat Research*, vol. 4 – *Collagen as a Food* (eds. T. R. Dutson & A. M. Pearson). AVI: Westport, 1987, pp. 251–66.

25. T. W. Schenz, M. A. Rosolen, H. Levine & L. Slade. In *Proceedings of the 13th NATAS Conference* (ed. A. R. McGhie). Philadelphia, PA, Sept. 23–6, 1984, pp. 57–62.

26. H. Levine & L. Slade. *Carbohydr. Polym.* **6** (1986), 213–44.

27. H. Levine & L. Slade. In *Food Structure – Its Creation and Evaluation* (eds. J. R. Mitchell & J. M. V. Blanshard). Butterworths: London, 1987, chap. 9.

28. L. Slade & H. Levine. *Pure Appl. Chem.*, in press.

29. H. Levine & L. Slade. *J. Chem. Soc., Faraday Trans. 1*, in press.

30. P. J. Flory. *Principles of Polymer Chemistry*. Cornell University Press: Ithaca, 1953.

31. F. W. Billmeyer. *Textbook of Polymer Science*, 3rd edn. Wiley-Interscience: New York, 1984.

32. F. Franks, M. H. Asquith, C. C. Hammond, H. B. Skaer & P. Echlin. *J. Microsc.* **110** (1977), 223–38.

33. A. P. MacKenzie. *Phil. Trans. Royal Soc. London B.* **278** (1977), 167–89.

34. J. D. Ferry. *Viscoelastic Properties of Polymers*, 3rd edn. John Wiley & Sons: New York, 1980.

35. F. Franks. *Biophysics and Biochemistry at Low Temperatures*. Cambridge University Press: Cambridge, 1985.

36. T. Soesanto & M. C. Williams. *J. Phys. Chem.* **85** (1981), 3338–41.

37. F. Franks. In *Water – A Comprehensive Treatise* (ed. F. Franks), vol. 7. Plenum Press: New York, 1982, pp. 215–338.

38. F. Franks. In *Chemistry and Technology of Water-Soluble Polymers* (ed. C. A. Finch). Plenum Press: New York, 1983, pp. 157–78.

39. N. A. Peppas & R. Khanna. *Polym. Engn. Sci.* **20** (1980), 1147–56.

40. H. E. Bair, G. E. Johnson, E. W. Anderson & S. Matsuoka. *Polym. Engn. Sci.* **21** (1981), 930–5.

41. J. E. Jolley. *Photogr. Sci. Engn.* **14** (1970), 169–77.

42. B. Wunderlich. *Macromolecular Physics, vol. 1 – Crystal Structure, Morphology, Defects*. Academic Press: New York, 1973.

43. J. F. Fuzek. In *Water in Polymers* (ed. S. P. Rowland), ACS Symp. Ser. 127. American Chemical Society: Washington, DC, 1980, pp. 515–30.

44. B. Wunderlich. *Macromolecular Physics, vol. 2–Crystal Nucleation, Growth, Annealing*. Academic Press: New York, 1976.
45. W. Borchard, K. Bergmann, A. Emberger & G. Rehage. *Progr. Colloid Polym. Sci.* **60** (1976), 120–9.
46. W. Borchard, K. Bergmann & G. Rehage. In *Photographic Gelatin II* (ed. R. J. Cox). Academic Press: London, 1976, pp. 57–71.
47. W. Borchard, W. Bremer & A. Keese. *Colloid Polym. Sci.* **258** (1980), 516–26.
48. P. Godard, J. J. Biebuyck, M. Daumerie, H. Naveau & J. P. Mercier. *J. Polym. Sci.: Polym. Phys. Ed.* **16** (1978), 1817–28.
49. M. Djabourov & P. Papon. *Polymer* **24** (1983), 537–42.
50. G. E. Wissler & B. Crist. *J. Polym. Sci.: Polym. Phys. Ed.* **18** (1980), 1257–70.
51. X. Jin, T. S. Ellis & F. E. Karasz. *J. Polym. Sci.: Polym. Phys. Ed.* **22** (1984), 1701–17.
52. S. Z. D. Cheng, M. Y. Cao & B. Wunderlich. *Macromolecules* **19** (1986), 1868–76.
53. J. Menczel & B. Wunderlich. *Polym. Prepr.* **27** (1986), 255–6.
54. B. Wunderlich. Personal communication, 1985.
55. R. L. Whistler & J. R. Daniel. In *Starch: Chemistry and Technology* (eds. R. L. Whistler, J. N. Bemiller & E. F. Paschall), 2nd edn. Academic Press: Orlando, 1984, pp. 153–82.
56. D. French. In *Starch: Chemistry and Technology* (eds. R. L. Whistler, J. N. Bemiller & E. F. Paschall), 2nd edn. Academic Press: Orlando, 1984, pp. 183–247.
57. B. Wunderlich. *Macromolecular Physics, vol. 3 – Crystal Melting*. Academic Press: New York, 1980.
58. A. Hiltner & E. Baer. *Polym. Prepr.* **27** (1) (1986), 207.
59. L. Mandelkern. *Polym. Prepr.* **27** (1) (1986), 206.
60. R. F. Boyer, E. Baer & A. Hiltner. *Macromolecules* **18** (1985), 427–34.
61. R. C. Domszy, R. Alamo, C. O. Edwards & L. Mandelkern. *Macromolecules* **19** (1986), 310–25.
62. F. D. Blum & B. Nagara. *Polym. Prepr.* **27** (1) (1986), 211–12.
63. A. S. Marshall & S. E. B. Petrie. *J. Photogr. Sci.* **28** (1980), 128–34.
64. H. F. Zobel. In *Starch: Chemistry and Technology* (eds. R. L. Whistler, J. N. Bemiller & E. F. Paschall), 2nd edn. Academic Press: Orlando, 1984, pp. 285–309.
65. J. W. Donovan. *Biopolymers* **18** (1979), 263–75.
66. C. G. Biliaderis, T. J. Maurice & J. R. Vose. *J. Food Sci.* **45** (1980), 1669–80.
67. K. Kulp & J. G. Ponte. *CRC Crit. Revs. Food Sci. Nutr.* **15** (1981), 1–48.
68. A. Hayashi & S. C. Oh. *Agric. Biol. Chem.* **47** (1983), 1711–16.
69. J. D. Ferry. *J. Amer. Chem. Soc.* **70** (1948), 2244–9.
70. S. H. Cakebread. *Manufact. Confect.* **49** (1969), 41–4.
71. J. M. G. Cowie. *Polymers: Chemistry and Physics of Modern Materials.* Intertext Publ.: New York, 1973.
72. G. E. Downton, J. L. Flores-Luna & C. J. King. *Indust. Engn. Chem. Fund.* **21** (1982), 447–51.
73. H. Batzer & U. T. Kreibich. *Polym. Bull.* **5** (1981), 585–90.
74. J. A. Brydson. In *Polymer Science* (ed. A. D. Jenkins). North Holland Publ.: Amsterdam, 1972, pp. 194–249.

75. W. P. Brennan. *Thermal Analysis Application Study No. 8*. Perkin Elmer Instrument Division: Norwalk, 1973.
76. D. R. Olson & K. K. Webb. *Organ. Coat. Plast. Chem.* **39** (1978), 518–23.
77. H. W. Starkweather. In *Water in Polymers* (ed. S. P. Rowland). ACS Symp. Ser. 127. American Chemical Society: Washington, DC, 1980, pp. 433–40.
78. I. W. Kellaway, C. Marriott & J. A. J. Robinson. *Can. J. Pharmaceut. Sci.* **13** (1978), 83–6.
79. J. M. Flink. In *Physical Properties of Foods* (eds. M. Peleg & E. B. Bagley). AVI: Westport, 1983, pp. 473–521.
80. T. S. Ellis, X. Jin & F. E. Karasz. *Polym. Prepr.* **25** (2) (1984), 197–8.
81. G. ten Brinke, F. E. Karasz & T. S. Ellis. *Macromolecules* **16** (1983), 244–9.
82. P. Moy & F. E. Karasz. In *Water in Polymers* (ed. S. P. Rowland). ACS Symp. Ser. 127, American Chemical Society: Washington DC, 1980, pp. 505–13.
83. C. Carfagna, A. Apicella & L. Nicolais. *J. Appl. Polym. Sci.* **27** (1982), 105–12.
84. S. Bone & R. Pethig. *J. Mol. Biol.* **157** (1982), 571–5.
85. K. Nakamura, T. Hatakeyama & H. Hatakeyama. *Polymer* **22** (1981), 473–6.
86. C. A. J. Hoeve & M. B. J. A. Hoeve. *Org. Coat. Plast. Chem.* **39** (1978), 441–3.
87. M. Scandola, G. Ceccorulli & M. Pizzoli. *Int. J. Biol. Macromol.* **3** (1981), 147–9.
88. E. C. To & J. M. Flink. *J. Food Technol.* **13** (1978), 551–94.
89. F. Franks. *J. Microsc.* **141** (1986), 243–9.
90. F. Franks. In *Properties of Water in Foods* (eds. D. Simatos & J. L. Multon). Martinus Nijhoff: Dordrecht, 1985, pp. 497–509.
91. S. Gaeta, A. Apicella & H. B. Hopfenberg. *J. Membr. Sci.* **12** (1982), 195–205.
92. Y. Mohajar, G. L. Wilkes, H. b. Gia & J. E. McGrath. *Polym. Prepr.* **21** (2) (1980), 229–30.
93. M. Froix. *Polym. Prepr.* **20** (1) (1979), 975–9.
94. J. V. Standish & H. J. Leidheiser. *J. Coat. Technol.* **53** (1981), 53–8.
95. A. Sfirakis & C. E. Rogers. *Org. Coat. Plast. Chem.* **39** (1978), 444–7.
96. H. W. Starkweather & J. R. Barkley. *J. Polym. Sci.: Polym. Phys. Ed.* **19** (1981), 1211–20.
97. D. J. Crofton & R. A. Pethrick. *Polymer* **23** (1982), 1609–14.
98. H. W. Starkweather. *Macromolecules* **15** (1982), 752–6.
99. M. L. Williams, R. F. Landel & J. D. Ferry. *J. Amer. Chem. Soc.* **77** (1955), 3701–6.
100. E. E. LaBarre & D. T. Turner. *J. Polym. Sci.: Polym. Phys. Ed.* **20** (1982), 557–60.
101. J. M. Barton. *Polymer* **20** (1979), 1018–24.
102. G. E. Johnson, H. E. Bair, S. Matsuoka, E. W. Anderson & J. E. Scott. In *Water in Polymers* (ed. S. P. Rowland). ACS Symp. Ser. 127. American Chemical Society: Washington, DC, 1980, pp. 451–68.
103. K. Hiraoka, M. Gotanda & T. Yokoyama. *Polym. Bull.* **2** (1980), 631–6.
104. B. C. Watson & Z. W. Wicks. *J. Coat. Technol.* **55** (1983), 59–65.
105. L. M. Robeson. *Macromolecules* **14** (1981), 1644–50.
106. J. M. Gosline & C. J. French. *Biopolymers* **18** (1979), 2091–2103.
107. P. Neogi. *Am. Inst. Chem. Engn. J.* **29** (1983), 829–39.
108. R. L. D'Arcy & I. C. Watt. In *Water Activity: Influences on Food Quality* (eds. L. B. Rockland & G. F. Stewart). Academic Press: New York, 1981, pp. 111–42.
109. I. C. Watt. *J. Macromol. Sci.-Chem.* **A14** (1980), 245–55.

110. A. Berens & H. B. Hopfenberg. *Stud. Phys. Theoret. Chem.* **10** (1980), 77–94.
111. H. B. Hopfenberg, A. Apicella & D. E. Saleeby. *J. Membr. Sci.* **8** (1981), 273–82.
112. R. J. Pace & A. Datyner. *J. Polym. Sci.: Polym. Phys. Ed.* **19** (1981), 1657–8.
113. P. P. Roussis. *Polymer* **22** (1981), 768–73.
114. J. S. Aspler & D. G. Gray. *J. Polym. Sci.: Polym. Phys. Ed.* **21** (1983), 1675–89.
115. C. A. J. Hoeve. In *Water in Polymers* (ed. S. P. Rowland). ACS Symp. Ser. 127, American Chemical Society: Washington, DC, 1980, pp. 135–46.
116. G. Ranade, V. Stannett & W. J. Koros. *J. Appl. Polym. Sci.* **25** (1980), 2179–86.
117. J. E. O. Mayne & D. J. Mills. *J. Oil Colour Chem. Ass.* **65** (1982), 138–42.
118. M. Karel. In *Concentration and Drying of Foods* (ed. D. MacCarthy). Elsevier Applied Science: London, 1986, pp. 37–51.
119. A. Apicella & H. B. Hopfenberg. *J. Appl. Polym. Sci.* **27** (1982), 1139–48.
120. M. Escoubes & M. Pineri. In *Water in Polymers* (ed. S. P. Rowland). ACS Symp. Ser. 127, American Chemical Society: Washington, DC, 1980, pp. 235–52.
121. F. Franks. *Cryo-Letters* **3** (1982), 115–20.
122. V. Stannett, M. Haider, W. J. Koros & H. B. Hopfenberg. *Polym. Engn. Sci.* **20** (1980), 300–4.
123. H. T. Oyama & T. Nakajima. *J. Polym. Sci.: Polym. Chem. Ed.* **21** (1983), 2987–95.
124. G. L. Brown. In *Water in Polymers* (ed. S. P. Rowland). ACS Symp. Ser. 127, American Chemical Society: Washington, DC, 1980, pp. 441–50.
125. G. R. Mauze & S. A. Stern. *J. Membr. Sci.* **12** (1982), 51–64.
126. S. A. Stern & V. Saxena. *J. Membr. Sci.* **7** (1980), 47–59.
127. W. J. Koros & D. R. Paul. *J. Polym. Sci.: Polym. Phys. Ed.* **19** (1981), 1655–6.
128. H. A. Iglesias, J. Chirife & R. Boquet. *J. Food Sci.* **45** (1980), 450–7.
129. J. Pouchly, J. Biros & S. Benes. *Makromol. Chem.* **180** (1979), 745–60.
130. P. Chinachoti & M. P. Steinberg. *J. Food Sci.* **51** (1986), 453–9.
131. P. Chinachoti & M. P. Steinberg. *J. Food Sci.* **51** (1986), 997–1000.
132. J. Biros, R. L. Madan & J. Pouchly. *Collect. Czech. Chem. Commun.* **44** (1979), 3566–73.
133. K. L. Petrak & T. J. Bumfrey. *Polym. Bull.* **3** (1980), 311–17.
134. A. Hiltner, H. Shiraishi, S. Nomura & E. Baer. *Midl. Macromol. Monogr.* **4** (1978), 249–59.
135. M. Karel & J. M. Flink. In *Advances in Drying* (ed. A. S. Mujumdar), vol. 2. Hemisphere Publ.: Washington, 1983, pp. 103–53.
136. S. R. Kakivaya & C. A. J. Hoeve. *Proc. Natl. Acad. Sci. USA* **72** (1975), 3505–7.
137. H. G. Burghoff & W. Pusch. *Polym. Engn. Sci.* **20** (1980), 305–9.
138. R. K. S. Bhatia. *J. Inst. Chem. (India)* **51** (1979), 225–32.
139. N. Kinjo, S. Ohara, T. Sugawara & S. Tsuchitani. *Polym. J.* **15** (1983), 621–3.
140. A. N. Bulygin, Y. L. Vinogradov, A. Y. Lukyanov, Y. I. Malko & A. A. Tager. *Vysokomol. Soedin.* **A25** (1983), 1020–4.
141. H. Bizot, A. Buleon, N. Mouhoud-Riou & J. L. Multon. In *Properties of Water in Foods* (eds. D. Simatos & J. L. Multon). Martinus Nijhoff: Dordrecht, 1985, pp. 83–93.
142. S. Z. Dziedzic & M. W. Kearsley. In *Glucose Syrups: Science and Technology* (eds. S. Z. Dziedzic & M. W. Kearsley). Elsevier Applied Science: London, 1984, pp. 137–68.

143. D. G. Murray & L. R. Luft. *Food Technol.* **27** (1973), 32–40.

144. A. P. MacKenzie. In *Microprobe Analysis of Biological Systems*. Academic Press: New York, 1981, pp. 397–421.

145. A. P. MacKenzie & D. H. Rasmussen. In *Water Structure at the Water–Polymer Interface* (ed. H. H. G. Jellinek). Plenum Press: New York, 1972, pp. 146–71.

146. D. Rasmussen & B. Luyet. *Biodynamica* **10** (1969), 319–31.

147. B. Luyet. *Ann. NY Acad. Sci.* **85** (1960), 549–69.

148. E. Maltini. *I.I.F.–I.I.R.–Karlsruhe* **1** (1977), 1–9.

149. Virtis Company. *Inc. Virtis SRC Sublimators Manual.* Gardiner, NY, 1983.

150. P. J. Flory. *Faraday Disc. Chem. Soc.* **57** (1974), 7–18.

151. J. R. Mitchell. *J. Texture Stud.* **11** (1980), 315–37.

152. E. E. Braudo, I. G. Plaschchina & V. B. Tolstoguzov. *Carbohydr. Polym.* **4** (1984), 23–48.

153. P. V. Bulpin, A. N. Cutler & I. C. M. Dea. In *Gums and Stabilizers for the Food Industry 2* (eds. G. O. Phillips, D. J. Wedlock & P. A. Williams). Pergamon Press: Oxford, 1984, pp. 475–84.

154. F. Reuther, G. Damaschun, C. Gernat, F. Schierbaum, B. Kettlitz, S. Radosta & A. Nothnagel. *Coll. & Polym. Sci.* **262** (1984), 643–7.

155. H. S. Ellis & S. G. Ring. *Carbohydr. Polym.* **5** (1985), 201–13.

156. M. J. Miles, V. J. Morris & S. G. Ring. *Carbohydr. Res.* **135** (1985), 257–69.

157. W. W. Graessley. In *Physical Properties of Polymers* (ed. J. E. Mark, A. Eisenberg, W. W. Graessley, L. Mandelkern & J. L. Koenig). American Chemical Society: Washington, DC, 1984, pp. 97–153.

158. D. G. Medcalf. In *New Approaches to Research on Cereal Carbohydrates* (eds. R. D. Hill & L. Munck). Elsevier: Amsterdam, 1985, pp. 355–62.

159. H. Krusi & H. Neukom. *Staerke* **36** (1984), 300–5.

160. T. E. Beesley. *Amer. Lab. May* (1985), 78–87.

161. M. Richter, F. Schierbaum, S. Augustat & K. D. Knoch. US pat. 3 962 465 (1976).

162. M. Richter, F. Schierbaum, S. Augustat & K. D. Knoch. US pat. 3 986 890 (1976).

163. E. E. Braudo, E. M. Belavtseva, E. F. Titova, I. G. Plashchina, V. L. Krylov, V. B. Tolstoguzov, F. R. Schierbaum & M. Richter. *Staerke* **31** (1979), 188–94.

164. J. M. Lenchin, P. C. Trubiano & S. Hoffman. US pat. 4 510 166 (1985).

165. G. A. Reineccius. Personal communication (1986). S. Anandaraman & G. A. Reineccius. *Food Technol.* **40** (1986), 88–93.

166. D. R. Meyer, US pats. 4 556 567, 4 556 568 (1985).

167. W. M. Green & M. W. Hoover. US pat. 4 161 545 (1979).

168. B. A. Cole, H. I. Levine, M. T. McGuire, K. J. Nelson & L. Slade. US pat. 4 374 154 (1983).

169. B. A. Cole, H. I. Levine, M. T. McGuire, K. J. Nelson & L. Slade. US pat. 4 452 824 (1984).

170. J. L. Holbrook & L. M. Hanover. US pat. 4 376 791 (1983).

171. A. E. Bevilacqua & N. E. Zaritzky. *J. Food Sci.* **47** (1982), 1410–14.

172. E. K. Harper & C. F. Shoemaker. *J. Food Sci.* **48** (1983), 1801–6.

173. M. L. Kahn & R. J. Lynch. US pat. 4 552 773 (1985).

174. J. Szejtli & M. Tardy. US pat. 4 529 608 (1985).

175. N. Nagashima & E. Suzuki. In *Properties of Water in Foods* (eds. D. Simatos & J. L. Multon). Martinus Nijhoff Publ.: Dordrecht, 1985, pp. 555–71.
176. G. W. White & S. H. Cakebread. *J. Food Technol.* **1** (1966), 73–82.
177. W. Vink & R. W. Deptula. US pat. 4 311 722 (1982).
178. R. Lees, *Confect. Product. Feb.* (1982), 50–1.
179. H. Ogawa & Y. Imamura, US Pat. 4 547 377 (1985).
180. D. H. Miller & J. R. Mutka, US Pat. 4 499 112 (1985).
181. M. L. Kahn & K. E. Eapen, US Pat. 4 332 824 (1982).
182. P. G. Keeney & M. Kroger, in *Fundamentals of Dairy Chemistry* (eds. B. H. Webb *et al.*), 2nd edn. AVI: Westport, 1974, p. 890.
183. E. R. Morris, A. N. Cutler, S. B. Ross-Murphy & D. A. Rees. *Carbohydr. Polym.* **1** (1981), 5–21.
184. S. Tsourouflis, J. M. Flink & M. Karel. *J. Sci. Food Agric.* **27** (1976), 509–19.
185. G. V. Barbosa-Canovas, R. Rufner & M. Peleg. *J. Food Sci.* **50** (1985), 473–81.
186. J. Chevalley, W. Rostagno & R. H. Egli. *Rev. Int'l. Choc.* **25** (1970), 3–6.
187. E. Fukuoka, S. Kimura, M. Yamazaki & T. Tanaka. *Chem. Pharmaceut. Bull.* **31** (1983), 221–9.
188. J. F. Gueriviere. *Indust. Aliment. Agric.* **93** (1976), 587–95.
189. T. M. Herrington & A. C. Branfield. *J. Food Technol.* **19** (1984), 409–35.
190. E. Maltini. *Annal. Ist. Sper. Valor. Technol. Prod. Agric.* **5** (1974), 65–72.
191. P. B. McNulty & D. G. Flynn. *J. Texture Stud.* **8** (1977), 417–31.
192. R. Moreyra & M. Peleg. *J. Food Sci.* **46** (1981), 1918–22.
193. V. N. Morozov & S. G. Gevorkian. *Biopolymers* **24** (1985), 1785–99.
194. E. A. Niediek & L. Barbernics. *Gordian* **80** (1981), 267–69.
195. N. Passy & C. H. Mannheim. *Lebensm.-Wiss. u.-Technol.* **15** (1982), 222–25.
196. M. Peleg & C. H. Mannheim. *J. Food Process. Preserv.* **1** (1977), 3–11.
197. P. L. Poole & J. L. Finney. *Biopolymers* **22** (1983), 255–60; *Int. J. Biol. Macromol.* **5** (1983), 308–10.
198. D. S. Reid. *Cryo-Letters* **6** (1985), 181–8.
199. N. Rosenzweig & M. Narkis. *Polym. Engn. Sci.* **21** (1981), 1167–70.
200. G. Tardos, D. Mazzone & R. Pfeffer. *Can. J. Chem. Engn.* **62** (1984), 884–7.
201. J. J. Wuhrmann, B. Venries & R. Buri, US Pat. 3 920 854 (1975).
202. K. S. Marsh & J. Wagner. *Food Engn. Aug.* **58** (1985).
203. F. Franks. *Cereal Foods World* **27** (1982), 403–7.
204. A. G. Walton, in *Nucleation* (ed. A. C. Zettlemoyer). Marcel Dekker: New York, 1969, p. 225.
205. A. P. MacKenzie, in *Freeze Drying and Advanced Food Technology* (eds. S. A. Goldlith, L. Rey & W. W. Rothmayr). Academic Press: New York, 1975, pp. 277–307.
206. K. W. Lang, Doctoral Thesis, University of Illinois, 1981.
207. M. Loncin, in *Freeze Drying and Advanced Food Technology* (eds. S. A. Goldlith, L. Rey & W. W. Rothmayr). Academic Press: New York, 1975, pp. 599–617.
208. J. Chirife, G. Favetto & C. Fontan. *Lebensm.-Wiss. u.-Technol.* **15** (1982), 159–60.
209. D. C. Thill, R. D. Schirman & A. P. Appleby. *Agron. J.* **71** (1979), 105–8.
210. S. C. Wiggans & F. P. Gardner. *Agron. J.* **51** (1959), 315–18.
211. F. Shafizadeh, G. D. McGinnis, R. A. Susott & H. W. Tatton. *J. Org. Chem.* **36**

(1971), 2813–18.
212. K. Ghiasi, R. C. Hoseney & E. Varriano-Marston. *Cereal Chem.* **60** (1983), 58–61.
213. P. L. Russell. *Staerke* **35** (1983), 277–81.
214. T. Fearn & P. L. Russell. *J. Sci. Food Agric.* **33** (1982), 537–48.
215. J. Longton & G. A. LeGrys. *Staerke* **33** (1981), 410–14.
216. D. B. Lund, in *Physical Properties of Foods* (eds. M. Peleg & E. B. Bagley). AVI: Westport, 1983, pp. 125–43.
217. D. B. Lund. *CRC Crit. Revs. Food Sci. Nutr.* **20** (1984), 249–73.
218. J. M. V. Blanshard, in *Polysaccharides in Food* (eds. J. M. V. Blanshard & J. R. Mitchell). Butterworths: London, 1979, pp. 139–52.
219. K. J. Zeleznak & R. C. Hoseney. *Cereal Chem.* **63** (1986), 407–11.
220. A. C. Eliasson, in *New Approaches to Research on Cereal Carbohydrates* (eds. R. D. Hill & L. Munck). Elsevier: Amsterdam, 1985, pp. 93–8.
221. A. Guilbot & B. Godon. *Cah. Nutr. Diet.* **19 (3)** (1984), 171–81.
222. L. Slade, R. Altomare, R. Oltzik & D. G. Medcalf, US Pat. 4 657 770 (1987).
223. J. L. Jane & J. F. Robyt. *Carbohydr. Res.* **132** (1984), 105–18.
224. M. Yamamoto, S. Harada, T. Sano, T. Yasunaga & N. Tatsumoto. *Biopolymers* **23** (1984), 2083–96.
225. A. Buleon, F. Duprat, F. P. Booy & H. Chanzy. *Carbohydr. Polym.* **4** (1984), 161–73.
226. M. Kowblansky. *Macromolecules* **18** (1985), 1776–9.
227. I. D. Evans. *Staerke* **38** (1986), 227–35.
228. S. Hizukuri. *Carbohydr. Res.* **147** (1986), 342–7.
229. Anonymous, in *Encyclopedia of Chemical Technology*, vol. 11, 3rd edn. Wiley & Sons: New York, 1980, pp. 711–19.
230. A. G. Ward & A. Courts. *The Science and Technology of Gelatin.* Academic Press: New York, 1977.
231. I. V. Yannas. *J. Macromol. Sci.-Revs. Macromol. Chem.* **C7** (1972), 49–104.
232. I. Tomka, J. Bohonek, A. Spuhler & M. Ribeaud. *J. Photogr. Sci.* **23** (1975), 97–103.
233. Anonymous. *Gelatin.* Gelatin Manufact. Inst. Amer.: New York, 1982.
234. D. A. Ledward, in *Functional Properties of Food Macromolecules* (eds. J. R. Mitchell & D. A. Ledward). Elsevier Applied Science: London, 1986, pp. 171–201.
235. W. F. Harrington & P. H. von Hippel. *Adv. Prot. Chem.* **16** (1961), 1–138.
236. P. L. Privalov. *Adv. Prot. Chem.* **35** (1982), 1–104.
237. J. A. Darsey & W. L. Mattice. *Macromolecules* **15** (1982), 1626–31.
238. J. Y. Chatellier, D. Durand & J. R. Emery. *Int. J. Biol. Macromol.* **7** (1985), 311–14.
239. D. Durand, J. R. Emery & J. Y. Chatellier. *Int. J. Biol. Macromol.* **7** (1985), 315–19.
240. T. Nishio & R. Hayashi. *Agric. Biol. Chem.* **49** (1985), 1675–82.
241. P. Reutner, B. Luft & W. Borchard. *Colloid Polym. Sci.* **263** (1985), 519–29.
242. M. De Brou & C. Den Tandt, US Pat. 3 930 052 (1975).
243. W. Borchard. Personal communication (1985).
244. I. Tomka. *Polym. Prepr.* **27** (1986), 129.

245. R. R. Leshik, N. A. Swallow, S. J. Leusner & D. J. DiGiovacchino. US Pat. 4 546 002 (1985).

246. N. R. Jones, in *The Science and Technology of Gelatin* (eds. A. G. Ward & A. Courts). Academic Press: New York, 1977, pp. 365–94.

247. N. J. Kalafatas, H. Rosenthal & G. A. Consolazio, US Pat. 3 927 221 (1975).

248. K. Hayashi, M. Washizawa & S. Yokoo, US Pat. 4 224 348 (1980).

249. T. V. Kueper & T. H. Donnelly, US Pat. 4 224 353 (1980).

250. J. L. Shank, US Pat. 4 341 810 (1982).

251. J. L. Shank, US Pat. 4 426 443 (1984).

252. J. L. Shank, US Pat. 4 528 204 (1985).

253. G. M. Brown, P. M. Bosco & R. L. Danielson, US Pat. 4 401 685 (1983).

254. P. M. Bosco & R. L. Danielson, US Pat. 4 407 836 (1983).

255. P. M. Bosco & R. L. Danielson, US Pat. 4 409 255 (1983).

256. J. Brown, P. E. Ellis & M. J. Draper, US Pat. 4 588 602 (1986).

257. P. H. von Hippel & T. Schleich, in *Structure and Stability of Biological Macromolecules* (eds. S. N. Timasheff & G. D. Fasman). Marcel Dekker: New York, 1969, pp. 418–574.

258. I. Tomka, US Pat. 4 360 590 (1982).

259. R. G. Morley & W. R. Ashton, US Pat. 4 346 120 (1982).

260. J. T. Fiscella, US Pat. 4 391 834 (1983).

261. R. G. Morley, US Pat. 4 427 701 (1984).

262. F. Franks, Personal communication (1984).

263. A. M. Kragh, in *The Science and Technology of Gelatin* (eds. A. G. Ward & A. Courts). Academic Press: New York, 1977, pp. 439–74.

264. P. D. Wood, in *The Science and Technology of Gelatin* (eds. A. G. Ward & A. Courts). Academic Press: New York, 1977, pp. 413–37.

265. J. A. Pace, US Pat. 4 325 761 (1982).

266. W. R. Ebert, US Pat. 4 428 927 (1984).

267. R. D. Cilento, C. Riffkin & A. L. LaVia, US Pat. 4 427 737 (1984).

268. R. C. Hoseney, K. Zeleznak & C. B. Lai. *Cereal Chem.* **63** (1986), 285–6.

269. A. L. DeVries. *Ann. Rev. Physiol.* **45** (1983), 245–60.

270. C. S. P. Sung & E. Pyun. *Polym. Prepr.* **27** (**1**) (1986), 78.

3

Cellular water relations of plants

A. DERI TOMOS

University College of North Wales, Dept of Biochemistry and Soil Science, Bangor, Gwynedd LL57 2UW

3.1 Levels of study

All life forms on earth are totally dependent on water. In plants it generally constitutes 80–90% of herbaceous tissues and over 50% of woody tissues. [1] In seeds and spores the content may drop to 20% or below although ultimately desiccation tends to kill even seeds as some residual metabolism is required to maintain viability. [2] Indeed, in the case of the so called 'recalcitrant' seeds (e.g. acorns) this minimum can be quite high. On the other hand the dormant stages of some plants (the cryptograms) can withstand total desiccation. The biophysics of these that allows such behaviour is far from understood. [3]

Water plays diverse physical and chemical roles in plants. Meidner & Sheriff [4] classify these into processes that involve structural, physical (such as translocation) and metabolic processes. The varied *physical* processes that involve water have been grouped into a class of phenomena that have been termed *water relations*.

The water relations of plants may be studied over a range of levels. These extend from the biophysical role of water at the molecular level to the global role of water in weather systems in agriculture and plant communities. At one extreme the focus is at atomic resolution, at the other the focus of resolution may be intercontinental. The often conflicting importance of water to agriculture and industry in areas of the world deficient in the commodity (not all of them poor by any means) has recently increased interest and effort towards an understanding of the role of water in plant life. [5] This review will deal with current developments and understanding of plant water at the resolution of the single cell. The study brings together results from a wide range of disciplines. This has advantages and disadvantages. The latter are mainly problems of communication – multidisciplinary volumes such as that of Franks [6] and this series will help to answer this. The advantages, however, are those gained when a problem is viewed from different directions. These wholly outweigh the disadvantages.

I propose to bring together many of the strands that currently are 'plant water relations' studies at the single cell level. If coverage is only superficial in parts, hopefully the references supplied will provide an introduction to the original literature in those areas.

Direct analysis at single cell resolution has only recently become practical in most cases, especially with higher plants, and so information gained from tissue and organ studies will be included when it appears justified to extrapolate information to single cell level. The water relations behaviour of whole plants will not be dealt with and only passing mention will be made to xylem, phloem and stomata as these have an extensive literature of their own (e.g. [7, 8]). Hopefully we are working towards a general understanding of water in plants that is beginning to allow whole plant phenomena to be related to cellular parameters. Considerable strides in this direction have been made recently with the application of new techniques in the studies of water flows across multicellular tissues. [9–16] Somewhere in the behaviour of individual cells is the full understanding of plant physiology; without an understanding at this resolution the plant remains a classical black box. A Laplacian outlook must be the order of the day! Ironically it may be that a major barrier of understanding will be that of tissue architecture at a multicellular level, which is a study that traces its origins to Henshaw (1661) and Hook. [17] The problem facing the plant physiologist may be smaller than that facing his 'animal' counterpart. Plant cells appear to possess relatively fewer cell types and combinations, and are often relatively large.

Indeed, much information regarding single plant cells is derived from 'giant-celled' algae such as *Valonia*, *Nitella*, *Chara* and *Acetabularia* whose relatively large cells allow manipulative techniques not generally possible for higher plant cells. [18] Although higher plants are not simply agglomerates of algal cells it must be said that at our current level of understanding we are not in a position to distinguish algal behaviour that may not apply to other plants. [19]

It is a quarter of a century since a seminal review by Dainty [20] set the foundations of modern plant cell water relations and a number of excellent reviews on the subject have appeared in the intervening period. [19, 21–26] While the excellent monograph by House [27] will continue to provide challenges for plant cell water relations research for many years to come.

Many reviews of the various aspects of water relations of tissue and organs that relate to single cells have also appeared. These include [28–36], (a list that is not meant to be exhaustive). Slatyer [37], Nobel [38], Kramer [1] and Baker [39] also include bibliographies and useful summaries of various aspects.

3.2 Plant cell structure

Plant cells range enormously in size, from the freshly divided cells of meristems to the 'giant' cells of the Characean algae (with dimensions measured in millimetres and a volume of 100 μl or more). In all cases the protoplast is bounded by a semipermeable lipid–protein membrane, the plasmalemma, which in many tissues, including those of some algae, can extend as the linings of intercellular pores (the plasmodesmata) linking the cytoplasm of a cell with that of its neighbour (figure 3.1). In higher plants and most algae the plasmalemma is surrounded by a cell wall (about 2–12 μm thick) of high tensile strength that allows the development of excess hydrostatic pressure within the cell. Some cells, e.g. many flagellate algae and the reproductive cells of some plant species as high as *Cycas* in the evolutionary scale, are effectively wall-less in which case negligible hydrostatic pressure is possible. [40, 41] In mature cells much (most in many cases) of the volume inside the plasmalemma is bounded by a second membrane, the tonoplast. These membranes delimit the vacuoles, of which cells may have one or many. The wall and these two membranes result in the cell being comprised of three distinct compartments. The wall space, the cytoplasm (between plasmalemma and tonoplast – often only 2–15 μm thick) and the vacuole. Other organelles occur within the bounds of the cytoplasm (nucleus, mitochondria, chloroplasts etc.) and although their collective or individual volume can be considerable at times, our current knowledge of their water relations is rudimentary. Figure 3.1 illustrates diagrammatically these compartments and the interrelationships between them. This review will consider the state of water in each compartment and the equilibria and flows of water between them.

When the wider behaviour of cells in their role of units within a tissue are considered, the adjoining cell walls are taken to form a continuum called the apoplast, the behaviour of which is determined by the behaviour of the

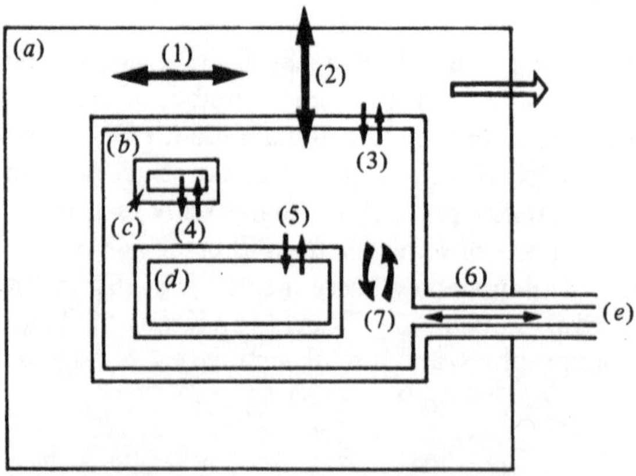

Fig. 3.1. Plant cell compartments and their hydraulic interactions: (*a*) cell wall; (*b*) cytoplasm; (*c*) organelles (e.g. mitochondria); (*d*) vacuole; (*e*) plasmodesmata. The arrows indicate the water pathways considered in the text: (1) parallel to wall fibrils; (2) normal to wall fibrils; (3) trans-plasmalemma; (4) trans-organelle membrane; (5) trans-tonoplast; (6) trans-plasmodesmata; (7) cyclosis. (The open arrow refers to apoplasmic flow.)

individual cell walls. [42] The existence of plasmodesmata linking adjacent cytoplasms has led correspondingly to the concept of an uninterrupted continuum of cytoplasm through a tissue, the symplasm. [43] The symplasm is sometimes defined as including the vacuole [43] and sometimes not. [31] Clearly if the symplasm is taken to include the vacuole then it does not imply a pathway uninterrupted by membranes. (It is important to be explicit when defining symplasm in the context of water flows. This will be especially important in considering solutes flow across highly vacuolated tissues.)

The quantitative description of cellular water relations requires the definition of relevant and measurable parameters. Despite over a century of development full agreement has still not been reached on several aspects. A brief outline of some of the basic considerations will be included here.

3.3 Thermodynamic basis of plant water relations

3.3.1 Osmosis

Current forms of analysis were introduced by Slatyer & Taylor [44] although plant material has contributed to our understanding of the physics of biological water since the pioneering experiments with osmosis of Pfeffer (1877) (see [45]). This early work demonstrated the osmotic behaviour of individual cells, including the semipermeable nature of their boundaries. Van't Hoff (1886) (see [45]) soon showed that Pfeffer's results for dilute solutions could be summarised by the equation

$$\pi = kmT \tag{3.1}$$

where π is the osmotic pressure of a dilute solution, T the absolute temperature, k a constant and m the molar concentration of the solute. (See the comments of Weatherley [28], Zimmermann & Steudle [23] and Passioura [26] for using osmotic *pressure* rather than osmotic *potential*.) Written in the form

$$\pi V = nkT \tag{3.2}$$

where n moles of solute are dissolved in volume V of solution, Van't Hoff recognised its resemblance to the equation of state for 'ideal' gases

$$PV = nRT \tag{3.3}$$

where P is the pressure, V the volume occupied by n moles and R is the gas constant. Van't Hoff subsequently formalised the relationship which now bears his name, although it was further modified by Morse (1905) (see [45]) who observed empirically that the use of molal rather than molar quantities improved its applicability.

$$\pi \approx (n/V)RT = RTc \tag{3.4}$$

where c is concentration.

In both plant and animal water relations it was soon realised that cells have an apparent osmotic 'dead space' and that the osmotic volume determined by quantitative analysis was somewhat smaller than the geometric volume observed under the microscope. To allow for this V was replaced by $(V - b)$. [46] (Here b is the non-osmotic volume.) Initially it was considered that b represented the geometric non-aqueous volume of the cell, but subsequent observations showed that this is not so, and that cells contain 'non-osmotic' water. Ponder [47] expressed the difference between the 'osmotic' and 'non-osmotic' water volumes by means of a ratio known as 'Ponder's R'

$$R = \frac{v^0 - b}{v_w^0} \tag{3.5}$$

where v^0 is the geometric volume of the cell and v_w^0 the volume of water. A discussion of values and basis of both b and R is given by House. [27] They certainly cannot be explained simply in terms of geometric parameters, however, but in terms of changes of osmotic coefficients of solutes with concentration.

The apparent equivalence of osmotic pressure and hydrostatic pressure as a driving force of water across a semipermeable membrane in the Van't Hoff equation is based on empirical observation. For example Dainty [20] cites the experiments of Mauro [48] and Robbins & Mauro [49] with artificial membranes as examples. Several attempts have been made to explain this, but as the mechanism of water transport across membranes remains uncertain the theoretical basis of this relationship is obscure. Ray [50] and Dainty [20] proposed models in which an osmotic pressure gradient results in a hydrostatic pressure difference at the mouth of a fine water filled pore, from which solute is excluded, crossing the membrane. In these models this hydrostatic pressure step is responsible for the observed behaviour of osmotic pressure and the apparent physical equivalence of the two. However, the occurrence of pores is still unresolved (see section 3.5.2). If water does indeed cross through water filled pores then several authors (see [27]) have pointed out that the permeability of water determined by hydrostatic or osmotic experiments will be quantitatively different from that determined by diffusional studies using radio-tracers. We shall return to this in sections 3.3.3 and 3.5.2.

It is to the second law of thermodynamics that we turn for most of our development of a quantitative description of water relations. In his investigations of the implications of this law Gibbs [51] called the work required to transport a mole of a material from a system to some reference point in its surroundings the *chemical potential* (μ) of that component. A substance tends to move from a region where its chemical potential is high to one where its chemical potential is low. Of crucial importance to our development of useful parameters for the description of cellular water relations is that at equilibrium the chemical potential of such a component is the same throughout the system. (For a biologically oriented description of the total free energy – now known as the Gibbs free energy – and its relationship to μ the reader is referred to Katchalsky & Curran [52]; or Nobel [38].) The useful starting point for the treatment of plant cell water and solutes is the following expression for μ of a component of a solution (see [38] for its development)

$$\mu_j = \mu_j^* + RT \ln a_j + \bar{V}_j P + z_j FE + m_j gh \qquad (3.6)$$

where μ_j is the molar chemical potential of component j in the solution, μ_j^* its molar chemical potential at the reference point in the surroundings, a_j is the activity of component j, $\bar{V}_j$ is the partial molar volume of the component, z_j the charge number of an ionic species, F the Faraday constant, E the electrical potential of the compartment, m the mass per mole of the component, g the gravitational constant and h the height above the reference point.

This expression conveniently separates each of the macroscopic components that influence the equilibrium point of component j; *viz.*:

$$\mu_j = \text{reference value} + \text{concentration/temperature} + \text{pressure component} + \text{electrical component} + \text{gravitational component}$$

The presence of the unknown reference component μ_j^* makes it impossible to quantify μ_j. (Although if necessary the reference point is defined as the chemical activity of a pool of pure water at 1 atm pressure at the temperature of the system.) However, in general the equation is used to determine the *difference* in μ_j between two points, $d\mu_j$, and an absolute value for μ_j^* is unnecessary as it cancels out when the chemical potential of a component at one point is subtracted from that at another.

In plant water relations this equation can have widespread application even when restricted to components of electrical neutrality (i.e. water and uncharged solutes) when $z_j = 0$. In this case the electrical component becomes zero. Similarly, at the resolution of the single cell or tissue, the influence of the difference in h from one point in a system to another is so much smaller than the range of the other components that it is conventionally ignored. (For comparison it can be shown from equation (3.6) that hydrostatic pressure of 1 MPa in an aqueous solution is approximately equivalent to 0.4 osmolal concentration and a 100 m difference in height. The first two values are of the order found in plant cells, the last is many orders of magnitude larger than the typical dimensions of cells and tissues considered here.)

Under these circumstances the following relationship can be obtained from equation (3.6)

$$\Delta\mu_j = \Delta[RT \ln a_j + \bar{V}_j P] \qquad (3.7)$$

The first term of equation (3.7), concerning temperature and concentration, may be related to the osmotic pressure, π, of a solution by the fundamental definition of the osmotic pressure of dilute solutions

$$RT \ln a_w = -\bar{V}_w \pi \tag{3.8}$$

where the subscripts refer to water. Therefore the chemical potential difference between water at two points in a system is given by

$$\Delta\mu_w = \Delta[-\bar{V}_w \pi + \bar{V}_w P] \tag{3.9}$$

If $\bar{V}_w$ is constant throughout the system this becomes

$$\Delta\mu_w = \bar{V}_w \Delta(P - \pi) \tag{3.10}$$

This thermodynamic description is defined as being under equilibrium conditions. Equation (3.10) can be used, for example, for the description of two bulk phases of aqueous solution at equilibrium across a rigid ideal semipermeable membrane. At equilibrium $\Delta\mu_w = 0$ and, since $\bar{V}_w$ is not zero,

$$\Delta(P - \pi) = 0 \tag{3.11}$$

If, as is conventional, the values of P are measured relative to atmospheric pressure, then P for the intracellular compartment is the *turgor pressure*. This is a parameter of central importance to the plant as we shall see.

A widely used quantity has been defined from this expression. This is the *water potential* (ψ – psi) of a solution. To arrive at this, equations (3.6) and (3.8) are combined, and the electrical term reduced to zero for neutral water. Thus

$$\mu_w = \mu_w^* - \bar{V}_w \pi + \bar{V}_w P + m_w g h \tag{3.12}$$

and since $m_w / \bar{V}_w = \rho_w$ (the density of water)

$$\psi = \frac{\mu_w - \mu_w^*}{\bar{V}_w} = P - \pi + \rho_w g h \tag{3.13}$$

If, as above, the gravitational component is also reduced to zero when the dimensions of the system under study are small, equation (3.13) not only defines ψ as a function of μ_w but also shows that at equilibrium

$$\Delta\psi = \Delta(P - \pi) = 0 \tag{3.14}$$

and

$$\text{turgor pressure} = \psi + \pi_i \tag{3.15}$$

Equality of ψ has replaced equality of μ_w as diagnostic of equilibrium. The popularity of this expression, rather than equation (3.6) in plant water relations research rests on the fact that P and π are more easily measured than the components of equation (3.6) (see appendix for methods).

Turgor pressure is not the only phenomenon that has been described in terms of water potential. It has been extended to describe the fluxes of water into plant cells and through plant tissues and organs. To describe such fluxes an equation is often used that is analogous to the laws of Fourier (for heat flow), Fick (for diffusion) or Ohm (for electric current, *viz.* $V = IR$):

flow = path conductivity × driving force

$$J_w = L_w \times (\psi_{wii} - \psi_{wi}) \tag{3.16}$$

where the subscripts i and ii refer to the water potentials at either end of the flow pathway for flux J_w of water, and L_w is the conductivity of the pathway (i.e. the reciprocal of the resistance term in Ohm's Law).

3.3.2 Hydraulic conductivity, reflection coefficient and solute permeability

Although widely used, several authors have strongly argued against the adoption of water potential to describe water relations. Useful in practice [21, 26], its use as a parameter must be accompanied by the demonstration that the assumptions implicit in its derivation are valid. Indeed Oertli [53] and Zimmermann & Steudle [23] conclude that the term is sufficiently in error under certain conditions as to be better discarded. Oertli [53] bases his argument on the observation that $\bar{V}_w$ can vary over a wide range in a cell depending on the interactions between water and solutes, or the binding of water to surfaces, i.e. from equation (3.13) $\Delta\psi$ may not be zero at thermodynamic equilibrium when $\Delta\mu_w = 0$. This objection also invalidates equations (3.10) and (3.11) under these circumstances. Oertli [53] argues that under certain circumstances $\bar{V}_w$ may even take on negative values, in which case even the direction of the water driving forces are incorrectly defined. Unfortunately it is, in practice, difficult to determine $\bar{V}_w$ in the compartments where it is likely to deviate most from its value of $1.805 \times 10^{-5}\ \mathrm{m^3\,mol^{-1}}$ at 20 °C in pure water (in the cell wall pores or the protein-rich cytoplasm, for example).

The objection exemplified by Zimmermann & Steudle [23] has a wider basis. Although water potential is, in practice, used even by its opponents, its use leads to a more general criticism of treatments of turgor and related phenomena of the type described above. Introduced to plant water relations by Slatyer & Taylor [44] and Dainty [20] (see also [54]), this is that such use of the second law of thermodynamics explicitly and erroneously assumes equilibrium of the components within the system. This is the case not only for water and solute fluxes (equation (3.16)) but also in the case of such apparently 'static' systems as the maintenance of a steady turgor pressure in a plant cell. A plant cell at constant turgor pressure is not at thermodynamic equilibrium with its surroundings since its membranes are never 'ideally' semipermeable to all solutes. Indeed, if they were, metabolic processes within cells would be starved of metabolites unable to cross the membrane. Let us reconsider the osmotic relations of such cells. The osmotic pressure of a solution is defined as the excess pressure that must be applied to a solution to prevent net flow of solvent across a rigid semipermeable membrane separating pure solvent from the solution. If the membrane is 'non-ideal', i.e. leaky to solutes as well as solvent, when net solvent flow is reduced to zero, net solute flow will continue. A balance pressure to bring about zero net *volume* flow,

$$J_v = \bar{V}_w J_w + \bar{V}_s J_s = 0 \tag{3.17}$$

(where $\bar{V}_s$ is the partial molar volume of the solutes and J_s their net flow across the membrane), will not be the same as the pressure required to bring J_w to zero. It is impossible to bring J_s, J_w and J_v all to zero simultaneously under these circumstances. Therefore constant turgor is not an indication of a true equilibrium state and strictly the theory of equilibrium thermodynamics does not apply to it.

However, this is not the only difficulty in applying a linear description of the type of equation (3.16) to flows across leaky membranes. The diffusing solute and solvent may interact with each other within the membrane in some way, i.e. the flow on one may not be independent of the flow of the other. (Indeed solute flows may well interact with each other, although this will be slight for dilute solutions and is generally ignored.)

These problems have been tackled in several theoretical ways that extend the usefulness of some of the concepts of conventional thermodynamics to systems that stray slightly away from equilibrium. One is by the use of kinetic or hydrodynamic theories involving frictional forces. [55, 56] Another, currently favoured in plant water relations, is the use of non-equilibrium (or irreversible) thermodynamics. [52] This traces its origins to the work of Rouss in 1801 on the electrical and osmotic behaviour of porous material which showed that the application of an electromotive force may produce not only a flow in charge across such a material, but also a non-conjugated flow of volume (now called electroosmosis). Conversely, the application of hydrostatic pressure was found to produce a non-conjugated flow of electricity (streaming potentials). The behaviour of thermocouples with regard to heat and electric flow described by Seebeck and Peltier subsequently demonstrated another system in which a link may occur between the driving forces of the flow of one quantity and the flow of another. In 1854 Kelvin pioneered the thermodynamic treatment of these phenomena that was used by Lord Rayleigh to describe a theory of sound. Onsager [57, 58] finally extended this to a set of equations, of a type called phenomenological equations, that express the linear dependence of all slow flows on all forces operating in a system. For a full development the reader is referred to Katchalsky & Curran. [52] However, the principle may be illustrated by the simplest case of the set of equations for the description of two flows (e.g. that of water, J_w, and of one solute, J_s).

$$J_w = L_{w,w} X_w + L_{w,s} X_s \tag{3.18}$$

$$J_s = L_{s,w} X_w + L_{s,s} X_s \tag{3.19}$$

where X_w and X_s refer to the 'Fourier–Fick–Ohm' driving forces for water and solute ($d\mu_w$ and $d\mu_s$ respectively) and $L_{w,w}$ and $L_{s,s}$ to the 'Fourier–Fick–Ohm' conjugating coefficients (e.g. the conductivity coefficient of equation (3.12)). $L_{w,s}$ and $L_{s,w}$ are not included in the 'Fourier–Fick–Ohm' treatment. These 'cross coefficients' represent the interactions of solvent and solute as they cross the membranes i.e. the effect of X_s on J_w and X_w on J_s respectively. If they have zero values, then the equations revert to those of Fourier, Fick and Ohm. In the examples of Rouss, Seebeck and Peltier, however, this was not the case. What, therefore, of the case of volume flows ($J_v = \bar{V}_w J_w + \bar{V}_s J_s$) in plants?

Use of a *dissipation function* allows the definition of the forces and fluxes in terms of the measurable parameters dP and $d\pi$ (see Reference 52; or, for an outline description, References 23, 56). Two equations are derived.

$$J_v = L_p \Delta P + L_{pD} \Delta\pi \tag{3.20}$$

$$J_D = L_{Dp} \Delta P + L_D \Delta\pi \tag{3.21}$$

where J_D is the velocity of the solute relative to the water flow (this, in effect, fixes a reference point for both flows), L_D is a 'straight' solute conductivity coefficient and L_p the 'straight' hydraulic conductivity. That dP and $d\pi$ can be varied independently (unlike $d\mu_s$ and $d\mu_w$ which are related by the Gibbs–Duham equation [52]) allows the designation of meaning to the linear cross coefficients. L_{pD} gives the volume flow for an osmotic gradient when dP is zero and L_{Dp} gives the difference in velocity of solute and water when the flow is driven purely by hydrostatic pressure. Zimmermann & Steudle [23] call these the 'osmotic coefficient' and the 'ultrafiltration coefficient' respectively.

At volume flow equilibrium, $J_v = 0$. From equation (3.20) this gives

$$dP = \sigma d\pi \tag{3.22}$$

where

$$\sigma = -L_{pD}/L_p \tag{3.23}$$

σ is termed the reflection coefficient. If it is included in equation (3.20) we obtain a useful equation to describe volume flow:

$$J_v = L_p(\Delta P - \sigma\Delta\pi) \tag{3.24}$$

(The reflection coefficient was introduced by Staverman in 1948 (see [59]), and is sometimes called the Staverman coefficient in animal studies. A more intuitive derivation of σ and L_{Dp}, not relying on a dissipation function, is given by Nobel. [38])

By combining the relationship $J_s = velocity_s \times \bar{c}_s$ (where $\bar{c}_s$ is the average concentration between the two compartments) with equations (3.20) and (3.21) an equivalent expression for J_s can be obtained:

$$J_s = (1 + L_{Dp}/L_p)\bar{c}_s J_v + \omega\Delta\pi_s \tag{3.25}$$

where

$$\omega \equiv \frac{\bar{c}_s(L_p L_D - L_{pD}^2)}{L_p} \tag{3.26}$$

ω is the coefficient of solute permeability.

At this point in the argument we encounter disagreement. Following Onsager [57, 58] all authors on plant water relations have accepted 'Onsager's Law' that deals with the symmetry of the phenomenological equations such as (3.18), (3.19) and (3.20), (3.21). [20, 23, 38, 54] This states that

$$L_{i,j} = L_{j,i} \qquad (i \neq j) \tag{3.27}$$

where the subscripts refer to generalised fluxes. This might be expected from Newton's third law of action and reaction. [38, 54] If this is the case $L_{pD} = L_{Dp}$, which signifies that in equation (3.25)

$$-L_{Dp}/L_p = \sigma \tag{3.28}$$

i.e.

$$J_s = (1 - \sigma)\bar{c}_s J_v + \omega \Delta\pi_s \tag{3.29}$$

We have, therefore, three independent parameters (L_p, σ and ω) all of which must be used to describe quantitatively osmotic phenomena in the presence of permeant solutes. [20] From equations (3.24) and (3.29) it can be seen that for a truly semipermeable membrane $\omega = 0$ and $\sigma = 1$. An often ignored property of these three parameters is that they are all concentration dependent. [23]

These parameters are the basis of much current work on plant cell water relations. However, attention must be drawn to the review of Hill [55] who presents arguments against some of the bases of irreversible thermodynamics especially the symmetry of the Onsager cross coefficients. While it is unlikely that major revision will be needed to our understanding of cell water relations in the near future [55] the chapter on water transport in Silver [56] makes stimulating reading in this area which is described as a 'mine field'.

In the meanwhile plant physiologists continue to use the parameters as defined by irreversible thermodynamics. In addition to hydrostatic and osmotically powered flows they also provide a basis for the description of other phenomena, *viz.* of electroosmosis, streaming potentials and ultrafiltration, the first of which we shall encounter in section 5.3.2. However we have in the process lost the general validity of the widely used parameter, water potential. That, from now on, must be used with care.

At $J_v = 0$, for example at steady turgor pressure, from equation (3.24) we obtain:

$$\Delta P - \sigma \Delta\pi = 0 \tag{3.30}$$

This will be different from equation (3.14) for all values of σ other than unity.

This illustrates the difficulties of using water potential as a parameter for describing both turgor pressure and the driving forces of water flow in plants. Its use is only justified when $\sigma = 1$. Even if this may in most cases approximate the case for flows across membranes in the presence of metabolites, in describing water flow through the apoplast ($\sigma \Rightarrow 0$), for example, the parameter is meaningless. [60]

The reflection coefficient gives an indication of the 'ideality' of the pathway/solute couple in question. Each solute and pathway will have its own independent value of σ. It is even possible to have negative values for σ. In these cases the pathway is more permeable to the solute than it is to the solvent and the Onsager cross coefficients speed the solvent rather than retard it.

3.3.3 *Membrane diffusive permeability*

Water molecules cross membranes in the absence of chemical potential gradients by diffusion. Such movement is a statistical process arising from the random independent movement of individual molecules and is generally estimated either by the use of labelled water molecules (of which $^1H_2O^{18}$ is considered the best [27, 61]) or by NMR techniques. [61, 62] The measured parameter is the membrane diffusive permeability, P_d.

The mechanisms of hydrostatic/osmotic water flow across membranes remains obscure. If it also is by a diffusive mechanism then L_p and P_d are related by the expression

$$\frac{L_p RT}{\bar{V}_w} = P_d \tag{3.31}$$

(see [20, 27 p. 184, 63]). However, the veracity of this relationship is not certain, and the apparent membrane diffusive permeability calculated using equation (3.31) from a measured L_p is called P_f. It is often observed that values of the P_f/P_d ratio are greater than 1. This has been used as an argument that hydrostatic/osmotic flows pass through fine water filled pores in the membrane. However, as estimation of P_d is far more prone to error due to the unstirred layers of water on either side of the membrane than is P_f, this conclusion is not fully accepted (see [20] for treatment of unstirred layers). We shall return to this when considering specific cases (section 3.5.2).

3.3.4 *Volumetric elastic modulus*

We are thus armed with a battery of parameters that we may attempt to measure in order to develop quantitative descriptions of cellular water relations. One other parameter of interest in dynamic systems remains. This is an empirical parameter that describes the elastic properties of the cell wall. It crucially deals with the dependence of cell volume on turgor pressure. Cell volume is a function of pressure. The relationship of which was defined by Philip. [64]

$$P = \varepsilon\left(\frac{V}{V_0} - 1\right) \tag{3.32a}$$

or

$$\varepsilon = V\frac{\mathrm{d}P}{\mathrm{d}V} \approx \frac{\Delta P}{\Delta V}V \tag{3.32b}$$

This modulus is a three-dimensional equivalent to Young's modulus. It predicts the fractional change in volume in response to a pressure change and thus describes the slope of plant cell volume/pressure curves. This is crucial not only for knowing the behaviour of cell volume as turgor pressure changes, but also for knowing the behaviour of π_i, intracellular π, since changes in cell volume at constant solute content will alter π_i in a way only predictable if the elastic modulus is known. As pointed out by Zimmermann & Steudle [23] the elastic modulus is a function of cell shape and size, as well as the specific rheological properties of cell wall material. They provide an equation relating the volumetric elastic modulus of long cylindrical cells (if the effect of the end walls may be ignored) to the Young's modulus of the wall in the radial (E_R) and longitudinal (E_L) direction, the Poisson ratios (σ_{RL} and σ_{LR}) relating the effect of a change in length on the diameter (and vice versa) and the dimensions of the cell, r = radius and d = the cell wall thickness.

$$\varepsilon = \frac{d}{r}\,\frac{2E_L E_R}{E_R(1 - 2\sigma_{RL}) + 2E_L(2 - \sigma_{LR})} \tag{3.33}$$

This relationship is not simple even for such a uniform structure as that of a *Chara* or *Nitella* internode or of a root cortical cell and few attempts have been made to quantify ε in this way. Most tissue cells will have shapes considerably more complex than this, including many that will be irregular.

Equations (3.32a) and (3.32b) on the other hand have been used successfully in conjunction with various methods of measuring cell and tissue water relations. It must be emphasised that they are defined in the absence of any external forces, such as a change in the pressure of a neighbouring cell on the cell under study, or on the mutual influence of cells in a tissue. The values of ε determined for tissue cells are not necessarily comparable with those of isolated cells. [23]

Equations (3.32a) and (3.32b) also only apply strictly to cells composed of materials that are ideally elastic; any departure from ideality due to *viscoelastic* properties and the value of ε will be a function of time in addition to the other parameters involved. A more generalised model of wall rheology is provided by the 'spring and dashpot' model (e.g. [65, 66]). Masuda and coworkers (e.g. [67]) describe stress relaxation as linear with log time after a lag period T_0, and express this as the equation

$$S = b \log\left(\frac{t + T_{\mathrm{m}}}{t + T_0}\right) + C \tag{3.34}$$

(see figure 3.2 for the identification of the variables). A mechanical model of such material comprising springs and 'dashpots' (hydraulic dampers) is also illustrated in figure 3.2. The outcome of this is that ε will be maximum if measured 'instantaneously' (i.e. within the lag period T_0) and subsequently will decrease as a function of b in equation (3.34). This model introduces the concept of a *relaxation spectrum* [65, 68, 69] which considers viscoelastic relaxation as a series of discrete events each with its own time constant. This behaviour is now recognised in studies on cell growth and wood technology but is slow to appear in other aspects of cellular water relations. The 'take-home message' is the strict necessity of reproducible conditions during mechanical extension. Different time scales of measurement will result in different portions of the relaxation spectrum being studied. Cell walls vary

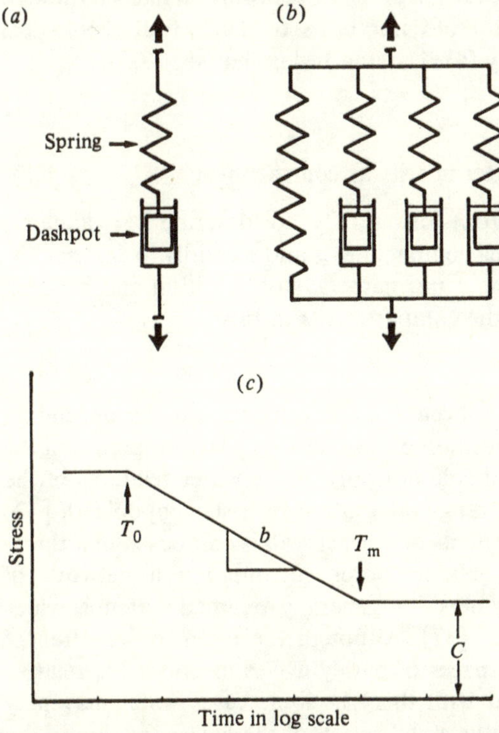

Fig. 3.2. Mechanical model of viscoelastic cell wall comprising springs and 'dashpots' (hydraulic dampers). (*a*) Maxwell element, (*b*) generalised wall (after [66]), (*c*) behaviour of strain relaxation of wall after applying a stress.

enormously in the relative shapes of their relaxation spectra, e.g. for *Nitella* rapid relaxations are most significant (the contents of the dashpots are more fluid) while for wood very much longer time scale extension predominates in the spectrum. [70]

In water relations studies the nature of the techniques to measure ε range from fairly rapid measurements with the pressure probe (less than 1 s) to much longer ones (minutes or hours) for techniques requiring long equilibration times. The values of ε measured rapidly have been termed the 'instantaneous' elastic moduli (ε_i), while those measured at longer time scales the 'stationary' elastic moduli (ε_s). [71] It is therefore important in water relations studies to know which type of parameter is being measured and whether it is relevant to the physiological property of interest (e.g. most cells *in vivo* do not undergo instantaneous turgor–volume changes).

Occasionally there appears in the literature a comment to the effect that thick cell walls indicate walls with a higher value of elastic modulus. If cell wall material were fully homogeneous such statements would be fully justified. However, not only can cell walls be composed of laminae of different chemical composition, but also the observations that have led up to the multinet hypotheses indicate that the orientation of cellulose fibrills – a likely (although non-proven) candidate for the molecular basis of much wall rheological behaviour – varies considerably from lamina to lamina (see [72]).

3.4 Physical environment of water in cellular compartments

Two sets of challenges confront us. Firstly to describe the aqueous environment of the individual cell compartments and secondly to describe the water relations properties of the water pathways both within and between them. Let us consider each of the compartments in turn.

3.4.1 The cell wall/apoplast

Some 5–40% of the water in cells occurs in the walls, the amount depending on the age, thickness and composition of the walls. [1] The properties of the wall material play an important role not only in the water relations of the individual cells but in the long-range integration of tissues (apoplast). [42] The bulk of the aqueous components of the cell wall is composed of a three-dimensional network of channels or pores surrounding a network of polysaccharide and protein polymers. It is generally assumed that most water is held there by surface tension. [37] (Although the recent observation of water droplets in intercellular spaces of barley leaves in only 67% relative humidity [73] is not consistent with this.) In some cases water may play important structural roles (e.g. the stabilisation of the $\beta(1 \rightarrow 3)$ xylan in the cell walls of some green algae [74]) and must not be treated simply as a 'universal inert filler'. [75]

While our knowledge of the biochemical relationship between the various polymers is far from complete, our information concerning the geometry of this aqueous continuum is even less so. Various proportions of the volume of cell walls *in vivo* are composed of water and the amount will depend on a range of factors. Preston & Wardrop [76] showed that the volumetric water content of the turgid cell walls of *Avena* coleoptiles exceeds 50%. Mature secondary walls, with their higher proportion of 'encrusting substances' would be expected to have a lower water content. Slatyer [37] presents data for pine wood flakes as an example of this. In this material Christensen & Kelsey [77] showed a water content of 0.33 g water per gram of solids at 100% relative vapour pressure. If the solids have a mean density of about 1.5 g cm^{-3} (see [78 table 3.1]) this is approximately 33% v/v. At the other extreme Gaff & Carr [79] suggest values of 1.5 g water per gram solid (corresponding to almost 70% v/v). Slatyer [37] comments on the paucity of data concerning the water content of living cell walls, this does not seem to have improved in the intervening period. Slatyer [37] suggests values of 10–100 nm for the diameters of the interfibrillar spaces, diminishing to 1 nm in heavily cutinised walls. Briggs, Hope & Robertson [80], citing the work of Preston, Nicolai, Reed & Millard [81] and Frey-Wyssling, Wyckoff & Mühlethaler [82], describe the wall as a basketwork of microfibrils 10–30 nm in diameter with spaces of the order of 10 nm between them. In considering the dimensions of the cellulose microfibrils Laüchli [42] suggests values of 1 nm for intermicellar and 10 nm for interfibrillar spaces. Considering the problems associated with electron microscopy of plant cell walls [83], however, more direct measurements are required of these spaces. Carpita, Sabularse, Montezinos & Delmer [84] demonstrated that only small macromolecules (radius 1.6–1.9 nm) could penetrate the cell wall of a variety of cell types. These workers predicted that hypertonic solutions would cause plasmolysis (contraction of the protoplast away from the cell wall) only if the solute were small enough to pass through the cell wall, and result in cytorrhysis (collapse of the wall) if the solute were too large to penetrate. They tested cells against a range of macromolecules. Miller [85] using similar techniques for the fern *Onoclea sensibilis* estimated pore diameters to be between 2.9 and 3.5 nm for prothalial cell walls and less than 0.8 nm for the intine. This approach has been criticised, however, by Tepfer & Taylor [86] who proposed that such a technique would dehydrate the walls faster than the macromolecules could penetrate them. This would cause shrinkage of the wall and of the dimensions of the pores. Using a preparation isolated from *Phaseolus vulgaris* hypocotyls and a gel filtration technique with much lower concentrations of the macromolecules they have shown that proteins of up to 60 000 dalton can penetrate substantial portions of the spaces of the walls (bovine serum albumin, 67 000 dalton, is totally excluded from the wall preparation). The apparent exclusion limit suggested is 5.6 nm for the pore diameter. However, this data refers to isolated material. In fully turgid cells the pores may have larger dimensions. Nevertheless they are a long way from reaching even the 10 nm frequently quoted.

The state of water within structures of such small dimensions is a matter of much interest and discussion. Some 150 years ago Poisson maintained that a density profile should exist in water adjacent to an interface with a solid. The dimensions of this profile, and its significance to the properties of water and solutes is still controversial. The considerations involved in the context of biological systems have been reviewed (e.g. [87, 88]). Some current models suggest that van der Waals, hydration and electrostatic forces make water within 3–5 nm of an interface behave differently from bulk water. This would suggest that very little of the water in the cell wall pores can be considered to be identical with bulk water. Indeed a growing amount of literature is appearing claiming that water in contact with macromolecules, or with other surfaces, is very different from bulk water [89, 90]. In the latter review Drost-Hansen hints that such 'vicinal' water may extend to more than 20 nm from surfaces and quotes Clegg [91] as arguing that nowhere within a (animal?) cell is one likely to be more than 5–10 nm from a 'surface' of some kind. This evidently would include all the water of the apoplast. Others have claimed that such water is quite normal, e.g. the self diffusion of water in biological gels indicates that most of it is in a normal bulk state. [92]

Discussing a number of reports on the viscosity of water in fine capillaries and thin films, Clifford [87] concludes the likelihood of an 8 nm thick layer of water at the capillary surface that is not sheared by forces causing liquid flow. On the other hand measurements of ion mobilities in a radio frequency (RF) field in 3 nm radius pores etched through mica are consistent with essentially normal bulk viscosity for 100 mol m^{-3} potassium chloride solution. [93] Clifford [87] comments in this case that viscosity may depend on flow rate (thixotropy) which was zero in these experiments. (The finest resolution for 'normal' water in a biological system appears to be the observation that water freezes normally in the 1.5 nm pore of the enzyme lysozyme (R. G. Bryant, quoted in [94 p. 388]).)

If the viscosity of water does increase dramatically in pores of less than 5 nm radius then bulk flow in the wall parallel to the cell membrane may behave with some 'unstirred' properties in which flow will be diffusional rather than by bulk movement. Clearly a consensus on the behaviour of water in fine, irregular pores is necessary to be able to predict water flow properties in cell walls.

This is in addition to the 'technical' problem of unstirred layers that extend many microns beyond solid surfaces. [20, 27] The entire cell wall must be considered an unstirred layer in this context, although presumably only normal to the plane of the membrane. Many attempts have been made to obtain the thickness of unstirred layers at membrane surfaces (see [27]) and in all cases the thickness of the layer is many times the thickness of any plant cell wall. House [27 table 4.1] lists some early estimates in which attempts were made to reduce the magnitude of the unstirred layer by rapid physical agitation of the medium bathing the membrane. Ginzburg & Katchalsky [95], for example, observed a decrease to a limiting value as the stirring rate during the experiment was increased from 0 to 2000 rpm. The limiting value was about 12 μm. The linear diameters of the interstices of the cell wall are orders of magnitude smaller than this. Indeed the entire walls of higher plants are typically in the order of 1–2 μm thick. (A 'classical' description of flow through macroscopic porous material may be found in Leyton. [96])

Little direct data regarding the macroscopic behaviour of water in plant cell walls is available. Some evidence suggests that within the walls the diffusion of small neutral molecules may not be similar to that in bulk water. Richter & Ehwald [97] made an estimate of the diffusion coefficient of sucrose in the wall free space of sugarbeet taproot tissue and found it to be between 6 and $9 \times 10^{-11}\ \mathrm{m^2\,s^{-1}}$ (at 25 °C), which is lower than its value in bulk water ($4.2 \times 10^{-10}\ \mathrm{m^2\,s^{-1}}$ at 20 °C [98]) even when considering the tortuosity of the interstices of the wall that complicates these estimates. Aikman, Harmer & Rust [99] estimate a tortuosity factor of 1.29 from electron micrographs of sugarbeet cell walls.

The bulk flow of water through the cell wall is central to the physiology of both growth (see section 3.5.3) and the transpiration stream. Slatyer [37], Steudle & Jeschke [11], Cosgrove [36] and others have all commented on the paucity of data concerning wall hydraulic conductivity.

Various measurements of cell wall L_p have been performed (table 3.1). For example Zimmermann & Steudle [100] using a modified pressure probe technique estimated a value for L_p of $6.9 \times 10^{-6}\ \mathrm{m\,s^{-1}\,MPa^{-1}}$ for the cell wall of the giant-celled alga *Nitella flexilis*. Similar values have been obtained for *Chara* (see references in [42]). Tyree [101] on the other hand estimated a value for maize root walls that is considerably lower ($14 \times 10^{-9}\ \mathrm{m\,s^{-1}\,MPa^{-1}}$). This was based on a previous measurement [102] of an intrinsic hydraulic conductivity coefficient of $0.14 \times 10^{-9}\ \mathrm{m^2\,s^{-1}\,MPa^{-1}}$ for *Nitella*.

The diffusive conductance (P_d) to water of surface cuticle layers (approx. $5 \times 10^{-7}\ \mathrm{m\,s^{-1}}$) has been shown to be pH dependent [103]. This is thought to relate to the dissociation of fixed carboxylic acid groups present. [104] It would appear that no equivalent measurements have been made of the corresponding possibility in more typical wall material.

The anisotropic structure of all walls makes it highly likely that water flow may be favoured in one direction over another. For example it is evident from the study of electron micrographs that the walls of mature cells have a multilaminate structure. (The apparent orientation of lignin with the phenyl propane rings parallel to the plane of the cell surface [105] indicates that this may be even more important in secondary thickened walls.) The distribution of non-cellulosic material (that may be crucial for microenvironment behaviour) is being tackled by such processes as immunological localisation,

Table 3.1. *Recent measurements of physiological turgor pressure (P), volumetric elastic modulus (ε) and hydraulic conductivity of cell membranes and walls of algae and fungi. Measured with the pressure probe. For previous data see [23]. (Here f(P) and f(V) are functions of turgor pressure and volume respectively; and L_{pt} and L_{pp} are the L_ps of the tonoplast and plasmalemma respectively.)*

Species	Physiological P (MPa)	$\varepsilon_i/\varepsilon_s$ (MPa)	$L_p \times 10^6$ ($m\,s^{-1}\,MPa^{-1}$)	Ref.
Algae				
(*a*) Membranes				
Nitella flexilis		8 ($P = 0.05$) (ε_s)		[71]
		20 ($P = 0.6$) (ε_s)		[71]
		20 ($P = 0.05$) (ε_i)		
		80 ($P = 0.6$) (ε_i)		
			14.1–19.2 ($P = 0$, perfusion)	[286]
Chara corallina				
(cell membranes)			0.8–1.5 (pH indep.)	[139]
			0.5–2.0 hydrostatic (conc. dep.)	[289]
			0.2–1.8 osmotic (conc. dep.)	[289]
		20.7	1.6–1.8 (pressure clamp)	[395]
(tonoplast, L_{pt})			3.0–10.0	[278]
(plasmalemma, L_{pp})			2.0–4.0 ($P > 0.04$)	[278]
			($L_{pt} > L_{pp}$; $P > 0.04$)	[278]
			($L_{pt} < L_{pp}$; $P \to 0$)	[278]
Halicystis parvula		0.05–0.2 (ε_i)	(pressure indep. 5 kPa < P < 90 kPa)	[71]
		0.1–0.2 (ε_i, P = 5–15 kPa)		[71]

		1.6	(ε_i, $P = 90$ kPa)			[71]
				0.08–0.2	(volume relaxation)	[71]
				0.1–0.25	($P > 20$ kPa) (turgor relaxation, $\varepsilon = \varepsilon_i$)	[71]
				0.5	($P = > 0$)	[71]
Valonia utricularis				0.035–0.09	(immobilised in gel)	[303]
Acetabularia mediterranea				0.14	(immobilised in gel)	[303]
	0.2–0.26 (oscillate)	2.9		0.12	($\Delta\pi_i$)	[228]
				0.09	($\Delta\pi_o$)	[228]
Lamprothamnium sp.				0.5	(*pressure clamp*)	[276]
				0.20		[276]
(*b*) Cell walls						
Nitella flexilis				13.2–19.9	($P = 0$, perfusion)	[286]
				6.9		[100]
				9.0		[102]
				3.6		[400]
				5.0		[401]
				1.8–3.5		[116]
Chara corallina		10	($P = 0.5$)	6.0–30.0		[278]
Fungi						
Phycomyces blakesleeanus	0.11–0.66			0.69		[220]

e.g. Moore, Darvill, Albersheim & Staehelin [108] have shown that whereas xyloglucan is uniformly distributed throughout the walls of suspension cultured sycamore callus, rhamnogalacturonan 1 is restricted to the middle lamella and the regions around the point of contact between adjacent cells.

Electrolyte diffusion in the wall is strongly influenced by the fixed charges present in the polymer matrix. [80, 106] This behaviour provides additional information regarding the water relations of the wall. Anions are preferentially excluded from a sizable proportion (e.g. 50% for *Chara* [107]) of the water free space (WFS) of the wall. For comparison a neutral solute, mannitol, diffuses freely in 92% of the WFS. [107] This volume which is under the influence of the predominantly negative fixed charge of the solid phase is the Donnan free space (DFS). Estimates for the concentration of charge in the wall range from 560 mol m^{-3} DFS in sugarbeet disk walls [80] to 1000 mol m^{-3} DFS in *Chara*. [109] Other data quoted in Walker & Pitman [106] suggest that the concentrations are lower in the walls of excised barley roots and leaf slices (300 mM in the latter), while Aikman *et al.* [99] reported 140 mmol per kilogram wet weight of sugarbeet wall. A relationship of wall exchange capacity with salt tolerance has also been observed.

Aikman *et al.* [99] showed that the diffusion coefficient of Rb^+ through sugarbeet cell wall material is $0.19 \times 10^{-9}\ m^2\ s^{-1}$ in comparison with a value of $2.1 \times 10^{-9}\ m^2\ s^{-1}$ for rubidium chloride in bulk water. Other values for diffusion coefficients in wall free space range from $8 \times 10^{-10}\ m^2\ s^{-1}$ for potassium chloride in *Nitella* to $(2\text{–}4) \times 10^{-13}\ m^2\ s^{-1}$ for Fe^{3+} in compressed barley roots (see [106] for references).

From exchange data this information has been used to estimate the p*K* values of the fixed anions in the DFS. Values of 2.8 and nearly 3 have been obtained for sugarbeet disks (see [80]). This is generally interpreted as corresponding to carboxylic acid groups of the galacturonic acid residues of pectin molecules. The charges on protein and glycoproteins are also likely to be significant [43] and the relatively large apparent free space for SO_4^{2-} in sunflower roots (57%) [110] is seen as an indication of this. The exchange data also suggest subtle differences between the DFS of *Chara*, where it has the properties of an electric double layer outside which the WFS behaves as bulk water (figure 3.3(*a*)) and of sugarbeet disks where the electric double layers overlap giving the impression of a more homogeneous Donnan phase (figure 3.3(*b*)). [80] The impression this presents is of the cellulose fibrills being coated by charged polymers surrounded by a diffuse double layer of ions forming the DFS, surrounded in its turn by WFS in which anions diffuse relatively easily (figure 3.3). Aikman *et al.* [99] predict such 'anion' channels from their kinetic data. Although Briggs, Hope & Robertson [80 p. 95] suggest that in sugarbeet walls the DFS extends to a certain extent throughout the holes (figure 3.3(*b*)), while in *Chara* the centres of the holes will be classical WFS (figure 3.3(*a*)), such behaviour is likely to be related to the geometric dimensions of the wall pores. It must be said that despite considerable advances in our understanding of basic wall biochemistry relatively little is known about the microarchitecture of the wall and any physiologically important effects related to this.

The Gibbs–Donnan properties of DFS space may prove to be of importance in several processes relating to the water relations of cells and tissues. Within the DFS they will result in increases in the local hydrostatic pressure of the cell wall solution. [80, 111] These hydrostatic pressures are the result of the increased osmotic pressure of the DFS caused by the Donnan distribution of ions. A relationship described above (equation (3.30)) will be of relevance. Thus the osmotic equilibrium set up by a 100 mol m^{-3} of fixed anions at equilibrium with 63.2 mol m^{-3} potassium chloride in the WFS will be 118.32 mol m^{-3} ($K^+ + Cl^-$) in the DFS. The difference of 51 mol m^{-3} (assuming osmotic coefficients of unity for K^+ and Cl^-) is equivalent to a pressure difference of 0.125 MPa. The actual magnitude of these forces in cell walls appears not to have been calculated, but the equivalent forces in ion-exchange resins are considerable. Ginzburg & Cohen [112] calculated pressures of 9.8 and 30 MPa for X4 and X12 cross linkages of the sodium form of the Dowex 50 cation exchanger in the presence of mmolar concentrations of the chloride salt of the respective cation.

This will have at least two notable effects on the wall. As argued by Ginzburg & Cohen [112] the pressure will tend to exclude *neutral* molecules

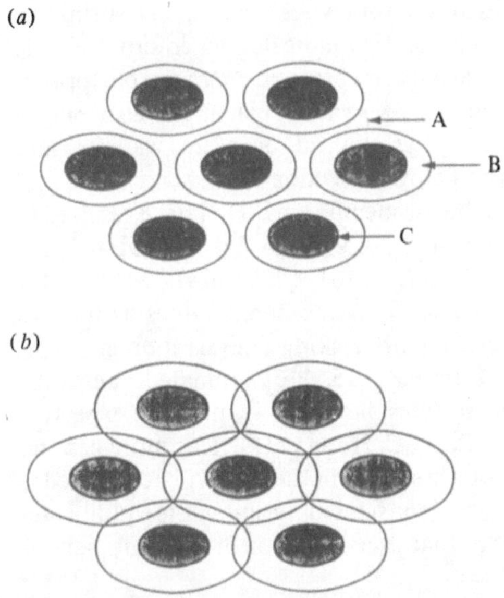

Fig. 3.3. Suggested configurations of solid material, Donnan and non-Donnan space in the cell wall. (*a*) Wall zone containing discontinuous DFS (*Chara* ?); (*b*) wall containing homogeneous DFS (sugarbeet?). A – WFS, B – DFS, C – solid material (e.g. cellulose).

from the DFS due to the pressure term in their chemical activities (equation (3.6)). The larger the partial molar volume ($\bar{V}$) the more significant will be the effect. Presumably this will influence the penetration of proteins into wall interstices (cf. [86]).

Secondly the pressure will be maintained by the development of stress between adjacent fixed charges. The closer the charges to each other the greater will be the tension. Use of IR spectroscopy shows that the pectin carboxyl groups are arranged parallel to the cellulose fibrills. [113] It is possible, therefore, that this distribution of charges orients the 'Donnan pressure' in specific directions. The high values suggested for resins, if applicable to cell walls, might exceed the stress on the wall caused by cell turgor pressure. It remains to be seen whether this could actually power growth at zero turgor pressure. Cleland (unpublished) has suggested that the stress yield threshold observed for expanding cells represents a critical extension of an elastic member before labile bonds are exposed. Internal pressures of the type discussed here could well 'pre-stress' such members and thus modify the yield threshold. A study of the interactions of the Donnan properties of walls and growth is overdue. The 'Donnan pressure' may also play subtle roles in modulating the sizes of the wall interstices, higher internal pressure opening up the wall structure. The dimensions of the apoplast pores may depend on the product of this 'Donnan pressure' and on the Young's elastic modulus of the polyvalent anion backbone. Tepfer & Taylor [86] showed clearly that walls are capable of shrinkage. A 25% solution of polyethyleneglycol (PEG 6000) dehydrated root walls resulting in a shrinkage of about 5%. If the behaviour is elastic this indicates an internal elastic modulus of 2 MPa (25% PEG 6000 has an osmotic pressure of approx. 1.25 MPa). Depending on the range of pK_a values pH might also play a role in modulating pore dimensions. A physiological role of this wall pressure in thermophilic bacteria is discussed below in section 3.5.1.

Measurements of σ for artificial cellulose membranes [114] have resulted in values of σ for sucrose of 0.075 (with an L_p of 21×10^{-6} m s^{-1} MPa^{-1}) for Visking dialysis tubing and 0.019 (with an L_p of 101×10^{-6} m s^{-1} MPa^{-1}) for Dupont wet gel. Ginzburg & Katchalsky [95] measured values as high as 0.163 for sucrose and 0.123 for glucose with Visking dialysis tubing.

In *Nitella* cell walls Tazawa & Kamiya [115, 116] estimated values for neutral sucrose of 0.25–0.5. For electrolytes, however, as might be expected, values nearer unity were claimed by Barry & Hope [117]. This more general situation, however, in which many of the solutes in the wall will be charged is more complex, but has still been considered using non-equilibrium thermodynamics. [20, 52] We find that the reflection coefficient here is described by the equation

$$\sigma = 1 - \frac{\omega \bar{V}_s}{L_p} - \frac{\bar{c}_s \varphi_w}{X t_1^0} \qquad (3.35)$$

(where ω is the solute permeability, φ_w is the volume fraction of water in the membrane, X the concentration of fixed charges in the membrane matrix, and t_1^0 the transference number of the counterions in free solution. [52 p. 167]) The reflection coefficient is dependent on the ratio of mean electrolyte concentration to fixed charges in the matrix. The dependence of the effective osmolarity difference across such a charged membrane on the true osmolarity difference is represented in figure 3.4 (taken from [118]). The concentrations of solutes of any kind in the wall space is an aspect of current interest (section 3.5.1).

We have here, however, a problem regarding the reflection coefficient. Being defined according to equation (3.35), it is open to the criticism referred to above [55] that in porous ('non-selective') barriers it can be shown that certain basic assumptions of irreversible thermodynamics break down.

A model system of similar dimensions and charge sign has been studied, in another context, by Hurtado & Drost-Hansen. [119] These workers

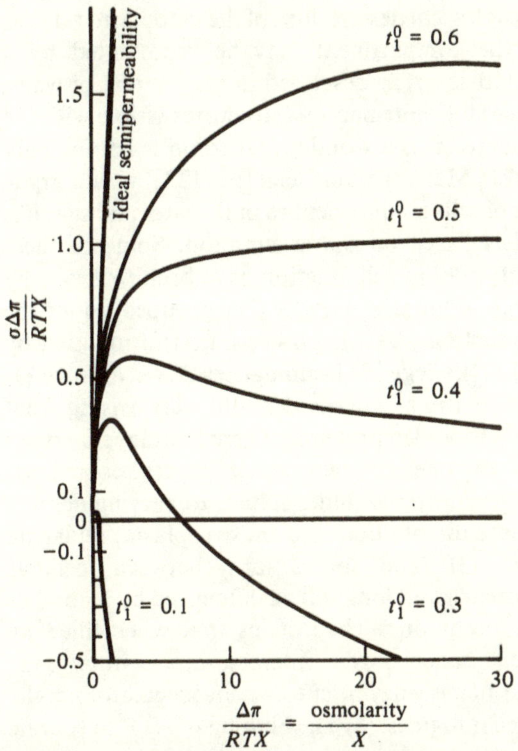

Fig. 3.4. Dependence of effective osmolarity (ordinate) on the true osmolarity (abscissa) across a charged membrane for various values of the transport number t_1^0 of the counterion. [118] Reproduced from the *Biophysical Journal* **2** (1962), 53–78 by copyright permission of the Biophysical Society.

measured the ionic selectivity of 14 nm pores through acid washed silica gel. The selectivity coefficients for potassium/sodium, potassium/magnesium and potassium/calcium all showed complex temperature dependence with maxima at about 15, 30 and 45 °C – independent of the nature of the accompanying anion: sulphate, iodide or chloride. The values ranged 1.3–1.65, 0.35–0.59 and 0.23–0.35 for potassium ions with sodium, magnesium and calcium respectively. Hurtado & Drost-Hansen [119] attribute this behaviour to vicinal water within the pore. Surprisingly the data was quantitatively similar to previous work reported by Wiggins [120] using silica-gel with much finer pore diameters (2.5 nm).

3.4.2 Cytoplasm/symplast

If the state of water in the wall space remains unclear, the equivalent situation in the cytoplasm is even more controversial. This is especially important as the controversy extends to the biochemical processes of this, the most important compartment of the living cell. In mature cells the cytoplasm is usually composed of a thin layer lining the cell wall and may contain as little as 5–10 % of the cell water. [1] Most physiological descriptions of the cytoplasm take, as a first approximation, that the compartment may be represented by a conventional dilute solution. However, as described in the context of walls, when all the membrane surfaces and proteinaceous structures visible with the electron microscope are considered it is a wonder that room is left for bulk water free of surface effects. [121] Many workers (e.g. [91, 122]) would argue that no water in the cytoplasm of cells is equivalent to bulk water and that it is misleading to construct models based on the assumption. Some of these alternative views on the role of water in cell function have been reviewed by Clegg [91]. The most thorough alternative view is that grouped under the heading of the association–induction (A–I) hypothesis, first formulated by Ling about 30 years ago (see [122]). Clegg [91] summarises this as follows. (1) Virtually all cytoplasmic water exists as polarised multilayers arising from fixed charges on surfaces (mainly extended proteins). These multilayers extend far beyond the dimensions of one or two water layers considered conventionally. (2) This structured water excludes solutes to varying degrees and is held responsible for solute distribution into the cytoplasm, replacing active membrane transport. (3) The interactions between cellular macromolecules (primarily proteins) and ions will be influenced by a number of factors. (4) 'Cardinal sites' occur on such proteins that when filled by specific molecules initiate cooperative interactions within the protein–ion–water system. Hormones, enzyme effectors and regulatory cyclic nucleotides are examples of such initiators. (5) The binding of ATP to specific cardinal sites generates the cooperative formation of the multilayers of water which, amongst other things, lead to a selective accumulation of potassium ions over sodium ions. Hydrolytic splitting of ATP leads in some undefined way to a degradation of the polarisation of water and its associated phenomena i.e. the cell moves to a lower energy state. While it is generally accepted that plant and animal cytoplasms are far from being simple solutions few plant physiologists would agree with the full blown A–I hypothesis. However, evidence against the 'simple solution' approach cannot be ignored. An example of this in a bacterial cell context is proposed by the work of Ginzburg & Ginzburg [123] using the extremely halophilic bacterium *Halobacterium marismortui*. They show that a high degree of selective discrimination for potassium over sodium ions in the cell is due to its internal water structure rather than any membrane action. They argue that 55% of the cell water has a translational diffusion coefficient of 10^{-12} m^2 s^{-1} (3 orders of magnitude lower than that in bulk water – H_2O^{18} – 2.35×10^{-9} m^2 sec^{-1} at 18 °C (see [27]).

3.4.3 Vacuoles

The bulk of the water in mature cells is found in a large central vacuole. Although these may in some cases contain other substances, the properties of water within them are generally considered to approach that of a 'dilute' solution. Reviews of vacuolar contents are available. [124, 125] Quantitatively the water relations of the vacuole reflect those of the rest of the cell. This is fortunate as it is by far the easiest compartment to analyse.

3.5 Interactions between compartments

Since water permeates all parts of the plant body it is not easy to delineate convenient blocks of phenomena and parameters that can be considered in isolation. Different techniques measure different collections of parameters, and different physiological problems cross the boundaries of definition of parameters. Following Dainty [21], however, I shall attempt to look at the interactions of the various compartments under the headings of 'static' and 'dynamic' processes. To these I shall add a third, which spans both headings and hence is better dealt with independently. The third heading will be the water relations of growth and movement.

The static processes are those in which volume flow (J_v) is not an important consideration either because it does not occur as in the case of steady turgor pressure, or because its rate is unimportant to the phenomenon under consideration. Examples of this will be osmotic and turgor adjustment and their implications.

3.5.1 Static systems ($J_v = 0$)

Turgor pressure Turgor pressure in plants is a direct legacy of the primeval cell's desire to sequester impermeant ions within a semipermeable membrane. [19, 121, 126, 127] It was perhaps inevitable that amongst the population of such ions one sign would predominate – proteins, nucleic acids, many metabolic intermediates, etc. generally carry a net negative charge at both physiological and environmental pHs. The conditions were therefore set from the start for the osmotic disequilibrium described by the Gibbs–Donnan effect. The magnitude of the osmotic disequilibrium will depend on the concentration of the indiffusable anion and the concentration and valency of the permeant electrolytes in the external solution. An extensive quantitative treatment of this is included in Briggs *et al.* [80] The net result is a higher osmotic pressure within the cell than outside, this results in a net flow of water into the cell which expands. As Hempling [121] reminds us, it is impossible for such wall-less cells to reach equilibrium and the cell volume will increase until cell lysis occurs. Hempling [121] calls this 'the eternal spectre of a watery death'!

Cells have countered this fate in two main ways. These are by solute transport and the development of a cell wall. A third option would have been to render the cell membrane effectively impermeable. This appears to have been done in the case of the membrane of water hardened trout eggs where $L_p = 3 \times 10^{-11}$ m s^{-1} MPa^{-1}. [128, 129] However, truly impermeable membranes would render all but dormant life impossible within them.

Wall-less cells have developed transport systems (notably the sodium ion extrusion pump [19, 126]) to export osmotically active materials against their electrochemical gradients out of the cell. Thus a volume equilibrium is reached due to the expenditure of metabolic energy. Walled cells have overcome destructive swelling by allowing the development of a hydrostatic pressure gradient to balance the osmotic pressure gradient across the plasmamembrane. Although this solves the problem at the cell surface it does not do so at the tonoplast which also generally separates two aqueous compartments, one with a much higher concentration of indiffusable anions than the other. Here transport mechanisms are still required to redress the balance, as in wall-less cells. [130, 131]

Strictly, changes in external π could also be counteracted by changes in σ [19] – but as this would in practice be accompanied by a corresponding change in ω for the solute in question it would result in the extracellular osmoticum flooding the cell. An effect that would be deleterious to at least parts of the cell. From these humble beginnings turgor pressure has developed to be a central and fundamental characteristic of plant cells. It allows an erect habit in non-woody plants [1], drives growth [132] and drives certain other mechanical processes such as stomatal action [133] and reversible tissue movement. [134]

In 'modern' plants the turgor and osmotic pressure levels of cells are not left to the whims of macromolecules and metabolism but appear to be controlled parameters even in a marine environment, where external water potential might be expected to remain constant. In the case of terrestrial plants, Coster [135] points out that they rarely enjoy a natural environment free from water stress for more than a few days. Under saline conditions the 'problem' is extended by the water potential of the environment being depressed additionally by the presence of considerable concentrations of salts.

Changes in the water status of plant cells brought about by changes in the external water potential have widespread metabolic consequences (see [136]). It is not surprising therefore that most cells possess homeostatic systems to counteract these effects. As we shall see in section 3.5.2 changes in extracellular water potential result in (more or less) instantaneous changes in cellular water potential i.e. unlike trout eggs the cell is not capable of isolating itself from changes in water potential.

An active water pump capable of using energy to maintain a water potential gradient across the cell membranes could also buffer changes in cell water potential (in the same way as active pumps maintain chemical activity gradients of many other solutes). As we shall see in section 3.5.2, although water potential gradients may be maintained across composite membranes in this way, the current consensus is that they do not play any major role in cell water relations although the mechanism by which contractile vacuoles achieve this remains to be determined. [137] In the absence of water pumps, or permeability barriers, therefore, these changes in cellular water potential cannot be avoided.

Osmotic adjustment The term 'osmoregulation' has been used to describe a whole range of phenomena involving changes of cell and tissue osmotic pressure. This is not surprising since until recently all methods of following both osmotic and turgor pressure modulation involved measuring bulk osmotic pressure. They therefore cover both 'passive' changes in cell osmotic pressure brought about by wall elasticity and 'active' transport or metabolically induced solute accumulation. New techniques, such as the pressure probe (see appendix), that measure turgor pressure directly should allow a dissection of the mixed bag of processes involved. Comment has been made on the rather lax use of the term 'osmoregulation' (e.g. [30, 40, 138, 412]). It would appear that in a few cases only (e.g. *Nitella*) is osmotic pressure a conserved parameter. The term should be retained for these examples. Most processes entail an adjustment of the osmotic pressure to allow regulation of the turgor pressure. The phrase 'osmotic adjustment' must be more descriptive of this process.

Mechanical osmotic adjustment Cells respond in two independent phases to changes in water potential. The first is a purely physical relaxation of the cell wall (according to equations (3.3.2a) and (3.3.2b)), the second is a transport (or metabolic) process aimed at recovering something of the original parameters of the cell. Since many of the analytical techniques rely on extrapolating turgor pressure from osmotic pressure measurements it has often been difficult to distinguish the relative importance of these processes. Use of the pressure probe, however, enabled Tyerman & Steudle [139] not only to describe the two processes independently but to use the data to calculate permeability coefficients, L_p and σ for *Chara* cells. The initial change in cell water potential involves a change in both osmotic and hydrostatic pressure. The former due to volume changes brought about by the latter. The relative proportions are a function of the elastic properties of the cell wall. The critical parameter here is the volumetric elastic modulus ε (equation (3.32)). A high value of this indicates a stiff, rigid wall while a low value indicates an elastic structure. Many authors have indicated that this is the first parameter under the control of the plant that acts in turgor or osmotic pressure adjustment. Figure 3.5 illustrates this for two model cells, the parameters of which are well within the observed physiological range. Both cells have initial osmotic pressures of 1 MPa and volumes of 100 pl in a solution of zero water potential (i.e. osmotic pressure = 0 and hydrostatic pressure = atmospheric). They differ however in that cell (*a*) has a volumetric elastic modulus of 10 MPa and cell (*b*) a volumetric elastic modulus of 1 MPa. Under these initial circumstances (assuming a reflection coefficient of 1) both cells will have a turgor pressure of 1 MPa. Each cell is now transferred to a solution where *P* remains atmospheric but π_0 is 0.5 MPa. It can be seen that with cell (*a*) ($\varepsilon \approx 10$ MPa) volume and internal osmotic pressure change relatively little in comparison with turgor pressure, while with cell (*b*) ($\varepsilon \approx 1$ MPa) turgor pressure changes relatively little in comparison with volume and internal osmotic pressure. Recent values of ε for higher plants, fungi and algae are given in tables 3.1 and 3.2. Additional earlier measurements are extensively listed in Zimmermann & Steudle [23] and Gutknecht *et al.* [19]

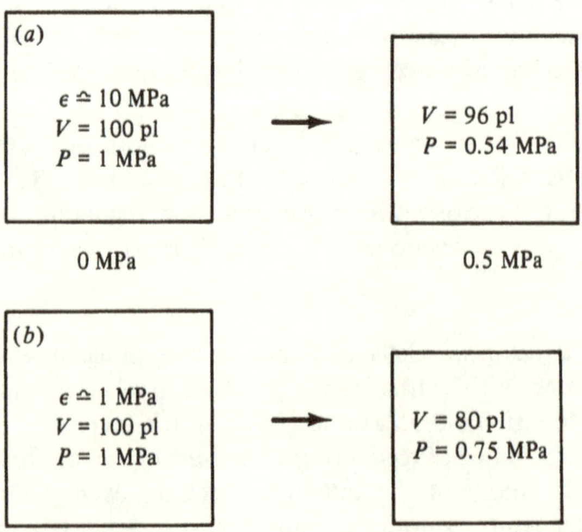

Fig. 3.5. Effect of volumetric elastic modulus, ε, on the shrinkage of a cell following a decrease in water potential.

The two extremes may be seen in the response of two different marine giant-celled algae. *Halicystis parvula* possesses a wall with a very low value of ε (0.06 MPa [140]) while that of *Valonia utricularis* is much higher (3 MPa at turgor pressures approaching zero, rising to 12–60 MPa at turgor pressures in the order of 0.4 Mpa. [141, 142]) (The physiological turgor pressures for these algae are 100 kPa for *V. utricularis* and 35 kPa for *H. parvula*.)

In the case of higher plants it has often been suggested that this parameter plays a role in water stress tolerance (see [32, 143]). Several studies using pressure bomb techniques (see appendix) suggest annual cyclic changes in bulk tissue ε that correlate with changes in water stress (see [32]). Increased stress is met by a decrease in the apparent value of ε. This should maintain turgor pressure at the expense of losing cell volume. Similar work suggests that tissue values of ε can also increase under environmental stimulus. This is harder to understand. Although we have little or no information of the molecular basis of ε, it is reasonable to assume that its value may be *decreased* within a preexisting structure by the breaking of bonds within the wall. An *increase* in ε on the other hand, at any turgor above zero, would have to involve inserting 'prestressed' bonds into the wall if it were to apply to the wall at turgor pressures below ambient. Levitt [144] has recently proposed such a process in his description of turgor recovery in wilted, excised cabbage leaves in the apparent absence of water uptake into the cells. He proposes that the cells actively contract by some unknown mechanism. He provides several hypothetical possibilities for this, including a role for lectins pulling the walls into folds. If such processes do occur, then a whole battery of new interpretations of cell and tissue water relations is opened up. As with so many such observations it will be essential to demonstrate such behaviour at the single cell level. (Although in studying the effect of pollutants at a single cell level with the pressure probe, Rygol & Lüttge [145] have indeed shown an increase in ε in individual cells.)

It is not clear from the literature whether the bulk of these seasonal changes in ε refer to the development of new cells, which, of course, can be produced with any wall elasticity. The question is whether such responses are brought about at the cell level, or whether the explanation lies in the heterogeneity of tissues.

While maintaining turgor pressure at the expense of loss of volume may be important in growing or moving tissues where critical turgor pressures need to be maintained, in other (non-growing) tissues the deleterious effect of volume change on tissue function may outweigh this. Tissue architecture is clearly a function not only of cell turgor but also of cell size and shape. By analogy, animal 'osmoconformers' (that correspond to plant cells with walls of low ε values) living in conditions of regularly changing water potential (brackish water environment) tend to swell up during periods immediately following a reduction in external salinity. This is disadvantageous and rapidly impairs body activities, including locomotory and food collecting mechanisms. [146]

Whether such a consideration is ever relevant to plants is unknown, but cannot, at present, be ignored. Lowering the value of ε may be a 'cheap' way of rapid turgor/osmotic adjustment but it is not without its potential problems to multicellular tissues. (It is unlikely that this argument applies to single cells.)

A 'solution' to the compromise between volume and pressure maintenance appears to have been adopted by certain intertidal algae. Such algae use very high turgor pressure under 'low salt' conditions as a buffer against loss of turgor pressure when the medium becomes salt. [19] Using the pressure probe Wienecke, Tomos & Wyn Jones [157] have measured turgor pressures in excess of 2.5 MPa in *Porphyra* in diluted sea water. However, whereas the protoplast had certainly swollen considerably relative to its state in full sea water, indicating a moderate value for the elastic modulus, the cell outline had hardly increased at all, indicating a very high value for the modulus. Thus a thick compressible wall within a stiff outer layer permits cell volume to be preserved despite changes in protoplast volume. Thus both volume and turgor pressure are partially 'buffered' by the cell wall behaviour. Such behaviour has not been sought or found in higher plants.

A critical feature of ε is that it is not constant, but is a positive function of pressure (i.e. cell walls are not Hookean). This 'strain hardening' [69] property has been shown using various techniques at both the tissue and the cell level. [23, 147–52] In some giant-celled algae and in some higher plants the elastic modulus also appears to be a function of cell volume. Zimmermann & Steudle [23] suggest that this may be another manifestation of the strain hardening effect in that tension (T) within the wall is related to turgor pressure (P) by the relationship

$$T = Pr/2d \tag{3.36}$$

where r and d are the cell radius and wall thickness respectively. Alternatively it reflects the varying ratio of areas of wall with differing rheological properties (e.g. end walls and sides).

Zimmermann & Steudle [23] indicate that such a relationship between ε and volume may even provide a model for a purely physical limitation of cell expansion growth. Although whether the relationship between wall elasticity and plasticity (section 3.5.3) is relevant is not known, this phenomenon is certainly worth further study. It is just as likely that the rheological properties that do limit growth at large volumes are reflected coincidentally in changes in elasticity.

Clearly the form taken by the turgor pressure dependence of ε may have subtle ecological significance that will become clearer as our understanding of the behaviour of materials in tissue architecture improves.

As discussed above (section 3.3.4), the rheological properties of viscoelastic structures are subject to a relaxation spectrum that may vary from cell to cell. This is reflected in wall elasticity by its having different values depending on the method of measurement. Zimmermann & Hüsken [71] showed that the stress (turgor pressure) dependence of ε in *H. parvula* differed greatly depending on whether an *instantaneous* elastic modulus (ε_i), or a long-term (*stationary*) elastic modulus (ε_s) was measured. In this study the authors claim that it is ε_i rather than ε_s that determines the rate of volume change following a rapid water potential change. It is difficult to see, however, how this can be the case for the longer-term field behaviour of cells. The measurement time dependence of ε in higher plants has been suggested by the pressure bomb measurements of Wenkert, Lemon & Sinclair [153] and by the hysteresis of measurements on the water relations of sea grass leaf tissue. [154] However, Brinckmann, Tyerman, Steudle & Schulze [151] showed a time *independence* of ε with *Tradescantia virginiana* leaf epidermal cells. Similarly, for red beet taproot it has been found that the turgor dependence of ε_i and ε_s are very similar using pressure probe methods. [155] Presumably this difference between algal and higher plant tissue parameters reflects differences in the relaxation spectrum of the wall material.

One aspect of the importance of ε to osmotic adjustment that has received some attention is the role of cells as 'water capacitors'. This involves the relationship between turgor/volume and water potential difference in tissues that undergo changes in water potential. [21] Water potential difference is given by

$$\Delta\psi = \Delta P - \Delta\pi \tag{3.37}$$

Within the cell, assuming that the Boyle–Van't Hoff law holds

$$\Delta\pi = \pi\Delta V/V \tag{3.38}$$

From equation (3.32b)

$$\Delta P = \varepsilon\Delta V/V \tag{3.39}$$

therefore combining equations (3.37)–(3.39)

$$\Delta\psi = (\varepsilon + \pi)\Delta V/V \tag{3.40}$$

This provides for the expression of 'cell capacity', C, defined as the increment change of volume per increment change of potential, i.e.

$$C = \Delta V/\Delta\psi = V/(\varepsilon + \pi) \tag{3.41}$$

This has been used, for example, to correlate changes in cell turgor during the diurnal crassulacean acid metabolism (CAM) cycle in leaf parenchyma of *Kalanchoë daigremontiana*. [15, 156] Here the nocturnal malic acid accumulation osmotically drives significant water storage in the leaf tissue. Changes in volume in the order of 2.3–10.7% v/v are observed. Care, however, must be taken when correlating cell capacity data derived from single cells and from whole tissues. Palta *et al.* [248] have recently shown that values for ε from whole sugarbeet taproots are much lower than those predicted from measurements on single cells using the pressure probe (see also [25]). The water storage capacity of the wall itself has been reviewed by Tyree & Jarvis [25] who consider it to be of little importance to the plant.

The cell wall elasticity governs the immediate response of the cell to changes in water potential. Following these changes most cells now use the option of modulating the relative contributions of the two components, osmotic and hydrostatic pressure. Evidently, following a change in water potential, P and π cannot both be conserved. In observing the behaviour of cells following such a change the apparent importance of each may be assessed (as we have seen in the case of cell wall rheology). To achieve this the osmotic and turgor pressures of the cells need to be measured. Until recently this has not been very easy for higher plants, and much of the early evidence comes from work on unicellular and giant-celled algae. However a considerable literature now exists to show that generally P is conserved at the expense of modulating π.

Biochemical osmotic adjustment Various responses are available to the plant cell following a change in water potential and the consequent adjustments to turgor and osmotic pressures due to the elastic properties of the cell wall.

(1) No further adjustment.
(2) Transport-induced or metabolic generation of cell solutes to recover initial *osmotic pressure*, and with it the original cellular environment for the enzymes, etc. of the cell (i.e. change the turgor pressure to the same degree as the change in the water potential).
(3) The same transport and metabolic processes as (2) but in order to recover the initial *turgor pressure* (i.e. alter the osmotic pressure to the same degree as the change in water potential).

Here again, for technical reasons, most detailed information refers to the responses of giant-celled algae. However a wide range of responses is seen. At one 'end' of the spectrum we find the intertidal alga *Porphyra* described above. Using the pressure probe Wienecke, Gorham, Tomos & Davenport [157] confirmed that the rhythm of changing extracellular water potential results in large changes in turgor pressure that are maintained without much further modification. In sea water turgor pressures of the order of 0.2–0.3 MPa are found. These reach values of over 2.5 MPa on dilution of the medium to levels experienced by the alga at low tide when they become immersed in freshwater from rainwater, streams or rivers. This alga appears to follow either strategy (1) or (2). (We currently have not the data to distinguish between two in this case.) Such behaviour is not restricted to marine/brackish water algae. The freshwater Characeae *Nitella* and *Chara* favour homeostasis of internal concentration over constant turgor pressure. [158, 159] Most algae studied, however, appear to favour strategy (3). These include *Ochromonas malhamensisi* [160, 161], *Chlorella pyrenoidosa* [162], *Chaetomorpha linum* (at 1 MPa) [163, 164], *Valonia utricularis* (at 0.15 MPa, depending on cell size) [141], *V. macrophysa* (at 0.15 MPa) [165], *Halicystis parvula* at (0.05 MPa)

[140], *Codium decorticatum* (at 0.23 MPa) [164, 166], *Griffithsia monilis* [167], *Ulva lactuca* [168], *Lamprothamnium succinctum* (at 0.9 MPa) [169, 170, 171], *Anabaena viriabilis* [172], *Synechocystis* [172] and *Chara buckellii* [173].

Bisson & Bartholomew [159] suggest that for plants growing in low water potential environments (soil or salt water) turgor regulation is favoured, whereas in freshwater, external osmotic pressure is so low that proportionately large changes in it will have little effect on turgor pressure. In this case osmoregulation will result in a reasonably effective turgor regulation. The degree of turgor maintenance in response to water or salt stress in higher plants was first reviewed by Hsiao *et al.* [30] In the case of salt stress total or partial turgor maintenance (measured in terms of increased osmotic pressure) has been described in many systems. More recently complete turgor maintenance has been shown at single cell resolution in long-term experiments for *Suaeda maritima* using the pressure probe. [174]

An increasing number of examples of turgor maintenance by increasing osmotic pressure in water stressed (other than salt stressed) plants has been reported. Maize roots [175] appear to regulate turgor pressure under water stress, as do soybean seedling hypocotyls. [176] In *Sorghum bicolor* plants in the field changes in tissue osmotic pressure closely mirrored changes in water potential [30] although in this type of experiment it is impossible to distinguish the wall induced changes from 'active' osmotic adjustment. Jones & Turner [177] looked further at this and found that since ε *increased* on drought pretreatment, the turgor adjustment was due to 'active' osmotic pressure adjustment. Fereres, Acevedo, Henderson & Hsiao [178] subsequently showed that the increase in leaf osmotic pressure was sufficient to maintain turgor pressure in *Sorghum* at a mean value of 0.5 MPa for 100 days of the growing season. Partial or complete turgor maintenance has been claimed for Orchard grass leaves (*Dactylis glomerata*) [179], soybean hypocotyl [176], pea root [180], sugarbeet leaves [181], apple leaves [182], *Hammada scoparia* [183], *Dryas integrifolia* [184], wheat [185, 186, 187], silver birch [188], *Zea mays* [178], several desert species [189], cotton [190, 191], sunflower [192], millet [193], rice [194] and sugarbeet taproot [413]. Other species showed no adjustment (see [34]).

Turner & Jones [138] comment on the limits of such osmotic adjustment. The commercial application of these studies in crop yield research is considerable. Readers are referred to reviews by Turner & Jones [138] and Morgan [34] for further details.

A decrease in cell water potential not associated with conventional 'stress' is that associated with growth. This will be dealt with specifically in a subsequent section, although the processes here are all relevant to this situation also.

Nature of perceived stimulus Cram [40] and Gutknecht *et al.* [19] comment on the basic principle of 'turgor' adjustment. The parameter to be adjusted (e.g. turgor pressure) is perceived and compared to a reference value. The difference ('error signal') is transduced and probably amplified to a control signal that influences an active process (such as transport). This causes a change in the original parameter to be adjusted. The feedback process continues until the 'error signal' is reduced to zero.

Gutknecht & Bisson [195] and Gutknecht *et al.* [19] list potential processes that could be the signals that relate the changes in cell water potential to the effectors of osmotic adjustment. Turgor pressure itself is not the only candidate, others are the absolute hydrostatic pressure, π_0, the ratio of π_i/π_0, the concentration of a specific solute and the partial pressure of dissolved gasses. These reviewers, however, list arguments to support turgor pressure as the primary signal in walled cells. Hastings & Gutknecht [196] and Bisson & Gutknecht [166] measured the effect of performing osmotic adjustment experiments with *Valonia ventricosa* and *Codium decorticatum*, respectively, under hyperbaric conditions (which alters absolute hydrostatic pressure without altering turgor pressure) and showed that in this species, at least, the absolute hydrostatic pressure has no role in addition to that of turgor pressure. They also showed that oxygen partial pressure had no effect. Use of the cell perfusion system with *Valonia* appears also to have ruled out sensing specific concentration gradients across the cell membranes. [19] Similarly these authors argue against changes in volume being important due to the large elastic modulus causing only small changes in this parameter. However evidence that concentrations of specific solutes do influence osmotic adjustment has been given recently for the wall-less unicellular alga *Poterioochromonas*. Here changes in calcium ion concentration brought about by osmotic shrinking and swelling appear to have a role in controlling isofloridoside metabolism. [197] Isofloridoside is the major regulated osmoticum in this species.

It is generally difficult to distinguish between volume changes and turgor changes as the conserved component. The proof of whether the turgor pressure is truly conserved may have to await the elucidation of the precise mechanism of turgor sensing. It is widely assumed that the volume homeostatis of wall-less algae (e.g. *Ochromonas malhamensis*) is volume rather than turgor sensitive simply due to the assumption that negligible turgor pressure is developed in such cells (e.g. [24]). Whether this is, in fact, the case remains to be shown since even membranes have a small, but measurable elastic modulus. An example that argues specifically for volume regulation is reported for another unicellular alga *Platymonas subcordiformis*. [198] This example has several noteworthy features. Rapid change of water potential by application of a hypoosmotic shock results in a swelling of the cells. Not all organelles respond equivalently. While the chloroplasts swell dramatically with resulting loss of internal structure, little change is seen in the nucleus. (The authors comment that the mitochondria were more or less unaffected, although the data presented does indicate swelling.) In the subsequent recovery period *cytoplasmic* volume is restored to its original value. *Total cell volume*, however, is not. The pressure differential across the plasmalemma/theca boundary is not conserved in this case.

The Donnan distribution of mobile ions across the tonoplast will tend, as noted above, to result in swelling of the cytoplasm at the expense of the vacuole due to the usual difference in protein concentration inside and outside the vacuole. Dainty [130], McNeil [199] and Leigh & Wyn Jones [125] have argued that this must require a controlled transport of solute across the tonoplast linked to a cytoplasmic volume regulation process.

The importance of water structure to macromolecule stability suggests the possibility that cells may be able to respond directly to the chemical potential of water. Such a process has recently been reviewed by de Meis. [200] In this system changes in the water activity in the cytoplasm lead to significant changes in the ΔG_0, the change in Gibbs free energy, of pyrophosphate hydrolysis. This is linked to calcium-ion-transport ATPase and cell energy transduction. Whereas the water activities considered are generally much lower than those thought to occur in a bulk compartment (e.g. 40% dimethyl sulphoxide (DMSO)) and are treated in the context of membrane surfaces, even relatively smaller decreases in water activity have measurable effects. For example, a 20% solution of glycerol (about 2 kmol m^{-3}, or 4.5 MPa osmotic pressure) reduces the K_m for inorganic phosphates of the phosphorylation of the calcium-ion-transport ATPase in *Rhodospirillum rubrum* at pH 6.0 from 1.5×10^{-3} to 0.8×10^{-3}. A 20% solution of DMSO reduces it to 7×10^{-6}.

The study of plant water relations is yet to reach this level of sophistication. In any case, as indicated by Hsiao *et al.* [30] changes in water potential of 0.5 or 1.0 MPa only affect the chemical activity of water to the extent of 0.4 and 0.7%. These authors consider such changes insignificant.

Nature of the solutes used for osmotic adjustment The nature of the osmotica used for turgor, volume or osmotic adjustment has been the subject of considerable interest over the last decade. It has been found that vacuole and cytoplasm respond in different ways in this respect. As no pressure differential can be established between them, unless the reflection coefficient of the tonoplast deviates from unity, cytoplasm and vacuole osmotic pressure must balance. The consensus is that whereas the vacuole of higher plants is relatively non-selective (Leigh & Wyn Jones [125] have chosen to use 'promiscuous'!) the cytoplasm of *all* living organisms has crucial limits on most osmotica. (This even applies to halophilic bacteria in which the cytoplasm contains very high concentrations of sodium chloride. In which case low concentrations of sodium chloride are 'toxic'.)

Although high concentrations of most solutes (especially inorganic ions) tend to inhibit cytoplasmic biochemistry a group of molecules have been identified that appear to be benign to such processes even at moderate concentrations. These are the 'compatible solutes' [201, 202] and include two main groups, (*a*) polyhydric alcohols and their derivatives and (*b*) small zwitterions typified by proline and glycinebetaine (N,N′,N″-trimethylglycine). [203] Potassium also appears to be 'tolerated' at relatively high concentrations and has also been referred to as a 'compatible solute' in halophylic bacteria. [201] The role of potassium ions in plant water relations has recently been reviewed by Hsiao & Läuchli. [204]

The characteristics of molecules that make them 'compatible solutes' are still unknown. Although Wyn Jones & Pollard [203] suggest that the property is related to the very small thermodynamic effect that they have on water structure. They speculate that with glycenebetaine, for example, the hydrophilic, hydrophobic, cationic and anionic interactions with water are finely balanced, and that with the polyols the nearest neighbour oxygen–oxygen distances correspond to those of oxygen in water. They point out a close identity of these molecules to the 'salting out' (macromolecule stabilising) compounds described by other workers (e.g. [205]). (See [33, 203] for further details.)

An aspect of cell osmotic adjustment familiar but far from understood is the action of the contractile vacuoles. A range of unicellular freshwater plants and animals possess this organelle that has the capacity of absorbing water from the surrounding cytoplasm and releasing it from the cell by exocytosis as a hypotonic solution. [137] This amounts to a water pump. The current perception of the mechanism appears to be that the vacuoles are filled osmotically by solute transport to generate an osmotic gradient. The solutes are then removed from the vacuole, leaving a hypotonic medium behind to be released by exocytosis. This is highly improbable (although see the description of a water pump across composite membranes in section 3.5.2). Water crosses membranes far more readily than the likely relevant solutes do even in such 'impermeable' membranes as those of the 'water hardened' trout egg. [27] Water would leave the vacuole simultaneously with the abstracted solutes. Some workers have commented on architectural features, such as the saccules of *Ochromonas danica* (see [40]) as being possibly significant.

The role of the cell wall in turgor control Cell wall properties are central to turgor pressure regulation in several ways. We have already seen how wall elastic properties influence cell turgor under conditions of changing water potential. In a subsequent section (section 3.5.3) we shall see that wall plastic properties play a role in determining turgor pressure under growing conditions.

A largely ignored or misunderstood aspect of the cell wall, however, has been the role of osmotic solutes contained within it. We must distinguish here between solutes held by Donnan forces and those not. Donnan solute distribution between DFS and WFS will not influence turgor pressure. While they will increase total osmotic pressure, the influence of the Donnan solutes on wall water potential will be exactly balanced by the Donnan hydrostatic pressure referred to above. Conversely the Donnan pressure in the wall will not influence turgor pressure as it, in turn, is exactly balanced by the osmotic pressure that causes it. (Any separation of these two phenomena will result in a perpetual motion machine pumping water down a pressure gradient.) That this simple relationship has become caught up in the mixture of different phenomena referred to as 'matric potential' is unfortunately and unnecessarily confusing. (This aspect is included in a wider critical discussion on matric potential by Passioura. [111])

'Non-Donnan' solutes, on the other hand, will influence cell turgor pressure. However, the levels of such solutes in walls have been difficult to assess. Indeed it is conventional to assume that apoplast solute concentrations are negligible (e.g. [206–9]). This assumption is based on observations that xylem sap pressed from material in the pressure bomb has a low osmotic pressure. It may be, however, that xylem sap is not representative of cell wall solution, i.e. that the apoplast is not osmotically uniform. It also appears that xylem sap osmotic pressure varies considerably from species to species.

Attempts have been made to measure cell wall inorganic ion contents by X-ray microanalysis (e.g. [210, 211]). These values, however, include both Donnan and non-Donnan components of the wall and are of little help in determining the relevant 'active' solute component.

Elution techniques are more likely to provide the relevant information, but in interpretation of such experimental results care must be taken to quantify material released from the Donnan distribution by the treatment or from inside the cell by changing the conditions of the wall. Bernstein [212] using castor bean, cabbage and sunflower leaves and Jacobson [213] using Venus' flytrap indicated that the osmotic potential in these plants might be in the range of 0.1–0.2 MPa. More recently Cosgrove & Cleland [214] using various forced perfusion techniques estimated that in etiolated pea internodes wall osmotic pressure is 0.29 MPa for apical segments and 0.18 MPa for basal segments. Similar results were obtained from stem regions of etiolated soybean and cucumber. This suggests not only that wall osmotic pressure may be considerable, but that stable gradients may occur within tissues. Based on conductivity and refractive index measurements before and after ashing Cosgrove & Cleland [214] suggested that 25% of the osmotic pressure was due to inorganic electrolytes and 75% to organic non-electrolytes with an average size similar to that of glucose. Using the pressure probe Leigh & Tomos [215] indicated that the turgor pressure of the cells of red beet taproot (under non-transpiring and well watered conditions when tension due to transpiration should be negligible) was 0.4–0.5 MPa lower than the full turgor expected under such conditions. Crude efflux analysis suggested that sufficient potassium ions might be present in the wall to account for the discrepancy. In *Kalanchoë daigremontiana* the turgor pressure measured directly was lower than the osmotic pressure. [156]

Also using the pressure probe, Clipson *et al.* [174] showed that the turgor pressure of leaf epidermal cells of the halophyte *Suaeda maritima* is only a small fraction of that expected from cell sap osmotic pressure. In 400 mol m^{-3} sodium chloride the cells of mature leaves of transpiring plants have sap osmotic pressures of approximately 960 mosmoles kg^{-1} (2.5 MPa) but turgor pressures in the order of 0.1 MPa. When wall hydrostatic tension is reduced by stopping transpiration by immersing the entire seedling in the hydroponic medium turgor pressure increases by only another 0.1 MPa (figure 3.6). The low turgor pressure can, therefore, only be due to wall osmotic pressure in the range of 2–3 MPa. By measuring sodium uptake and transpiration rates Clipson & Flowers [216] have estimated xylem sap concentrations under similar conditions and suggest values of 113 mol m^{-3} (equivalent to 0.6 MPa) and 58 mol m^{-3} (0.3 MPa) for night and day conditions, much lower than suggested for the leaf apoplast. Other estimates of xylem sodium concentrations for halophytes vary. Rozema, Gude, Bijl & Wesselman [217] calculated mean values between 1.3 and 15.1 mol m^{-3} for four species of plants without secretory glands grown in 200 mol m^{-3} sodium chloride. Ownbey & Mahall [218] on the other hand calculated higher values for *Salicornia virginica* in 210 and 530 mol m^{-3} sodium chloride (19.2 and 37.9 mol m^{-3} respectively).

While halophytes growing in high concentrations of salt are probably not good models for general descriptions of apoplast water relations they clearly show that apoplast osmotic pressure cannot be ignored.

(At the other end of the spectrum, however, using the criteria outlined

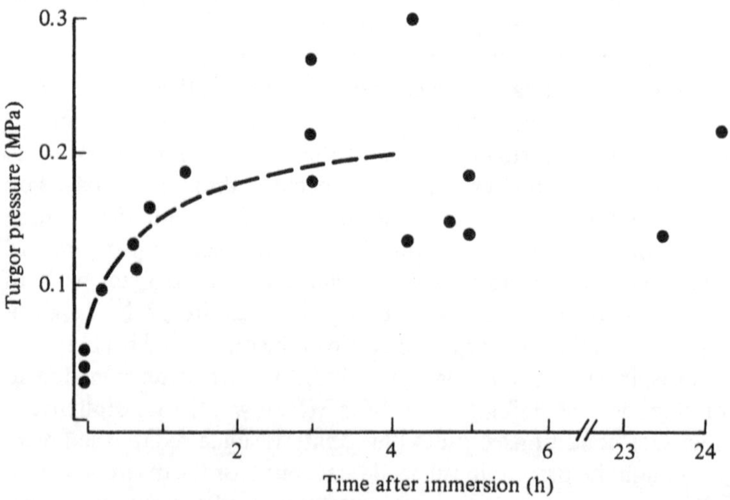

Fig. 3.6. Increase in turgor pressure of leaf epidermal cells of *Suaeda maritima* following immersion in its hydroponic bathing medium to stop transpiration. (Cell osmotic pressure, 1.4 MPa; hydroponic medium osmotic pressure <30 kPa.)

above, pressure probe experiments on wheat seedlings suggest that the apoplast of leaf epidermal cells of seven-day-old plants growing in Hoagland's solution contain negligible osmotically significant solutes. Leaf cell turgor pressure measured with the pressure probe (table 3.2) corresponded to their osmotic pressures.)

In cells with very low water potentials, such as salt adjusted halophytes or zeromorphs, it makes a great deal of sense to have high solute concentrations in the wall. If the wall water potential were maintained at a low level purely by transpiration pull, events such as occasional rainstorms that would stop transpiration and allow the cells to approach full turgor would probably result in the cells bursting. A similar situation is seen if salt adjusted *Suaeda maritima* leaves are vacuum infiltrated with dilute solutions or distilled water. Here the cells simply burst. [219] In the field the thick wax cuticle prevents rainwater getting in to the leaves and the predominance of an osmotic rather than a hydrostatic water potential gradient through the plant protects the leaf cells from excess pressure. It is paradoxical that the wax coat of halophytes may well have more to do with keeping water out of the plant than to keeping it in. It emphasises, however, that too *high* a turgor pressure may be just as much a physiological problem as one which is too low. This may also be the case in sugarbeet taproot cells during sucrose accumulation. The stability of osmotic potential gradients within the apoplast is of great interest in this context. Presumably a small energy input at some transport site will be sufficient under many conditions due to diffusion being such a slow process over distances of magnitude much larger than a single cell. Water movement down an osmotic water potential gradient will also be slow in the wall owing to the low value of the reflection coefficient in this pathway (see [12, 13, 30]). In sections of apoplast influenced by significant transpiration flow the situation is likely to be more complex.

It has been suggested that control of cell turgor pressure by modulation of the wall osmotic pressure can be much more effective than modulation of the protoplast osmotic pressure. [40] The rate of change of solute concentration in any compartment will be the flux rate into or out of that compartment *divided by the compartment volume*. Since the cell wall has a much smaller volume than the protoplast, a solute flux across the plasmalemma will result in a far faster change in wall osmotic pressure than would take place in the protoplast. Turgor pressure is determined by the differential osmotic pressure, hence its behaviour during the plasmalemma flux will be largely determined by the wall parameter. Observations consistent with this are presented in figure 3.7, which illustrates turgor adjustment in red beet tissue under two different conditions. These are (*a*) an 'open' case where excised tissue is bathed in a large external volume [410] and (*b*) a 'closed' case where freshly cut tissue was placed under silicone oil. The latter resembles the situation *in vivo* with a limited external space, while in the former the apoplast is in close diffusive contact with a large bulk volume.

Table 3.2. *Recent measurements of physiological turgor pressure (P), volumetric elastic modulus (ε) and hydraulic conductivity of cell membranes of higher plants. Measured with the pressure probe. For previous data see [23]. (Here f(P) and f(V) are functions of turgor pressure and volume respectively;* L_{pen} *and* L_{pex} *are endosmotic and exosmotic* L_p*s.)*

Species	Physiological P (MPa)	ε (MPa)	$L_p \times 10^6$ (m s^{-1} MPa^{-1})	Ref.
Mesembryanthemum crystallinum				
(bladder)	0.1–0.4[a]			[402]
(stems)	0.3			[392]
(petiole)	0.19 (at 10 wks; at 16 wks diernal rhythm with CAM)			[392]
(leaves)	0.04			[392]
Capsicum annuum				
(fruit)				
(giant cells)		0.1 ($P = 0.1$ MPa) ($\varepsilon = f(V)$)	5.8 (influenced by pollutants)	[145, 299]
		27.0 ($P = 0.41$ MPa)		[145, 299]
(normal cells)		0.1 ($P = 0.22$ MPa)	0.21	[145, 299]
		0.6 ($P = 0.4$ MPa)		[145, 299]
Oxalis carnosa				
(bladder cells)				
upper	0.17	0.19–1.7 ($\varepsilon = f(P)$; $\varepsilon \neq f(V)$)	0.38	[403]
lower	0.25	1.8–16.6 ($\varepsilon = f(P)$; $\varepsilon = f(V)$)	2.30	[403]
Tradescantia virginiana				
(epidermal intact)	0.35	4.0–36.0	0.02–1.1 (E_a = 50–186 kJ mole^{-1})	[9]
	0.41	4.0–24	0.014–0.55	[404]
		0.3–35 ($\varepsilon = f(P)$)	0.24–1.34 ($L_p = f(P)$)	[151]
(epidermal strips)		1–10 ($\varepsilon = f(P)$)	0.064 ($L_p \neq f(P)$; $L_p \neq f(\pi_0)$)	[154]
(subsidiary intact)	0.24	3.2–20.1	0.022–0.35	[404]
(mesophyll cell)		0.64–1.4	0.036–0.06	[404]
Chenopodium rubrum				
(parenchyma)	(0.41)	1.2–3.7	0.07–0.20	[147]
(cell suspension)		1.1–4.8 ($\varepsilon = f(P)$)	0.05–0.19	[147]
(gel embedded)		2.1–3.2	0.10–0.14 (no unstirred layer)	[147]

Kalanchoë daigremontiana						
(mesophyll)		1.3–12.8		0.2–0.88		[156]
Suaeda maritima						
(epidermal immature)	0.25–0.4				(independent of π_i)	[174]
(mature)	0.05				(independent of π_i)	[174]
	0.03–0.49	1.4–20.4	($\varepsilon = f(P)$)	0.03–0.43		[148]
Rhoeo discolor						
(epidermal)	0.16	4.3		0.18	(increased by ABA)	[275]
Triticum aestivum (root)						
(mature cortex)	0.68	7.6	($\varepsilon = f(V)$ or age)	0.12	($L_{pex}/L_{pen} = 0.9$)	[10]
	0.31–0.5	1.5–30.5	($\varepsilon = f(P)$)	0.08–0.22		[11]
	0.3–0.74[a]	3.0				[16]
(elongating cortex)	0.65	1.7–2.1[a]		0.45		[16]
(mature epidermal)	0.55	4.5	($\varepsilon = f(V)$ or age)	0.12	($L_{pex}/L_{pen} = 1.0$)	[10]
(mature hair cell)	0.44	4.4	($\varepsilon = f(V)$ or age)	0.10	($L_{pex}/L_{pen} = 1.1$)	[10]
(leaf epidermis)	0.9–1.2					[405]
Lolium temultenum						
(expanding leaf cells)	0.5[c]					[351]
Sinapis alba						
(seedling stem)	0.45–0.52[c]					[352]
Beta vulgaris						
(leaf epidermis)	0.4					
(leaf vein parenchyma)	0.4					
(leaf mesophyll)	0.43					
(leaf vein epidermal)	0.34					
(petiole epidermis)	0.41–0.59[b]					
(tap root)	0.6–0.8[b]	3.0–10.0	($\varepsilon = f(P)$)	0.21		[411]
Soybean (hypocotyl)						
(elongating epidermis)		0.38		6.1		[12]
(elongating cortex)		2.7		2.7		
(mature epidermis)		0.33		4.4		
(mature cortex)		4.4		2.8		
Maize leaves (midrib)						
(adaxial epidermis)	(0.68)	4.8		0.4		[13]
(adaxial parenchyma)	(0.69	10.3		0.3		
(mid epidermis)	(0.75)	3.4		1.4		

Table 3.2 (*continued*)

Species	Physiological P (MPa)	ε (MPa)	$L_p \times 10^6$ (m s^{-1} MPa^{-1})	Ref.
(mid parenchyma)	(0.75)	6.7	2.0	
(adaxial epidermis)	(0.68)	3.0	1.5	
(adaxial parenchyma)	(0.69)	4.1	2.5	
Pisum sativum (*epicotyl*)				
(epidermal)		1.2–20.0	0.02–0.2	[268]
(cortical)		0.6–21.5	0.04–0.9	[268]
	0.46			[406]
	0.62			[336]
Elodea densa				
(upper epidermis)	0.6–0.8	1.0–15.0 ($\varepsilon = f(P)$)	0.78 ($P > 0.4$ MPa) ($L_p = f(P)$)	[150]
Apple fruit				
(unripe)		2.5–6.0 ($\varepsilon = f(P)$; $= f(V)$)	0.3–0.34	[152]
(ripe)		8.8–18.2 ($\varepsilon = f(P)$; $= f(V)$)	0.1–0.11	[152]
Salix babylonica	0.51–0.93		(aphid stylet)	[373]

[a]Function of nutrients.
[b]Function of transpiration rate.
[c]Independent of growth rate.

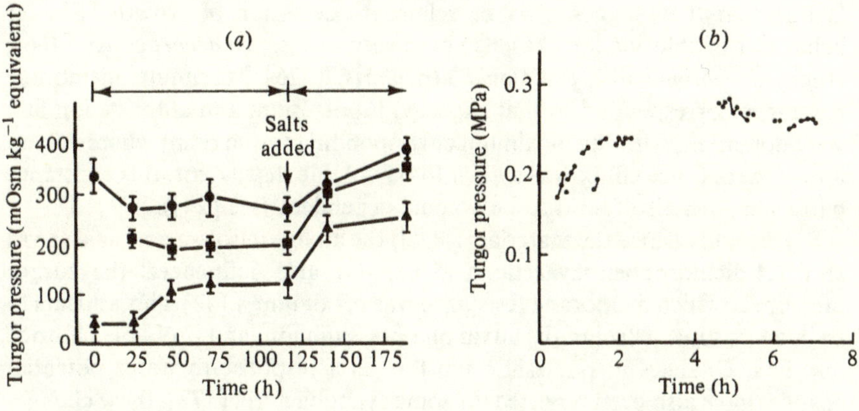

Fig. 3.7. The time scale of turgor adjustment in excised red beet root tissue. (*a*) 'Open system'. [410] (*b*) 'Closed system'.

Cosgrove, Ortega & Shropshire [220] have recently demonstrated the presence of solutes in a separate wall compartment between the cuticle and plasmalemma of the giant-celled sporangiophore of *Phycomyces blakesleeanus*. This compartment contains a solute concentration equivalent to an osmotic pressure of between 0.5 and 1 MPa. Cosgrove *et al.* [220] suggest a turgor control mechanism along the lines outlined above.

An exotic osmotic modulating mechanism has been suggested for the cell walls of heat resistant bacterial endospores. [221, 222] Here it is suggested that the Donnan pressure (of some 3 MPa) set up in the peptidoglycan polymer is directed towards an expansion of the wall into the protoplast. The result of this is to dehydrate the protoplast by reverse osmosis. Other workers in this field have suggested that ionic changes may result in the contraction of the entire wall to produce the dehydration. [222]

Such an active role of the wall has yet to be seen in any eukaryote, although it is reminiscent of the discussion of changing values of the elastic modulus noted above.

Turgor sensitive responses Reviews of early observations of turgor dependent processes are included in Zimmermann [24] and Gutknecht *et al.* [19] As such work is dependent on knowing cell turgor pressure most of the processes described to date are from algal systems. Probably the most widely studied system is that of potassium ion transport across the plasmalemma of *Valonia* species. Gutknecht [223] used perfused cells to show that potassium ion influx decreased significantly when turgor pressure was raised from 0 to 0.1 MPa. Potassium ion efflux on the other hand increased only slightly. Zimmermann, Steudle and coworkers [22, 224, 225] subsequently extended this work to higher turgor pressures, and also showed a cell volume dependence of the function of potassium ion efflux and turgor (a feature that they suggest may be related to cessation of growth). Similar behaviour of chloride ions has also been seen in *Codium decorticatum* [166], *Halicystis parvula* [140], *Valonia utricularis.* [226] Maximum membrane resistance corresponded to that expected if potassium and chloride ion flux were dominant, with the maximum corresponding to the point where net ion flow is zero. Since efflux was shown to be volume dependent, this resistance parameter was also found to be volume dependent (see [24]).

Surprisingly (since the material is algal) the auxins indole-acetic acid (IAA) and 2,4-dichlorophenoxyacetic acid (2,4-D) also influenced the turgor pressure at which membrane resistance was maximum. [142] This appears to be due to a direct effect of the auxin on potassium ion (and probably chloride ion) flux. Changes in potassium ion flux as a response to water potential changes have also been reported for some cyanobacteria [172], these changes precede the longer-term osmotic adjustment brought about by accumulation of low molecular weight organic solutes. [172]

Zimmermann & Beckers [227] have subsequently shown the turgor pressure stimulation of the generation of action potentials in *Chara corallina.* The resting membrane potential of about −160 mV is lost in less than 1 s when pressure pulses are injected into the internode with the pressure probe. Similar action potentials have been observed in the giant-celled marine green alga *Acetabularia mediterranea.* [228] In this alga turgor pressure is not held constant but regularly oscillates (at 23 °C) between 0.22 and 0.26 MPa with an interval from 16 to 28 min. The turgor pressure is generally rising, but at a crucial value an action potential occurs and the turgor drops fairly rapidly to the lowe.: end of its range, whence it begins again to rise. Mummert [229] measured the amount of chloride ion released during these action potentials and concluded that they are sufficient to account for the turgor drop (about 4% of the cell content). The relationship between chloride ion flux and turgor control is not simple, however, as turgor adjustment was occasionally seen in the absence of an action potential, and action potentials that were not followed by turgor adjustment were common.

In higher plants such experiments are technically more demanding, although much interest has been given recently to turgor sensitive processes here also. Several workers have observed an influence of external osmotic pressure on the transport of certain solutes across cell membranes.

Marre, Lado, Rasi Caldogno & Colombo [230] suggest that water stress increased auxin induced hydrogen ion excretion from pea stems. Conversely, Cleland [231] demonstrated that auxin-stimulated acid efflux from *Avena* coleoptiles was inhibited by external mannitol concentrations, while mannitol increased hydrogen ion excretion in the absence of hormone. (Cleland [231] comments that since the plasmalemma ATPase responsible is totally insensitive to auxin at incipient plasmolysis, *in vitro* experiments to detect activity on broken membranes may not be successful.) More recently Van Volkenburgh & Boyer [232] while studying the effect of water stress on cell growth have suggested that the proton efflux associated with acid wall loosening in maize leaves may also be turgor dependent. Similarly acidification of the medium by bean leaf tissue is also modulated by mannitol. [233, 234]

Smith & Milburn [235] provided evidence to suggest that the loading of phloem in *Ricinus communis* may be controlled by turgor pressure. Enoch & Glinka [236, 237] have shown a turgor dependent efflux of potassium ions from carrot tissue. Using excised seed-coats and cotyledons of developing seeds of *Pisum sativum* and *Phaseolus vulgaris* [238] several groups [239–44] have shown that sucrose and amino acid uptake and efflux by the cells in these tissues are influenced by the osmotic pressure of the bathing medium (modulated with mannitol). Wolswinkel and colleagues [240, 244] suggest that the basis of this is a turgor dependent sucrose and amino acid transporter in the tissues.

Net efflux from the seed-coat is reduced by 250 and 400 mol m^{-3} mannitol solutions relative to that in 100 mol m^{-3} mannitol. Conversely net uptake of labelled sucrose and valine by the seed is stimulated by a high extracellular osmotic pressure. [244] Similar data has been obtained for uptake into leaf segments [234, 245] and wheat endosperm slices. [246]

Working with excised sugarbeet tissue Wyse, Zamski & Tomos [247] showed two components to sucrose uptake into the taproot cells; one, saturating at an external sucrose concentration of about 20 mol m^{-3}, the other being non-saturable. The saturable component only occurs in the presence of a high external osmotic pressure due to the non-penetrating solute, mannitol. The penetrating solute, ethylene glycol, is ineffective. The saturating component is inhibited by both *p*-chloromercurylbenzoyl sulphonate (PCMBS) (an inhibitor of active sites containing thiol groups) and carbonylcyanide-*m*-chlorophenyl hydrozone (CCCP) (an uncoupler of oxidative phosphorylation) indicative that it is an active process.

Reinhold, Seiden & Volokita [245] and Hagege *et al.* [234] for leaf tissue and Wyse *et al.* [247, 414] for sugarbeet taproot report evidence that active proton extrusion, which promotes the proton motive force to power other solute uptake, may be the fundamental turgor pressure sensitive event. In the case of *Vicia faba* leaf fragments [234], *Senecio mikanioides* leaf segments [245] and excised sugarbeet taproot tissue [247] a maximum acidification of the external medium is observed at the turgor pressure levels that correspond to those found in intact plants. As the turgor of the excised tissue approaches 'full turgor' acidification is almost totally abolished (figure 3.8).

Palta, Wyn Jones & Tomos [248] have followed the turgor pressure of intact sugarbeet root cells during the day–night rhythm of transpiration in sugarbeet (figure 3.9) and find that if the acid efflux behaviour of the excised tissue were to be manifested in the intact plant, then the range of *in vivo* pressures would result in considerable changes in the wall pH. Experiments to see if this is indeed the case are currently underway in our laboratory.

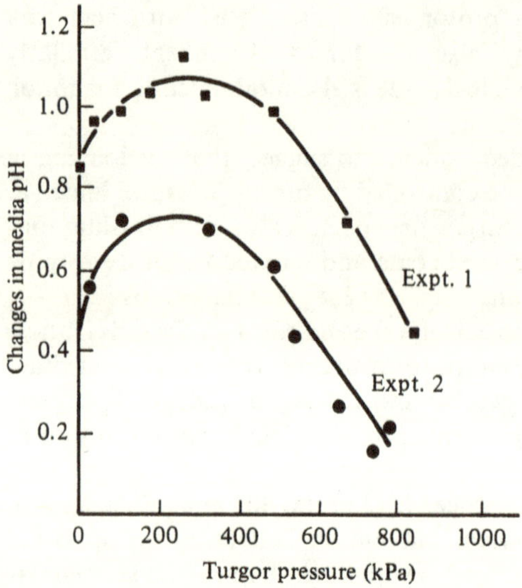

Fig. 3.8. Effect of bathing medium osmotic pressure on acid efflux from excised sugarbeet root tissue. The results from two typical experiments are shown. [247]

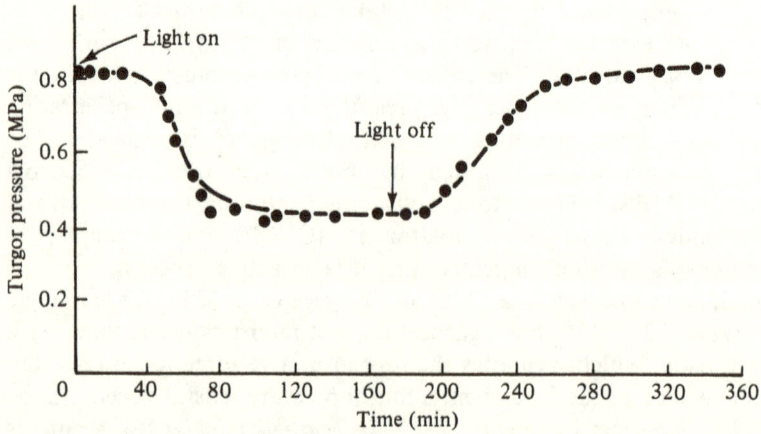

Fig. 3.9. Turgor pressure in intact sugarbeet root cells under illumination and darkness to simulate day/night conditions. [248]

If the control of proton efflux (presumably controlled by a proton/ATPase) does prove to be turgor sensitive in many situations then it may well prove central to the long-range integration of cells and tissues as it appears to be involved in so many other processes from growth to solute transport. Chalmers, Coleman & Walton [249], using carrot suspension cells suspended in solutions with a range of osmotic pressures, have employed an electrochemical technique (a ferricyanide–ferrocyanide couple) to show that a transmembrane redox reaction of these cells also appears to be turgor sensitive.

These turgor sensitive processes may be only a special case of a wider range of mechanosensory mechanisms in plants, including the rapid movements of carnivorous plants, a review of which may be found in Bentrup. [250]

Turgor sensing mechanisms The wealth of evidence for turgor (or volume) dependent processes asks the question of the nature of the mechanism of sensing the stimulus. Some of the general concepts involved have been reviewed extensively by Gutknecht *et al.* [19] and Zimmermann. [24, 251] The former propose a model by which compression of the plasmalemma against the pores between the cellulose microfibrills of the wall causes local stretching and tension in the membrane. This tension is the sensed quantity. The latter propose an electromechanical model for the membrane. In this the membrane is under considerable compression due to both the enormous electric field across it (of the order of $10^7\,V\,m^{-1}$) and the hydrostatic mechanical compression. Thinning of the membrane can occur as a result of direct compression or as a result of stretching of the membrane as the wall extends elastically. (An important aspect of the electromechanical model is that it clearly shows that electrical and mechanical phenomena are interrelated in this field.) The membrane is surprisingly compressible. In algal cells, animal cells and in those of bacteria values for the elastic compressive modulus of the order of 1–8 MPa have been obtained at physiological pressures. [252, 253] However, the hyperbaric experiments on osmotic adjustment in giant-celled algae referred to above [19, 166, 196], that show no effect of absolute pressure in a physiologically significant range, strongly suggest that it is not by straight compression that turgor is sensed.

The degree of thinning due to stretching of a membrane will be dependent on the elastic modulus of the cell wall (a wall with an infinite value for ε would allow no stretching at all).

Zimmermann [24] suggests other candidates for turgor pressure sensors. Charged polyelectrolyte structures, such as those found in cell walls, are potential sites for electromechanical transduction. Piezoelectricity is yet another electromechanical process that could be used by the cell, and such a phenomenon could be related to a permanent dipole moment within a compressible matrix. It is noteworthy, therefore, that proteins in membranes are very rich in alpha-helical sequences (40%) that are particularly strong dipoles. [254] Each peptide bond has a dipole of 3.63 debye, one twist of a helix approximately 13 debye, and a helix scanning the membrane a dipole of approximately 130 debye.

Figure 3.10 illustrates the arrangement of such helices in the transporter bacteriorhodopsin of bacterial purple membranes. [255] Current work on transport proteins from higher organisms suggests that such arrangements of alpha helices normal to the plane of a membrane may be common

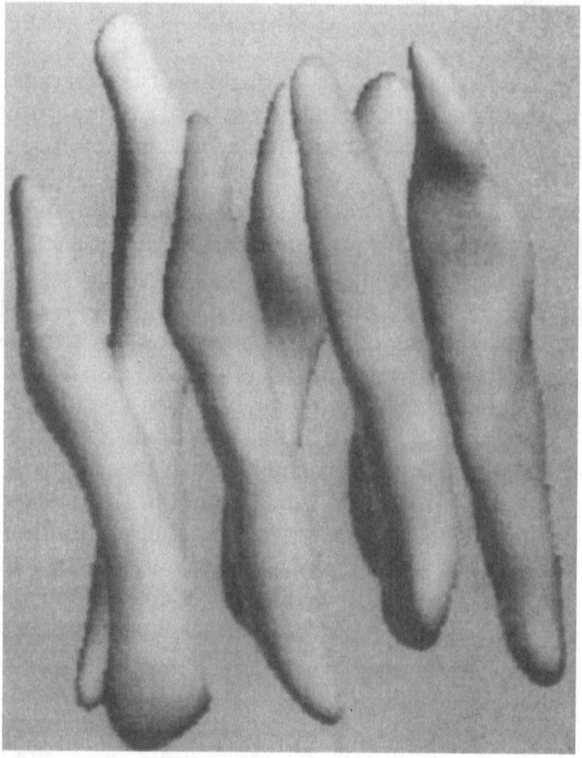

Fig. 3.10. A computer simulation of the alpha-helical portions of bacteriorhodopsin, a membrane protein. The helices lie slightly off normal to the membrane. [255] Reproduced from *Scientific American* **250** (February 1984), 56–66 with permission.

components. If this proves to be the case then transport and turgor sensing properties could reside in the same molecule.

One physical membrane process does appear to have been shown to be influenced by turgor pressure in the relevant range. Zimmermann and colleagues [253, 256, 257] have detected the presence of mobile charges in the membranes of *Valonia utricularis*. This phenomenon is shown by measuring the capacitance of the cell membranes following pulse charge experiments. Increase of turgor pressure from 5 kPa to 200 kPa increases the translocation rate constant of these charges by a factor of 2, and their total surface density by 30%. Something of the nature of the charges may be extrapolated from the fact that they (reversibly) disappear when the pH is lowered from 8.2 to 4 or 5, presumably due to protonation of anionic groups. [253] The protein synthesis inhibitor cycloheximide slows down the translocation rate, but leaves the concentration of charges unchanged. Büchner, Rosenheck & Zimmermann [257] interpret this as indicating that the charged moieties are coupled to, but not part of, a carrier protein. Such mobile charges are yet to be related to any 'macroscopic' transport process and also are yet to be demonstrated in another species. Based on the hyperbaric experiments of Hastings & Gutknecht [196] it is widely assumed that the most likely site for a turgor sensor is in the plasmalemma or cell wall. The response of the mobile charges to pressure changes, however, is found in both tonoplast and plasmalemma [257]. Either the interpretation of the data is incorrect or this is evidence that the tonoplast is capable of acting as the site of the sensor.

Lucas & Alexander [258] have indicated an important feature of some of these experiments and theories of turgor sensing mechanisms. In studying the control of bicarbonate and hydroxide ion transport in *Chara corallina*, they found an effect following plasmolysis of the cells but not following cytorrhysis (when the cells were air dried to reduce turgor pressure). The authors emphasise that cytorrhysis is a far more common phenomenon in nature than plasmolysis (cf. [208]) and that the 'turgor' effects on hydroxide ion transport observed was an artefact of plasmolysis. They point out that, in general, the membrane will be compressed against the wall at all times, even at reduced (or even zero) turgor pressure. This would tend to invalidate the stretching model of Gutknecht *et al.* [19] and the purely compressive component of the electromechanical model. [23] While this may well be true for water stress and the diurnal changes in turgor due to transpiration pull [14, 246], it is not true for the situations where assimilates or ions in the cell wall are controlling turgor. Here changes in direct membrane compression against the wall will occur. This will be the case for the proposed roles of the turgor sensing mechanism in the control of assimilate uptake etc.

Negative turgor pressures A question that is raised in several contexts is that of the existence of 'negative pressure' in cells. That hydrostatic tension equivalent to a 'negative pressure' of many MPa can be set up in a body of water, even in the presence of dissolved gases, has been demonstrated empirically many times (see [7] for a review). In general these studies are in the context of xylem sap. Degrees of tension can be unequivocally shown for the annulus cells of some fern sporangia that use the rapid release of the tension by cavitation as a mechanism of powering spore dispersion (see [7]). The distribution of such a characteristic amongst 'conventional' tissue cells is not clear. Claims of 'negative' turgor based on measurements with the pressure bomb have been questioned by Tyree [260] who suggests that the associated measurements of cell sap osmotic pressure are underestimates due to dilution with apoplasmic water on extraction.

However, there appears to be no reason why 'negative' turgor cannot be established in cells as long as the cell walls are sufficiently rigid to prevent collapse (cytorrhysis). As recently argued by Lucas & Alexander [258] and Oertli [208] under ordinary conditions, the hydrostatic pressure in the cell wall will always be lower than that in the protoplast. The plasma membrane will always be pressed firmly against the wall and there is no requirement of evoking adhesion of the membrane to the wall. (The exception will be under fluctuating saline conditions where natural plasmolysis might occur.) Sufficiently strong cells adjacent to apoplast containing water under tension will therefore be expected to develop 'negative' turgor pressures if the internal osmotic pressure is not high enough. Oertli [208] argues that this will be the case in sclerophyllous cells of xeromorphic plants. This should be amenable to test using the pressure probe.

Tyree & Richter [261] indicate that evidence for 'negative' pressure may be obtained independently of the assumptions of sap osmotic pressures from pressure/volume curves. They show that 'negative' turgor pressures of −0.2 MPa or more will produce significant curvature of the water potential isotherm at about −2 MPa. Thus 'negative' turgor could contribute up to 10% measured water potential without being detected.

The determination of turgor pressure, negative or otherwise, from indirect measurements of osmotic or water potentials is complicated by another feature of the cell wall – its solute content.

Uniformity of cells Tomos & Wyn Jones [131] have recently considered some of the implications of cell to cell inhomogeneity in the context of transport processes and have come to the conclusion that the general tacit assumption of some degree of uniformity within anatomically homogeneous tissue is likely to be correct, although this remains far from proven. This consideration is important when an attempt is made, as in this review, to draw on observations gained from measurements on tissues or organs to build up an understanding of single cell water relations. The whole is certainly more than the sum of the parts, the question is how uniform are the parts?

When considering water relations we are greatly helped by the high conductivity of membranes to water which is generally taken to mean that a cell is in water potential equilibrium with its immediate wall. This local equilibrium is discussed further by Molz & Ferrier. [262] Similarly the assumed high conductivity of the wall is taken to mean that adjacent cells are not likely to be too far from hydrostatic pressure equilibrium, although this may not be so over long distances (see section 3.4.1). As we have seen deviations of the reflection coefficient of membranes from unity will have an influence on these assumptions. More importantly standing osmotic gradients within the apoplast will certainly influence uniformity.

Current techniques of X-ray microanalysis do not appear to be able to measure cell contents with sufficient precision to allow definitive descriptions of the osmotic pressures of cells within tissues. At worst, some such data presented totally ignores the improbable osmotic consequences of the reported solute distribution. The problems of allocating osmotic roles to ions visualised in the X-ray microprobe cannot be overestimated.

The use of the pressure probe allows us at least to look at turgor pressure uniformity between adjacent cells, and leaves to others the problem of drawing up osmotic pressure balance sheets. Figure 3.11 illustrates the range of turgor pressures measured around stomatal complexes from two different plants of *Tradescantia virginiana*. It can be seen that uniformity exists within the epidermal cells (to within the practical resolution of the equipment). The range of turgor pressure of 23 kPa (which was within the experimental error in this case) corresponds to a difference in osmotic pressure of only 8 mosmoles kg^{-1}. The stomatal subsidiary cells represent another tissue and have another turgor pressure. Figure 3.12 illustrates the turgor pressures of successively deeper cells in the pea internode. [214] Here again turgor pressure uniformity is strongly suggested. On the other hand apparently uniform tissues may certainly have dramatic turgor pressure differences from cell to cell. Following the application of 25 mmol m^{-3} abscisic acid (ABA) to wheat roots for 24 h, adjacent cells in the cortex developed dramatically different turgor pressures. [263] Indeed, the epidermal cells in the tissue

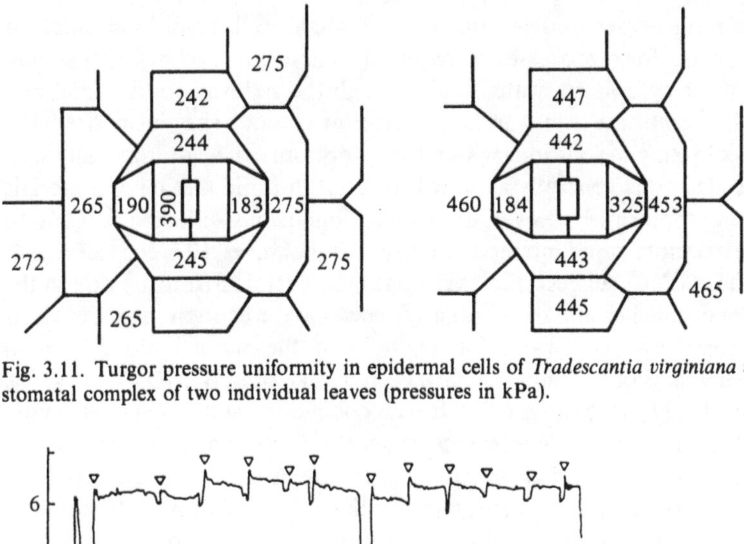

Fig. 3.11. Turgor pressure uniformity in epidermal cells of *Tradescantia virginiana* around the stomatal complex of two individual leaves (pressures in kPa).

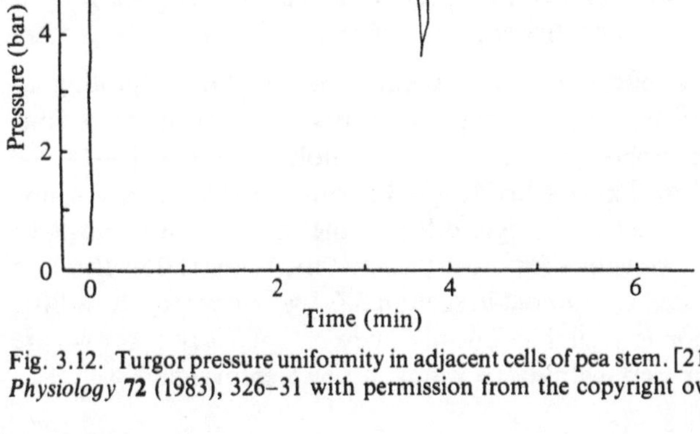

Fig. 3.12. Turgor pressure uniformity in adjacent cells of pea stem. [214] Reproduced from *Plant Physiology* **72** (1983), 326–31 with permission from the copyright owner.

appear to have consistently lower turgor pressures than the underlying cortical cells [10, 263], although this was not seen by Steudle & Jeschke. [11]

Turgor uniformity between adjacent cells is probably due to the presence of plasmodesmata, although at present only theoretically derived values are available for plasmodesmatal hydraulic conductance. This is considered in section 3.5.2. If the symplast as a whole does represent a single uniform hydrostatic unit, then in the presence of external osmotic pressure gradients either corresponding internal gradients must exist (or, more unlikely, gradients of reflection coefficient) or complex circulatory systems will occur within the apoplast/symplast geometry. Such a situation is reminiscent of the original symplast hypothesis of Münch [264] before it was restricted to phloem transport. Current developments with the pressure probe, involving the measurement of osmotic pressures from material obtained from single cells of known turgor should solve this question.

3.5.2 Dynamic situations ($J_v \neq 0$)

In considering water movement in our single cell resolution analysis information on four processes is required. These are cycling of the bulk solution of the cytoplasm, water flow through the cell wall (both radial and tangential), membrane transport and water flow through the plasmodesmata.

The bulk cycling of solutions within the cytoplasm can be dramatically seen in many cells and is generally assumed to result in rapid mixing of materials within the cytoplasm. However, at cellular dimensions diffusion is likely to result in even more rapid mixing of diffusable substances. [27, 265] Lüttge & Higinbotham [266] suggest, however, that cytoplasmic tortuosity due to the presence of organelles and membrane systems may be enough to give cyclosis a role. Streaming rates vary, for example in the slime mold *Physarum polycephelum* rates of 1.35 mm s^{-1} have been observed. In *Nitella*, 50 $\mu m\ s^{-1}$ is more typical. [27] It is often difficult to reconcile the static image of highly organised endoplasmic reticulum seen under the electron microscope with the rapid cyclosis seen in the light microscope. Little attention has been paid to cytoplasmic streaming in the study of cellular water relations although the observations that metabolic inhibitors influence water transport through root cortical sleeves (i.e. without a functional endodermis) have been presented as possible evidence of a role at a tissue level. [267]

Tissue pathways Much effort has been spent, however, in attempting to describe water flow through tissues, especially roots, in terms of the various pathways; apoplast, symplast and vacuole to vacuole pathways. [9–14, 31, 268] Molz & Ferrier [262] have provided a blueprint for interpreting whole tissue behaviour on the basis of single cell parameters. The problem is to provide useful operational values for those parameters. Volume flow through cell wall material has been considered in section 3.4.1 where we saw how little quantitative information is available. The parameters that are best known are those that describe the permeability to water of the membranes.

Membranes and plasmodesmata Membrane permeability to water can be measured in two different ways. Firstly as a derivative of L_p, in which case it is the permeability to water driven by a hydrostatic or osmotic gradient. This filtration, or osmotic, permeability (P_f) is calculated from L_p by the equation:

$$L_p RT/\bar{V}_w = P_f \tag{3.42}$$

which is based on the assumption that water diffuses across membranes, i.e. the membrane has no water filled pores. We shall return to this definition later in considering evidence for and against such pores. In general, however, with bulk flow we are concerned with L_p rather than P_f.

L_p has been estimated by a number of techniques (see appendix). Transcellular osmosis, perfusion and plasmometry have provided several values of L_p in special cases [19, 23], but most values of L_p for both algae and higher plant cells have now been provided by the use of the pressure probe (tables 3.1 and 3.2).

The pressure probe is proving to be a useful tool in this respect, being relatively (but far from totally) free from uncertain assumptions. One problem is the definition of precisely which pathway is being measured. In the majority of cases the tip of the pressure probe is inserted through both plasmalemma and tonoplast. When pressure relaxation curves are initiated they may result in water flow from or to the vacuole by various pathways. The only certainty is that water must cross the tonoplast. It remains uncertain whether much flow by-passes the plasmalemma and wall through the plasmodesmata. The remaining water crosses the plasmalemma into the wall. The problem is not fatal to the technique since regardless of pathway the L_p measured for the cell membranes by the pressure probe represents the relevant value for the cell *in situ* (Steudle, personal communication).

The assumption that plasmodesmata are conductive to water is central to the symplasmic concept of cytoplasmic continuity. Recent experiments to determine the molecular exclusion limit of the pores by injecting peptides and dyes would suggest that the pores are certainly open to bulk flow (see [269, 270]). The precise hydraulic conductivity of the pores, however, is not so easily established. Applying Poiseuille's equation to the pore [271] is made difficult by not knowing the precise dimensions of the pore *in vivo*. It is certainly smaller than the dimensions apparent from transmission electron micrographs. The molecular weight cut-off in *Elodea* leaves for injected peptides is equivalent to a molecular size of about 1.6 nm. [272] Graham, Clarkson & Sanderson [273] (quoted in [31]) have estimated plasmodesmatal hydraulic conductivity by measuring water uptake by mature zones of the root in which they claim that the pores are the only route of flow to the stele. Assuming a pore radius of 5 nm (corresponding to the apparent dimensions of the desmotubule in electron micrographs) and a viscosity of the cytoplasm of twice that of pure water, the flux rate of

2.5×10^{-8} m s^{-1} predicted a conductivity of 2.6×10^{-7} m s^{-1} MPa^{-1}. As this is within the range of membrane L_p values, the authors suggest that the symplasmic pathway is unlikely to provide the predominant pathway for water flow across the root cortex. Moreover, they suggest that cytoplasmic viscosity is likely to be higher than the twice that of pure water used (see also the discussion on water filled pores in section 3.4.1). Also, as noted above it would appear that a radius of 5 nm is an overgenerous estimate of pore dimensions. Evidence against plasmodesmata influencing the values of L_p gained by the pressure probe are: (1) that relaxation times of ligatured *Chara* cells are proportional to the total surface area of the cells, despite the fact that plasmodesmata are restricted to the end walls only [274]; (2) pressure probe measurements of the L_p of single suspension cultured cells of *Chenopodium rubrum* result in values similar to those of the mesophyll cells of the intact plant from which they derive [147]. This strongly suggests the absence of significant plasmodesmatal flows in the mesophyll cells. (3) L_p values of epidermal cells of *Tradescantia virginiana* for which each of the adjacent epidermal cells have been punctured and plasmodesmata presumably closed are substantially similar to those with intact neighbours [219].

On the other hand dramatic decreases of apparent L_p with time in the same cell have been observed on rare occasions (e.g. [11, 268, 275]). These could be explained in terms of the blocking of initially conducting plasmodesmata. It seems highly likely, however, that the drop in turgor generally associated with the penetration of a cell by the pressure probe may well block any plasmodesmata, possibly by sweeping endoplasmic reticulum into them (see [269]), although the recent series of microinjection experiments to measure plasmodesmatal conductance for solutes [270, 272] would suggest that at least some remain open. This may well require more careful use of the pressure probe to avoid such losses of turgor pressure on insertion. This can be achieved if the probe is pressurised in a cell and pushed immediately into the underlying cell without reducing pressure. (Shackel & Brinckmann [14] have recently developed a technique by which a glass fibre is used to seal the probe tip until it is in position in the vacuole.)

Zimmermann & Steudle [23] and Gutknecht *et al.* [19] list values of hydraulic activity L_p for a selection of giant algae. More recent data are shown in table 3.1. These fall within the range of 2×10^{-8}–5×10^{-6} m s^{-1} MPa^{-1} and are dependent not only on the species but on several conditions noted below (notably turgor pressure). Wendler & Zimmermann [276] comment that the L_p values of freshwater algae tend to be higher than those of sea water algae, although this may only represent the influence of high osmotic pressures on the parameter (see p. 242). Recently Wendler & Zimmermann [277, 278] using improved methods of recording and processing data have resolved the turgor pressure relaxation of cells of *Chara corallina* into two phases which they interpret as representing the conductivities of the plasmalemma and tonoplast respectively. Assuming that a range of potentially complicating factors such as a true three compartment model for the system, unstirred layers, solvent drag, non-linearity of force/flow and wall elastic effects are not involved they show that the L_ps of the membranes are different. The plasmalemma has a value for L_p of (2–4) × 10^{-6} m s^{-1} MPa^{-1} while the range for the tonoplast is (3–10) × 10^{-6} m s^{-1} MPa^{-1}. While confirming a difference between the membranes these data contradict previous suggestions that the conductivity of the tonoplast is so high as to have little effect on the pressure relaxation. [279]

The significance of these values of hydraulic conductivity for both algae and higher plants is that water crosses membranes relatively easily. House [27] illustrates this by the work of Collander [280, 281] with values of the P_f/P_s ratios (P_s is the solute permeability) for a range of neutral solutes. The extremes are 3.4 for methanol and 34 800 for glycerol. Values in the order of 10^3 may be calculated for P_{K_+} and P_{Na_+} and 10^6 for P_{Cl_-} for *Chara* membranes from the solute permeability data of Lannoye, Tarr & Dainty. [282] In the absence of considerable flows relatively small water potential gradients will be sustained across membranes, i.e. cells will be close to hydraulic thermodynamic equilibrium with their walls. This can be visualised by the observation that half times of turgor pressure relaxation for higher plant cells following pressure pulses applied by the pressure probe are measured in terms of seconds rather than minutes. [149, 283]

Factors influencing membrane L_p

Turgor pressure Several parameters and conditions have been found to influence L_p. The L_p of both algal (*Chara* and *Valonia*; Zimmermann & Steudle [100, 141, 284]) and higher plant (*Elodea* [150]) cells increases several fold as the turgor pressure is reduced towards zero. Following these unexpected observations Zimmermann & Steudle have argued that the response is due to the pressure dependent solute fluxes as discussed above (see [23]). The precise mechanism remains unclear but Zimmermann & Steudle propose that under conditions where active flux of solute (J_s^{act}) is a significant proportion of J_v, L_p will be overestimated by the conventional calculation (equation (3.44)). Fiscus [285] derived the equation:

$$\frac{dP}{dJ_v} = \frac{1}{L_p} - \frac{J_s^{act}RT}{J_v^2} \qquad (3.43)$$

to describe the apparent resistance of cell membranes to water flow during salt transport. Use of the expression

$$J_v = L_p\Delta\psi = L_p(\Delta P - \Delta\pi_i) \qquad (3.44)$$

tacitly neglects the second term in this equation. Zimmermann & Steudle [23] argue that this may be the basis for the apparent rise in L_p at low turgor pressures. They suggest that if J_s^{act} is taken into account then true L_p would become constant over the whole pressure range. In general links between water and solute flows remain poorly understood.

At low turgor pressures the increase in L_p brings it into the range of the L_p estimates for the cell wall. At this point the wall hydraulic conductivity becomes rate limiting in cellular water uptake. [141, 286]

Osmoticum concentration Low turgor pressures may be produced experimentally by increasing external osmotic pressure. This in itself may influence L_p in two ways. Not only is L_p dependent on concentration by definition [23] but also it has been proposed that dehydration of the membrane by osmotica results in changes in L_p. [284, 287–9].

Hormones The effect of hormones on the cell water permeability has been an issue of debate for some time. ABA, a hormone increasingly associated with stress effects, including water stress, has been claimed to increase root permeability in tomato, sunflower and maize, decrease it in soybean and *Phaseolus*, and have both effects consecutively in barley. Other reports show no effect (see references in [275]).

ABA is known to influence potassium ion fluxes in membranes [290] and this effect must contribute at least partly to these observed changes in permeability. However, some attempts have been made to observe changes in L_p at a cellular level. Glinka & Reinhold [291, 292] using excised carrot root tissue showed that ABA decreased the time required for water equilibration. Although far from conclusive, this is consistent with an increase in cell membrane L_p. More directly, using transcellular osmosis Ord, Cameron & Fensom [293], showed a 15% increase in L_p of *Nitella* membranes as the result of ABA treatment. Although the physiological role of higher plant growth substances in these giant algal cells must be questioned. Using the pressure probe Eamus & Tomos [275] showed an apparent 2–3-fold increase in the L_p of *Rhoeo discolor* epidermal cells. On the other hand, also with the pressure probe, Jones [294] showed no effect of ABA on the L_p of the root cortical cells of wheat up to 1 h after application. In this last case, the conductivity of a single cell before and after hormone application could be measured, which avoided errors in L_p due to inaccuracies in determining cell dimensions. Conversely plasmolytic measurements of the L_p of onion epidermal cells indicate a decrease in L_p on ABA application [295], although as noted above L_p at plasmolytic pressures may not be representative.

Auxin has also been claimed to have an effect of hydraulic conductivity of hypocotyl tissue. [296] Although Dowler, Rayle, Cande & Ray [297] found no such effect on the permeability of pea segments to tritium labelled water. Zimmermann & Steudle [225] showed that auxin, by interacting with a turgor pressure sensitive potassium ion flux across the membrane of *Valonia* cells, has an indirect effect on the water permeability. Clearly a consistent pattern is yet to emerge. At a single cell level the physiological significance of the small changes in L_p suggested by those studies that do show changes are hardly likely to influence the cell's water relations significantly. However, the accumulated effect across a tissue may well influence the relative importance of tissue water pathways that may be of very similar conductivities.

Other effects Other reports have described assorted properties of L_p. One, yet to be repeated and extended, is the remarkable work of Cailloux [298] who appears to have shown that different portions of the hair of a root epidermal cell possess different hydraulic conductivities, depending on whether the vacuole or cytoplasm underlies the area of plasmalemma in question.

The L_p of the *Chara* internode wall is independent of pH in the range 7.2–11.1. [139] This indicates that both the acid and alkaline bands of the *Chara* membrane will have the same conductivity. L_p is not influenced by the metabolic inhibitor, potassium cyanide [9], or by the cyclosis inhibitor cytochalasin B. [289] Some benzene derivatives decrease L_p after 2–6 h treatment, while mercury chloride has no effect. [299]

Using the pressure probe Steudle & Wieneke [152] showed that the L_p of apple fruit cells decreases slightly from 3×10^{-7} to 1×10^{-7} m s^{-1} MPa^{-1} as the fruit ripens. Finally, Weisenseel & Schmeibidl [300] showed that the red/far red light ratio has an influence on the water permeability of *Mougeotia*. The bases of these changes are unknown.

Diffusive permeability The second definition of membrane water permeability is as the permeability of the membrane to labelled water (P_d) defined according to the equation:

$$J_w^* = P_d \Delta c_w^* \tag{3.45}$$

where J_w^* and Δc_w^* are the flux and concentration step of labelled water across the membrane concerned. This corresponds to the permeability of self diffusion of water and hence is called the diffusive permeability.

Using cell perfusion techniques Gutknecht [301] measured P_d for the vacuole to medium pathway of *Valonia ventricosa* as 1.22×10^{-6} m s^{-1}. By perfusing the cells with sea water Gutknecht was able to remove the cell membranes, allowing measurement of the P_d of the cell wall and associated unstirred layers. Thus a value of 2.36×10^{-5} m s^{-1} could be calculated for the protoplast P_d alone. It is assumed that this value of P_d represents the diffusion of water across the membrane independent of the existence of any water filled pores. If, as noted above, the same pathway is taken by hydraulic or osmotic water flow, then $P_d = P_f$. This relationship has been used widely in attempts to determine whether both pathways are, in fact, the same. Examples of P_d for a number of plant cells are given in table 3.3. Unlike those of ions, the water permeabilities of biological membranes are of similar orders of magnitude to those of artificial lipid bilayers, highly suggestive that the lipid components of membranes contribute significantly to the pathway of water.

Table 3.3 *Values of P_d for plant membranes. The range of values for animal cells* ([27] *p.* 156) *is given for comparison.*

Species	P_d ($m\,s^{-1} \times 10^6$)	$P_d \bar{V}_w/RT$ ($m\,s^{-1}\,MPa^{-1} \times 10^6$)	Technique	Ref.
Elodea densa	300	2.22	NMR (paramagnetic ion)	[304]
Chlorella vulgaris	21	0.16	NMR (paramagnetic ion) (pulsed gradient)	[305]
Hedera helix	300	2.22	NMR (paramagnetic ion)	[306]
Nitellopsis	4.4	0.03	D_2O diffusion	[407]
Nitella mucronata	25	0.19	D_2O	[281]
Valonia ventricosa	2.36	0.02	Perfusion corrected for unstirred layers	[301]
Allium cepa	2.03	0.015	T_2O	[408]
(Animal cells	0.21–64	1.56×10^{-3}–0.47		[27])

Since the work of Hevesy, Höfer & Krogh [302] on frogskin permeability many workers have reported that $L_pRT/\bar{V}_w(P_f)$ is consistently greater than P_d (see [20]). The general conclusion from this has been that water must pass through water filled pores in the membrane. As pointed out by Dainty [20], however, this is far from certain. Whereas the measurements of L_p are relatively free of the influence of unstirred layers either side of the membrane [303], measurement of P_d by tracer exchange techniques will invariably considerably underestimate P_d. Dainty [20] argues that an unstirred layer some 10 μm thick on one side of the membrane would have an apparent P_d equivalent to the value of $L_pRT/\bar{V}_w$ estimated for the membranes of *Chara australis*, *Nitella translucens* and various red blood cells (about $2 \times 10^{-4}\,m\,s^{-1}$). Clearly unstirred layers may well exceed this value by an order of magnitude, House [27] quotes values from a minimum of 12 μm for the rapidly stirred medium bathing Sylvania we get to as high as 350 μm for isolated cornea of rabbit in an unstirred medium. Since quantitative measurement of the unstirred layers is generally difficult this approach does not appear particularly fruitful.

Paradoxically, some values of P_d determined by NMR techniques [62] appear to be considerably higher than P_f from the same tissue. Stout and colleagues found values of $3 \times 10^{-4}\,m\,s^{-1}$ for *Elodea*. [304] This value has been questioned by Steudle *et al.* [150] who measured P_f values 2 orders of magnitude lower for *Elodea* with the pressure probe. Steudle *et al.* suggest that the required use of moderate concentrations of manganese ions in the NMR technique somehow interferes with the membrane permeability. Stout and colleagues measured P_d values of $2.1 \times 10^{-5}\,m\,s^{-1}$ for *Chlorella* [305] and $3 \times 10^{-4}\,m\,s^{-1}$ for cell membranes of ivy bark. [306] The former value was obtained from two NMR techniques, one of which was performed in the absence of a potentially interfering exogenous paramagnetic ion.

Fettiplace & Haydon [61] tentatively argue that for artificial lipid bilayers there is considerable evidence against the presence of water channels, citing (1) a similarity between results from osmotic and isotopic exchange techniques (i.e. $P_d = P_f$); (2) that the specific conductance of membrane material is effectively that of alkanes saturated with water and not obviously consistent with the presence of water channels; (3) that the activation energy for transfer across the bilayer (E_a) is about twice that for self diffusion; and (4) that there is no unequivocal spectroscopic evidence for water within the chain region of the bilayer. Fettiplace & Haydon further consider that the rate limiting factor is probably that of solubility and diffusion of water in the chain region of the bilayer rather than a barrier in the surface of the bilayer.

For biological membranes, on the other hand, the situation is less clear. Fettiplace & Haydon [61] review some of the data available for erythrocytes. For the results from mammalian erythrocytes they claim that unstirred layers cannot account for a significant difference between P_d and P_f ($P_f/P_d = 3.3$). Also by comparison with the activation energy for water transport (E_a) in artificial membranes (45–119 kJ mol^{-1}), the activation energy for intact erythrocytes (16.4 kJ mol^{-1}) is close to that for self diffusion of water (19.3 kJ mol^{-1}). Most convincing is that treatment of human erythrocytes with the thiol reagent PCMBS dramatically reduces the P_f/P_d ratio to 1.1 and increases the activation energy to 48.2 kJ mol^{-1}. [307] For non-mammalian erythrocytes, however, the lower P_f/P_d ratio of 1.5 and an activation energy of 48 kJ mol^{-1} is suggestive of the absence of pores.

Such detailed measurements are not available for plant cells, especially those of higher plants. Dainty & Ginzburg [287] measured a value of 35.6 kJ mol^{-1} for E_a for *Nitella translucens* membranes using transcellular osmosis and Weigl [308] a value of 26 kJ mol^{-1} for corn roots using isotope exchange. Similar values were obtained for beech leaves by Tyree & Cheung [309] using a pressure bomb, although in this case it is difficult to identify the hydraulic resistances measured. Tomos *et al.* [9] using the pressure probe found somewhat higher values for the membranes of *Tradescantia virginiana* epidermal cells (69 and 110 kJ mol^{-1} for two cells with the most complete data). Commenting that E_a values higher than those for bulk solution do not necessarily mean the absence of pores from the membrane, Price & Thompson [310] suggest that values of E_a of up to 54–67 kJ mol^{-1} might be found for pores on the basis that this is the activation energy of free diffusion of water molecules in ice. If this were the case most of these values would still be consistent with the presence of pores. Clearly caution must be used in attempting to interpret these data. House [27] discusses the problems of drawing conclusions from E_a values alone. Another factor, for example, being changes in membrane fluidity brought about by change in temperature. [311]

The temperature dependence of membrane L_p will result in temperature having an effect on the relative importance of the various pathways of water through tissues as presumably the effect of temperature on bulk flow through apoplast and plasmodesmata will be much smaller. Indeed this may prove a useful way of contributing to our understanding of water flow pathways. Various short and long-term effects of temperature on water uptake by whole roots are known (see [1]). How these relate to cellular function remains unclear, although as temperature affects water movement through dead roots far less than through living ones it is probable that membrane effects are involved.

Flows across composite membranes House [27] and Zimmermann & Steudle [23] review and discuss the implications of water movement across more than one membrane in parallel. Such systems do not necessarily behave even qualitatively as single membranes. Since the wall/plasmalemma/tonoplast series is the path of water flow from outside into the vacuole this is clearly of physiological relevance.

Two phenomena, in particular, are predicted if two membranes in parallel possess different reflection coefficients. The first is a dependence of L_p on the direction of water flow (rectification), the second is an apparent active transport of water across the system.

Rectification with respect to water flow (in non-symmetrical endoosmotic and exoosmotic L_p values, $L_{pen} \neq L_{pex}$) was first observed in experiments involving shrinking and swelling of protoplasts. [312, 313] More recently the phenomenon has been observed several times in experiments on transcellular osmosis [116, 288, 314] and with the pressure probe. [315] Two explanations have been proposed for this behaviour. Since the effect is influenced by the osmotic concentration of the medium used (e.g. L_{pen}/L_{pex} rising from 1.19 in $20\,\mathrm{mol\,m^{-3}}$ sucrose to about 2.1 at $200\,\mathrm{mol\,m^{-3}}$ sucrose) it has been suggested that the effect is largely due to the dehydration of the membrane. [287, 288, 315]

In addition to this, however, Rabinowitch, Grover & Ginzberg [316] and Zimmermann & Steudle [23] believe that since polarity is observed in some cases even at zero concentrations another factor must also be playing a role. This factor they suggest to be due to two membranes in series with different reflection coefficients. Kiyosawa & Tazawa [279], however, showed rectification in tonoplast-free *Chara corallina* cells. In this case, two elements of differing properties may exist within the one membrane, e.g. the two surfaces. [23]

The second phenomenon is an active water pump. In general terms such a pump has been treated with scepticism since the high passive hydraulic conductivity of membranes would be expected to continually 'short circuit' such a pump. [1] (This would also negate any water pumping by electroosmosis. [20]) Ginsburg & Ginzburg [317] and Ginsburg [318], however, describe such a property for composite membrane systems. Looking at a whole tissue phenomenon, isoosmotic flow across the cortex of roots, they developed a model that would, indeed, have the properties of an active water pump dependent on the expenditure of metabolic energy. Meyer, Sauer & Woermann [319] describe such a system in detail. Theirs is a model system in which two artificial membranes are arranged in series. The two outer bulk phases are the same (isoosmotic). The compartment between the two membranes contains a solute for which the reflection coefficient for the two membranes is different. Although the osmotic potential gradient from the central compartment to each outer compartment is the same, the different linking forces due to the different reflection coefficients will drive water flow preferentially in one direction, hence the water pump. As the solute diffuses out of the compartment with time, however, the water flow would slow and stop. If fresh solute can be continually generated in the central compartment (Meyer *et al.* used the action of invertase on sucrose to generate glucose and fructose) the water flow will continue. Meyer *et al.* extended this process by adding sodium chloride to the central chamber also. This was dragged along in the water flow. The net result being that the transport of sodium chloride was powered by the splitting of the sources. Furthermore, when ion exchangers were used as membranes, sodium and chloride ion movements were not equal and a potential difference was set up across the membrane (the streaming potential). Clearly this is a very different pump from that generally considered under active solute transport and yet the conditions required are those that could be expected to occur in cells, e.g. plasmalemma/tonoplast.

Katou and Furumoto [320, 321] have recently proposed a related model based on the wall and plasmalemma as two components in a thermodynamic water pump transporting water into cells.

Reflection coefficients The reflection coefficients of a range of non-electrolytes have been measured in *Chara* by transcellular osmosis [288] and for *Chara*, *Nitella* and *Tradescantia virginiana* epidermal cells with the pressure probe. [139, 154, 289, 315] Values are listed in table 3.4. The coefficients are influenced by solute concentration [289] but not by pH. [139]

Although defined differently, membranes also have reflection coefficients towards electrolytes. No attempts to measure these in intact cells appear to have been made (cf. subcellular organelles below). Following comments that the relatively high permeability of the potassium ion in membranes signifies that the ion is not likely to be an important osmoticum, Hastings & Gutknecht [322] have argued that when the whole system is considered the effective reflection coefficient is unity. Thus the potassium ion can indeed act as a major cytoplasmic osmoticum. [204]

Osmotic properties of organelles

Nucleus House [27] summarises the early work on the osmotic behaviour of the nucleus (from animal tissues) and comments that despite the occurrence of the pore-complex in the membrane *in vivo* isolated nuclei swell

Table 3.4. *Cell and organelle reflection coefficients. (The amino acids represented are electrically neutral at the pH 6.2 buffer used.* [326])

Osmoticum	*Chara corallina* [289]	*Chara corallina* [288]	*Nitella translucens* [288]	*Nitella flexilis* [315]	*Tradescantia virginiana* [154]	*Pea* chloroplasts [331, 333][a]	*Potato* mitochondria [326][a]
D-Ribose							1.01
D-Xylose							0.97
D-Lyxose							1.00
D-Arabinose							0.98
D-Glucose							1.00
D-Galactose							0.97
D-Mannose							1.01
Sucrose				0.97	1.04	1.00	1.00[a]
Ethyleneglycol					0.99	0.40	0.25
Glycerol					0.93	0.63	0.44
Erythritol						0.90	0.71
Adonitol						1.00	0.98
Mannitol	1.02				1.06	1.01	0.99
Sorbitol						1.00	1.02
Urea				0.91	1.06		
Formamide	0.99	1.00		0.79	0.99		
Acetamide				0.91	1.02		
Dimethyl formamide	0.76						
Methanol	0.38	0.30	0.50	0.31	0.15	0.00	0.07
Ethanol	0.40	0.27	0.44	0.34	0.25		
Propan-1-ol	0.24	0.22	0.40	0.17	−0.58		
Propan-2-ol	0.45			0.35	0.26		
Butan-1-ol	0.14						
2 Methyl-propan-1-ol	0.21						
Acetone	0.17						
Glycine						0.03[b]	0.98
L-Alanine						0.05[b]	1.02
L-Threonine							0.96
L-Phenylalanine							0.99
L-Methionine							1.00
L-Cysteine							0.97

[a]Calculated relative to sucrose, taken to have $\sigma = 1$. [b]At $\pi < 0.1$ MPa, above this concentration $\sigma \Rightarrow 1.00$.

when presented with a hypotonic solution of electrolyte or non-electrolyte. (This behaviour of the pore-complex is reminiscent of that discussed above for plasmodesmata.) He comments furthermore that the response is poorly understood and that transport processes of the nucleus are generally poorly understood. Guard cell nuclei certainly change shape during stomatal movement. For example, in *Vicia faba* the organelles are oval when the stomata are closed and rounded when stomata are open. Also, in *Artemisia rotundifolia* they are rounded when the pore is closed and crenated when it is open (quoted in [8]). Whether such behaviour is related to the osmotic properties of nuclei remains to be seen.

In *Platymonas subcordiformis* no change in nuclear volume is observed on rapidly changing the external osmotic pressure of the cells, despite large changes in total cell volume. [198]

Mitochondria Here again the earlier work is summarised by House. [27] (Relatively little data are available for plant mitochondria specifically.) In summary, mitochondria appear to behave osmotically but with a significant non-osmotic volume. This non-osmotic volume ranges from 40 to 80% of the physical volume of the organelle, contrasting with only 30–40% that can be accounted for by 'solid' material (Tedeschi & Harris [323]). Bentzel & Solomon [324] characterised two aqueous compartments, one comprising 70% of the volume, apparently accessible to sucrose, not surrounded by a semipermeable membrane and hence non-osmotic, and a second comprising 30% of the volume with 50% apparent osmotic volume. House [27] indicates that there is no clear correlation between the size of these volumes and any anatomical compartment. (Such a conclusion is reminiscent of a similar controversy regarding the identification of osmotic volumes in the alga *Dunalliela*. [316, 325])

Mitochondrial shrinking and swelling as a function of its metabolic state is a well known phenomenon, and yet one that, according to House [27], is not so easily explained.

The reflection coefficients of potato tuber mitochondria towards a homologous series of alcohols, sugars and amino acids are given in table 3.4. [326]

Chloroplasts Chloroplasts behave like osmometers in the presence of various osmotica. [327] Nobel [328] showed that the volume of pea chloroplasts is a linear function of π_0^{-1} with an apparent non-osmotic volume of some 17 al per chloroplast (volume *in vivo* varies from 29 to 35 al depending on illumination). The non-osmotic volume is therefore 51% for dark and 41% for illuminated chloroplasts. Presumably this reflects the extensive membrane and other protein development. Gross & Packer [329] indicated that isolated granal vesicles also behave as osmometers. The work of Tolberg & Macey [327] and Gross & Packer [329] indicated that granal membranes respond differently from others in the organelle in their response to sodium chloride or sucrose as osmoticum. The granal membranes seem impermeable to both while some other membrane within the organelle is accessible to sodium chloride but not to sucrose.

Like that of mitochondria, the volume of chloroplasts also responds to changes in metabolic state. On illumination the volume decreases owing to a flattening of the chloroplasts. The slope of the volume/π^{-1} curve indicates that in the dark the organelles contain more osmotically active particles. Indeed it is known that illumination causes efflux of potassium and chloride ions from the chloroplasts (see [330]). A full explanation of chloroplast shrinking and swelling is still awaited, however. [38]

Wang & Nobel [331] have measured the reflection coefficients, σ, of the chloroplast membranes of *Pisum sativum* to a series of alcohols of different lipid/water partition coefficients using packed organelle volumes (corrected for interstitial fluid). They also showed the expected qualitative relationship between the two components.

Nobel & Cheung [332] and Nobel [326, 333] also investigated the apparent reflection coefficients of pea chloroplasts to certain amino acids (which they considered to be neutral at the pHs used in the experiment). They showed that at low concentrations the values of σ were very low (e.g. up to an osmotic pressure of 0.1 MPa, glycine has a σ of 0.03). At higher concentrations they rose, approaching a value of unity above 0.35 MPa osmotic pressure. They interpreted this to indicate the presence of a saturatable carrier for the amino acid, saturating at about the equivalent of 0.35 MPa osmotic pressure of glycine. They subsequently used this to show that certain amino acids compete for the same carrier. (Serine is transported by a carrier distinct from the one for glycine and alanine. [332]) Nobel [333] further used this technique to study the temperature dependence of σ in chill resistant (pea) and chill sensitive (bean) plants, showing a decrease in the value for glycerol from 0.48 (bean) and 0.62 (pea) at 1 °C to 0.03 at 28 °C. Whereas the curve for pea was more or less linear for the range, the curve for bean chloroplasts showed a dramatic discontinuity between 10 and 12 °C. It was suggested that this is due to a phase change in the membranes. Nobel & Wang [334] also showed an effect of ozone on the chloroplast σ for erythritol, and suggest that an important early effect of ozone on plants is an increase in chloroplast permeability.

3.5.3 *Hydraulic machines*

In ways reminiscent of machines powered by heat-generated steam pressure, plant cells are capable of performing mechanical work driven by osmotic pressure derived from chemical reactions (e.g. solute transport or metabolism). The most important of these processes is that of cell expansion growth.

Growth and cellular water relations The water relations of growing cells have recently been reviewed by Tomos [335] and Cosgrove. [336] Here again our understanding is dominated by thorough studies on the cells of giant algae, notably *Nitella*, although much information regarding the same parameters for higher plants has been obtained in the last four or five years.

Cell expansion is largely equivalent to the uptake of water into the protoplast. The uptake rate can be governed by three main components. These are: (1) the rate at which wall expansion will allow water entry; (2) the rate at which water can cross the membranes into the cell under the driving forces (pressure or osmotic) prevailing about the cell; and (3) the maintenance of the osmotic pressure of the cell by solute uptake or generation by metabolic processes.

The first is described by considering the wall expansion properties to have the behaviour illustrated in figure 3.13. Above a certain threshold (yield) value (Y) the material of the wall extends linearly with increasing stress (turgor pressure). Such behaviour characterises what is known as a 'Bingham substance' and is described by the equation

$$\frac{1}{V_0}\frac{\mathrm{d}V_0}{\mathrm{d}t} = \phi(P - Y) \tag{3.46}$$

where ϕ, the gradient of the line in figure 3.13, is the extensibility coefficient. However, as first formulated by Lockhart [337] and popularised by Ray, Green & Cleland [338] growth and turgor pressure are related in another way also.

Water flow into cells is governed by the hydraulic conductivity of the cells

$$J_\mathrm{v} = L_\mathrm{p}A(\sigma\Delta\pi - P) \tag{3.47}$$

i.e.

$$J_\mathrm{v}/V = (\mathrm{d}V/V)\mathrm{d}t = \text{Growth} = L_\mathrm{p}(A/V)(\sigma\Delta\pi - P) \tag{3.48}$$

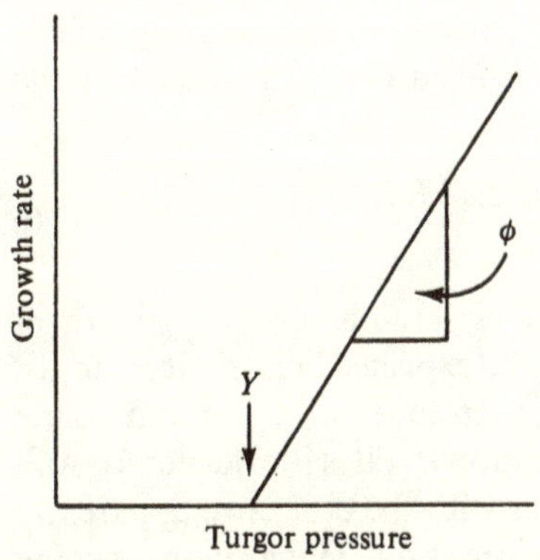

Fig. 3.13. Idealised behaviour of cell wall growth as a function of cell turgor pressure.

$L_p(A/V)$ is often abbreviated to a transport coefficient L. Lockhart [337] combined these relationships to give the expression

$$\text{Growth rate} = \frac{\phi L}{\phi + L}(\sigma\Delta\pi - Y) \tag{3.49}$$

(see also Ray *et al.* [338] and Cosgrove [339]). From these relationships it can be seen that growth rates may be controlled by either the membrane property L or the wall properties ϕ and Y. Cosgrove [340] clearly sets out the behaviour of turgor pressure as the ϕ/L ratio varies. Can we determine which of these parameters is the predominant one in growth? For the geometrically simple *Nitella* cell data for both L and ϕ are available. Using the growth data of Green, Erickson & Buggy [341] and values of L_p, A and V from various measurements [21], Cosgrove [339] quotes values of about $2.8 \times 10^{-4}\ \text{s}^{-1}\ \text{MPa}^{-1}$ for ϕ and calculates values of $2 \times 10^{-2}\ \text{s}^{-1}\ \text{MPa}^{-1}$ for L. Thus ϕ is two orders of magnitude smaller than L signifying that the wall properties are the limiting parameters in the growth of this cell. *Nitella* cells would be expected to grow according to equation (3.46). For *individual* higher plant cells the situation is not so easily defined. L_p values are similar (see table 3.2), but values of ϕ of these cells are difficult to estimate. Cosgrove [336] argues that observed growth rates of pea stem cortical cells would result in such small changes in the driving force of water uptake $(\sigma\Delta\pi - P)$ that water transport cannot be limiting. Several workers have shown that wall extensibility measured by stress/strain analysis of 'isolated' wall material correlates well with growth rate (see [16, 335, 342, 344]). Although the relationship between such measurement and ϕ remains obscure [345–6] these data also support the proposition that wall rheology rather than membrane hydraulic conductance is critical at this resolution.

Much of the current work on this aspect puts to one side the problem of uptake of osmoticum into a growing cell. Lockhart [337] considers both the situation where no solute uptake occurs and where osmotic adjustment is perfect. It is the latter that has been extended by recent workers. In plants with high osmotic pressures in their growing cells the amount of solute required to maintain π_i can be considerable. Growth at constant π_i is described by the relationship [347]

$$\frac{dV}{dt} = \left(\frac{\text{Net solute uptake} + \text{Net solute synthesis}}{-\pi_i}\right)\frac{RT}{1} \tag{3.50}$$

This aspect has been considered especially by studies on the growth of halophylic plants. [347] Table 3.5 illustrates the expected behaviour of turgor pressure associated with different relative magnitudes of L_p, ϕ and solute uptake. Constant turgor pressure at different growth rates is evidence for wall properties being the major limiting parameter with the other two apparently playing negligible roles. Constant turgor pressure at different growth rates has been demonstrated using tissue averaged methods. [348, 349, 350] Using the

Table 3.5. *The influence of L, ϕ and solute fluxes on water potential equilibrium, turgor and cellular osmotic potential in growing cells* [335]

	Water potential	Turgor pressure	Osmotic pressure
$L \gg \phi$			
Solute flux			
(i) Not limiting	Equilibrium	Constant	Constant
(ii) Limiting	Equilibrium	Lowered	Lowered
$L \leqslant \phi$			
Solute flux			
(i) Not limiting	Non-equilibrium	Lowered	Constant
(ii) Limiting	Non-equilibrium	Lowered	Lowered

pressure probe, similar constant turgor pressure at varying growth rates has been observed for individual cells of wheat roots [16], *Lolium temulentum* leaves [351] and phototropism in mustard seedlings. [352] The observations for *Lolium* and mustard represent fairly rapid changes in growth rate. Constancy of turgor under these circumstances is in contradiction to a result using psychrometric methods with bean leaves. [353] When rapid growth is initiated with white light a decrease in turgor pressure is observed. However, as Y also decreases $(P - Y)$ remains unaltered in this case.

Turgor pressure could remain constant if a limiting hydraulic conductivity was accompanied by a balancing increase in cell osmotic pressure. In this case a limiting conductivity would be overcome by increasing the driving force of water into the cell. Such could well be the case if highly sensitive, turgor dependent, uptake processes occurred in the membranes and suitable solutes were available in the apoplast. Due to the known, high hydraulic conductivity of higher plant membranes, however, this is unlikely to be crucial at cellular resolution. However, that turgor is, in fact, maintained under growing conditions at all is strongly indicative that water and solute fluxes are kept equivalent. Turgor sensitive processes must be prime candidates for such a system.

Although the behaviour of a single higher plant cell appears to be predictable from these considerations, that of tissues is not so. The hydraulic conductance of a path from source (soil or xylem vessel) to sink (the expanding cell) is not simply the membrane L_p, as in *Nitella*, but also includes the tissue pathways (apoplast or symplast) alluded to above. Here we find a disagreement based largely on values for the hydraulic conductance of the apoplast. Arguments have been put forward to support both the position that the hydraulic conduction of these pathways is growth limiting (e.g. [35]) and that it is not so. [36] The solution to this argument lies in determining the longitudinal hydraulic conductivity of cell wall material *in vivo* and in determining the distributions of non-Donnan solutes in the cell wall, that

Table 3.6. *Tissue and cell extensibility and stress yield thresholds for expanding plant tissues.*

Species	Y (MPa)	P (MPa)	Cell ϕ ($MPa^{-1}\,s^{-1}$)	Tissue ϕ ($MPa^{-1}\,s^{-1}$)	Ref.
Nitella	0.2		2.8×10^{-4}		[341]
Avena coleoptyle	0.6				[132]
Pea stem	0.2	4.6			[406]
	0.29	0.6		2.22×10^{-5} (in water)	[336]
				6.67×10^{-5} (in auxin)	
Betula pendula	0.071	0.07–0.45		3.61×10^{-6} (in dark)	[344]
				12.8×10^{-6} (in light)	
	0.182 (on water stress)				[362]
Acer pseudoplatanus	0.25	0.25–0.55		11.7×10^{-6} (in dark)	
				12.2×10^{-6} (in light)	
Sunflower	0.25				[350]
Maize	0.4				[360]
Bean leaves	0.2–0.4 ($Y = f(age)$)	0.32–0.58			[353]
Soybean stem	0.44	0.53		95×10^{-6}	[409]

appear currently to be confusing attempts to measure water potential gradients in tissues.

From an experimental point of view the situation is made more complicated by a further feature of these parameters. It is highly unlikely that ϕ and Y are constants in practice (changes in L_p have been considered above). The clearest illustration of this is the work of Green *et al.* [341], again with the giant-celled alga *Nitella*, the growth rate of which we have suggested should be described by equation (3.46). These workers showed that growth rates could indeed be altered by changing the external osmotic pressure by altering turgor pressure. However, with time the growth rate recovered to its previous rate without any further change in turgor pressure. Clearly such behaviour is not consistent with equation (3.46). Either ϕ or Y must be changing. Green *et al.* [341] showed that while ϕ appears to remain constant the yield threshold appeared to be under feedback control so that the crucial parameter $(P - Y)$ in equation (3.46) was regulated (cf. [353]). However, there is a minimum value below which Y will not fall, and the completeness of recovery of $(P - Y)$ tends to diminish as the turgor is brought closer to this point. The decrease of Y appears to be under metabolic control as it is inhibited by metabolic inhibitors, the reversal of the effect when turgor pressures are increased, on the other hand, appears to be a non-metabolic strain-hardening process.

Van Volkenburgh & Cleland [354] have reviewed data available for higher plants. They point out that in some systems the growth rate *is* determined simply by turgor pressure. [355–8] Of those that are not they conclude that both ϕ and Y may be modified citing the promotion of leaf expansion of bean leaves by white light as being accompanied by an increase in ϕ [349, 359–61], and the decrease in growth rate of water stressed sunflower leaves as being accompanied by changes in both ϕ and Y. [350] More recently Taylor & Davies [343–4, 362] compared the growth rate in the leaves of tree species that tolerate shaded environments to different degrees (birch – shade intolerant, and sycamore – shade tolerant). The yield threshold remained constant for each species under light and darkened conditions; ϕ, on the other hand, increased three fold for birch on illumination. No changes in ϕ were observed for sycamore. The conclusion is that sycamore growth is governed by changes in turgor pressure while that of birch is regulated in addition by wall extensibility. The result of this is that whereas sycamore leaf expansion is slowed by decreasing turgor brought about by illumination-stimulated transpiration, birch leaves adapt to the lower turgor pressures of a fully sunlit environment. This example provides a clear ecological role for the various combinations of cellular parameters. Values of Y and ϕ from the literature are shown in table 3.6.

Mechanical movement Reversible movement of plant cells and organs has been reviewed by Hill and Findlay. [134] Growth and reversible movement share many parameters of cellular water relations. Indeed, in some cases the two may be superimposed and may lead to confusion of interpretation. An example of this is described by Hill & Findlay in the case of *Drostera* tentacle movement. [363, 364] True reversible movement cannot involve plastic effects in the wall.

It is generally assumed that all examples of movement are associated with changes in cell volume brought about by turgor pressure changes in specialised motor cells. Changes in elastic modulus could bring about movement, but the problems of *increasing* values of ε in turgid cells discussed above suggest that this might not be expected. However, differences in elastic properties between different parts of the motor cell (or tissue) may certainly be related to movement. The classic example is that of the stomatal guard cell in which the wall lining the pore (the ventral wall) is taken to be far less extensible than the dorsal wall opposite. Also the ventral wall is taken to be far more extensible in an axial direction than it is in the radial direction. This conclusion is supported by the distribution of cellulose microfibrills within the various parts of the wall. [365, 366]

With the exception of the work of Meidner & Edwards [367] no direct measurements of turgor pressure have been reported on any mobile systems although changes in pressure are generally assumed to be responsible for a range of such processes. Hill & Findlay [134] review data for pulvinal movement in various plants, of which *Mimosa* is probably the most familiar, including the rapid curling of *Drostera* tentacles, stamen filament and stigma lobes, the traps of *Dionaea* and *Utricularia*, and the dramatic pollination mechanism of the trigger plant *Stylidium*.

Hill & Findlay [134] conclude that, in general, the movements follow a slow 'pumping up phase' in which cellular metabolic energy increases the osmotic pressures of cells, and hence their turgor pressures. The subsequently stimulated movement is brought about by a release of solutes from the cell down electrochemical gradients. Water follows osmotically. It can be seen, therefore, that the processes that bring about the actual movement are passive. For some of the more rapid movements this calls for remarkable changes in membrane properties. In *Stylidium*, for example, where a column holding the stamens and style (some 12 mg in weight) moves through some 200–300° in 10–25 ms, the motor cells are charged by import of potassium chloride, the concentration in the anterior cells being possibly as high as 600 mol m^{-3}. [368] This is the active process.

The discharge process requires the motor cells to produce maximum torque within 1–2 ms. Findlay & Findlay [369] estimate that a decrease in length (volume) of 5% in this time would be needed to account for the angular movement. Hill & Findlay [134] consider the implications of this for three alternative mechanisms. In the first they assume that osmotic adjustment to $0.95\pi_i$ occurs instantaneously, water then follows osmotically. In this case the cell membranes would have to have an L_p value of 2.8×10^{-3} m s^{-1} MPa^{-1}, some 3 or 4 orders of magnitude faster than the range given above. Although not considered by Hill & Findlay [134] this is also orders of magnitude higher than the few available values of wall L_p.

In their second case they assume that the reflection coefficient, σ, drops instantaneously to zero. In this case, assuming a positive turgor, all inwardly directed driving forces for water are lost and water is driven out by hydrostatic pressure. Here again L_p values of 1.4×10^{-4} m s^{-1} MPa^{-1} would be required.

They conclude that neither situation is likely and that the movement cannot be due to 'conventional' osmosis. They suggest the operation of a third alternative, electroosmosis. Assuming physiologically possible values for membrane potential differences and resting potentials for potassium and chloride ions they calculate that an L_p of 10^{-6} m s^{-1} MPa^{-1} and a membrane conductance of 138 S cm^{-2} would account for the observations. The L_p is now within the physiological range, the conductance values are still rather high. For comparison the conductance of the electroplax innervated membrane of the electric eel, *Electrophorus*, rises from 0.1 S cm^{-2} to only 10 S cm^{-2} when stimulated. [370] To allow for this Hill & Findlay [134] suggest that the L_p may well be an order of magnitude higher and this would bring the conductance values to within a meaningful range. We see from table 3.2 that this is indeed justified. Hill & Findlay [134] suggest that the mechanism for this change is the sudden opening of 0.6 nm pores over 0.1% of the surface of the motor cell. Clearly the membranes of these cells have remarkable properties.

3.5.4 *Long range transport*

Transport of water through the plant as a whole is beyond the scope of this review. Xylem transport has been thoroughly reviewed by Zimmermann. [7] Phloem transport is far more complex and is far from being as well understood. A stimulating review has recently been provided by Lang [371] (see also [372]).

One feature of xylem physiology can be related to the cellular parameters discussed here. A crucial water relations parameter of this cell is that it has a wall that can withstand considerable tension within the lumen without buckling, and that it can withstand this tension without drawing air through the cell wall. This latter feature is due to the fine nature of the water filled interstices of the wall. This property is widely discussed quantitatively in terms of capillary tension, e.g. Nobel [38] who calculates that channels of effective radius of 5 nm could withstand a tension of 30 MPa. It is likely that such values are only useful in that they provide an estimate of possible tensions. A fascinating suggestion, however, has been discussed by Zimmermann [7] to the effect that the walls may possess pores of specific sizes that allow air penetration at specific tensions. He calls these 'designed leaks' that act as safety devices within the xylem.

The water relations of phloem are even less well understood. In the context of this review, however, the remarkable use of aphid stylets to measure phloem turgor pressure is noteworthy. Pressures of up to 1.0 MPa were measured in the sieve tubes of *Salix babylonica*. [373] Equally remarkable is the apparent success of the 'phloem needle' approach of inserting a 25 gauge hypodermic needle into tissues. Pressures of 0.7–2.4 MPa were found at various heights in a red oak tree [374] and 0.03–1.0 MPa were found in the squirting cucumber, *Ecballium*. [375] The apparent influence of turgor pressure on phloem loading and unloading has been referred to above.

3.6 Concluding remarks

The mass of information presented here clearly tells us a great deal about the water relations of plants at single cell resolution. The recent development of new techniques is beginning to allow an integrated understanding of individual cells and their interaction in the behaviour of uniform tissues. However, the answers to four basic questions appear still to have eluded us. Firstly in order to link cell behaviour into tissue behaviour accurate measurements of cell wall hydraulic conductivity and geometry are required. Although the apoplast pathway is clearly important for solute flows across tissues it appears that apoplast, symplast and vacuole-to-vacuole pathways are remarkably finely balanced. Secondly the hydraulic relations of the plasmodesmata are very poorly known. Here, indeed, prediction of symplasmic behaviour with regard to water flow appears impossible. Thirdly, and probably most importantly, comes the apparently unresolved question of the state of water in the cytoplasm. Although fully blown A–I models of the type proposed by Ling (e.g. [122]) are not widely accepted, some deviation from solution ideality within the often minute volume of the cytoplasm between membranes and macromolecules must occur. Elsewhere [131] the importance of linking water and solute relations studies together for their mutual benefit has been argued. While this approach is fine for the vacuole and experiments in which a bulk external phase is available (i.e. fresh or sea water for single celled plants, or a bathing medium for excised sections of higher plants) it is more questionable in the case of the cytoplasm and the wall. In the case of the wall it would appear that this approach may be valid with certain justifiable modifications (based mainly on the, presumably, rigid arrays of the wall). The cytoplasm, however, must remain something of a *terra incognita* in this respect. Finally, we have the puzzle of the turgor sensing systems and mechanisms. So many plants appear capable of some form of turgor adjustment yet we know very little about what is being sensed and by what. If turgor pressure really is the effector we shall find a novel class of biophysical/biochemical interactions that may not only clarify the basis of an important phenomenon of plant physiology but also be of considerable biotechnological value.

Appendix

Techniques of measuring cell-related water relations parameters

Various techniques of studying the behaviour of the parameters outlined above applying to single cells, tissues and organs have been developed. A summary of the various techniques mentioned in this review is given here. Those techniques that provide information regarding cell parameters directly are given more emphasis than those that provide only tissue averaged values.

A3.1 Plasmometry

The use of the osmotic behaviour of individual cells, observed under the microscope, extends back to the work of Pfeffer, long before the nature of the semipermeable membrane was understood. More recently a major proponent of the use of plasmometry in plant water relations science has been the group of Stadelmann (e.g. [376]). (For a more recent introduction to the literature see [377].)

The technique is based on the phenomenon that when plant tissues are placed in a range of hypertonic media of non-penetrant solutes water is drawn out of the cells by osmosis. At external concentrations above that for loss of turgor pressure one of two events may occur. Either the wall will fold in on itself (a process called cytorrhesis) or, if the osmoticum can penetrate the cell wall, the plasmalemma lining the wall will pull away from the wall and become visible within the outline of the wall (plasmolysis). The precise cytorrhetic point is not easily observed under the microscope. The point at which the protoplast pulls away from the wall in plasmolysis, on the other hand, is often easily discernible in vacuolated cells. (Especially if they contain coloured pigments.) This point is called the point of incipient plasmolysis and corresponds to the point at which the differential hydrostatic pressure across the cell wall (turgor pressure) reaches zero. At this point from equation (3.30) it can be seen that $\sigma\pi$ of the cell must equal $\sigma\pi$ of the medium used to bring about incipient plasmolysis. If the reflection coefficients of all the osmotica are unity then $\pi_i = \pi_0$. (Indeed as long as the effective reflection coefficient of the bathing medium is the same as that of the vacuolar sap $\pi_i = \pi_0$ at incipient plasmolysis. This could be achieved by using an osmoticum of similar composition to the vacuolar sap (but see note on composite membranes, section 3.5.2).)

Two basic variants of the technique are employed. The most frequently used is the 'method of incipient plasmolysis' based more or less directly on the outline description above. [376] The second variant is called the 'plasmometric method' and was first used by Höfler. [378] It depends on being able to measure accurately the volume of the protoplast at an external osmotic pressure higher than that of the plasmolytic point. If several such measurements are made on the same cell at different osmotic concentrations then extrapolation to the plasmolytic point is possible. In practice this is done for cylindrical cells in which the protoplast pulls away from the cell wall at the two ends of the cell. [376] The technique works well for cells of simple geometry and has many advantages over the method of incipient plasmolysis. Following *rates* of change of volume after osmotic perturbation allows an estimation of a value for L_p for the cell membranes. By choosing suitable geometry (e.g. cylindrical cells) this may be performed quite accurately. Equation (3.24) is then used to calculate L_p from cell geometry and shrinkage/swelling kinetics. A word of warning in interpretation of results obtained by the technique has been given by Zimmermann and colleagues who have shown that in some cases for both algae (e.g. *Valonia utricularis* [141, 284]) and higher plants (*Elodea densa* [150]) L_p is a function of turgor pressure and that at low pressures L_p is increased by more than an order of magnitude over its value at 'physiological' external osmotic pressures. Naturally all plasmolytic measurements are made at zero effective turgor pressure. In addition they point out that L_p, σ and ω may be dependent on the concentration of the osmoticum outside the cell. At high concentrations they believe that the exposed membranes are dehydrated to an extent that influences their L_p.

Having determined π_i at zero turgor pressure it is now possible to attempt an estimate of the turgor pressure of the cell in its *in vivo* state. The principle is to find the external concentration at which the cell volume is the same as when under the natural conditions of the cell or tissue. At this point the turgor pressure will be the same as that under natural conditions, and can be determined from equation (3.22) by subtracting the osmotic pressure of the outside osmoticum from that of the cell at that volume. The value for π_i in the turgid state will be less than that of the plasmolytic point (π_g) but since the product $\pi_i V$ will be constant (as long as V is corrected for non-osmotic volume) π_i can be calculated from π_g and the difference in volume. This, in principle, also allows the calculation of a 'standing' elastic modulus for the cell from equation (3.32) or (3.33).

A feature of the technique is that a normal distribution of plasmolytic points within a population of cells may be interpreted in two ways. It reflects either a real heterogeneity of osmotic pressures at ambient turgor pressures or that although ambient turgor and osmotic pressures are constant a microheterogeneity of wall elastic modulus results in cell to cell variation at zero turgor.

A3.2 Pressure bomb

The robustness of the pressure bomb has made it a very widely used instrument in field studies of *whole plant water* relations. [379] It provides an indication of the value of some *tissue averaged* parameters. The device allows the water potential of an organ to be manipulated by altering the hydrostatic pressure of the symplast and vacuole [206, 209, 380] and permits the estimation of π_i, turgor pressure and ε. These are obtained from the analysis of a pressure/volume curve, in which the behaviour of symplast/vacuole volume as a function of applied pressure is observed. Similar data may be derived from similar pressure/volume curves obtained during air drying of a tissue. At each point of the curve tissue weight is plotted against tissue water potential measured psychrometrically. [381] This second method overcomes a potential source of error of the bomb method which is generally used with the assumption that apoplasmic wall osmotic potential is negligible. In some cases at least this assumption is in error. [174, 214, 215] Measuring water potential directly measures both wall hydrostatic tension and osmotic potential.

A3.3 Transcellular osmosis

A now classical way of measuring the hydraulic conductivity, L_p, and reflection coefficient of the membranes of cylindrical giant algal cells is the method of transcellular osmosis, developed by Osterhout [382] and by Kamiya & Tazawa [383] and used by Dainty and colleagues [107, 287, 288, 384] and the school of Kamiya & Tazawa (Osaka/Tokyo school). This work has contributed considerably to our understanding of the water relations of algal cells and membranes.

It involves the bathing of *Nitella* or *Chara* cells in a double chambered vessel, one chamber of which contains water, the other a solution of osmoticum. The cell is sealed into a hole separating the two chambers so that each end is bathed in a different solution (figure 3.14). The cell now behaves as a double semipermeable membrane as the water flow must cross the cell

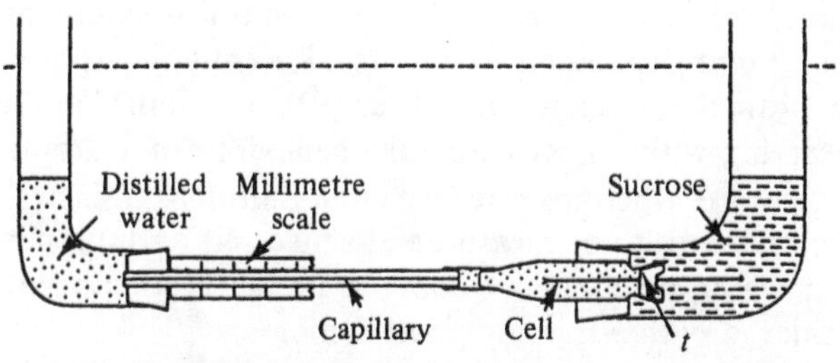

Fig. 3.14. Transcellular osmosis apparatus. [107] Reproduced from *Australian Journal Biological Science* **12** (1959), 395–411 with permission.

membranes twice. The effect of various osmotica on the osmotic behaviour of the membranes can now be studied.

This technique produced useful estimates of σ (table 3.4) and L_p ([23] and table 3.1). A surprising observation was that when the cell was arranged asymmetrically the values obtained for L_p were different in each direction. [288] These authors suggested that this was due to differential dehydration of the membrane by the external osmoticum resulting in a decreased L_p.

A3.4 Perfusion techniques

Because their size allows the necessary manipulation giant algal cells lend themselves to another manipulative technique that has contributed considerably. Early work on the use of perfused cells has been reviewed by Gutknecht *et al.* [19] and by Gutknecht & Bisson. [195] This provides a unique way of controlling membrane current and voltage as well as turgor, $\Delta\pi$ and specific ion concentrations, and probably represents the most complete method of studying plant cell solute and water relations available. (Recently Spyropoulos [286] has used this technique to calculate values for the Onsager cross coefficient L_{pD}, see equation (3.20).) Unfortunately it appears that it is impractical to attempt a similar treatment of regular higher plant tissue cells.

A3.5 Freezing point and psychrometric methods

A group of highly used techniques for providing *tissue averaged* information is based on the literal interpretation of equations (3.11) and (3.14). Since the vacuolar volume often comprises a major proportion of the cell and tissue the osmotic pressure of extracted tissue sap (or homogenised tissue) is taken to approximate π_i. Measurement of external water potential allows cell turgor pressure to be calculated by difference.

$$\text{Turgor} = \pi_i + \text{External water potential} \quad (3.51)$$

Indeed this approach can yield good results under certain fixed conditions in which there is no doubt as to external water potential (cells bathed in a solution of known osmotic pressure for example [259]). Osmotic pressure can be measured by freezing point depression techniques. Recently, equipment has been developed to allow the measurement of samples of volumes in the nanolitre range by observing with a microscope the behaviour of a droplet submerged in oil in an accurately thermostatted well (the Clifton osmometer). This will allow osmotic pressures to be measured at single cell resolution in some cases. Measuring water potential including a pressure component, however, cannot be achieved with a freezing point osmometer.

A revolution in whole plant water relations research occurred with the introduction of practical and robust thermocouple psychrometers for the accurate measurement of the vapour pressure of aqueous samples and tissues. These techniques work on the principle that the rate of evaporation of water from a droplet of water is proportional to the chemical potential (and hence water potential) of the water in the air around the droplet. Evaporation cools the water remaining in the droplet. The rate of cooling is therefore a function of the chemical potential of water in the air. If the temperature of the droplet can be accurately followed this rate of cooling can be measured. This is achieved by positioning the droplet on a thermocouple junction. An early but thorough review of the theory and techniques available may be found in Brown & van Haveren. [385] Two types of thermocouple psychrometer are in common use *viz.* the type devised by Spanner [386] which depends on the Peltier cooling effect to form a pure water droplet on the thermocouple, and the type devised by Richards & Ogata [387] in which pure water is held in a small loop about 0.5 mm in diameter that is wrapped in the thermocouple junction (see [388] for a theoretical consideration of apparatus dimensions).

Instruments such as the Wescor dewpoint system, using Spanner type psychrometry [389] (see [390] for a critical review) are used widely in laboratories studying plant water relations and can measure samples down to a few microlitres in volume. Boyer & Knipling [391] suggested that such non-equilibrium techniques might be in error (of up to 30% for some woody species such as rhododendron leaves) if the path of water flow from the thermocouple junction to the water surface being measured included a significant resistance to the diffusion of water vapour, and proposed an improved technique by which the 'wet-loop' psychrometer of Richards & Ogata [387] was 'charged' with an aqueous solution of identical chemical potential to the sample to be measured. At such a 'null point' no water flow would occur and no error would be incurred as a result of unknown resistances in the sample. It can be shown that the influence of the resistance on the apparent potential is linear and measurement of thermocouple output with two or more solutions on the loop allows extrapolation to the value corresponding to zero evaporation. This method is called the isopiestic technique and, although not as convenient to use as the widely employed Spanner type, allows very accurate and reliable measurements of the chemical potential of water in biological samples to be obtained.

It is important to note that the heat generated by respiration of living tissues interferes with these measurements. One way of accounting for this is to use a dry thermocouple in addition to a wet one in order to measure sample heat output. Various methods of conducting this heat away, or even stopping respiration by the use of nitrogen in the sample chamber have been suggested (see [388]).

The main use of these instruments in water relations research is to measure the chemical potential of water in extracted samples and also in intact tissues, where values are taken to represent the state of water on the surfaces of water evaporation, i.e. the cell wall/apoplast.

A3.6 Pressure probe

The only direct method of measuring turgor pressure is based on the insertion of a microcapillary attached to a pressure measuring device into individual cells. This allows not only measurements of the cell parameters but also the continuous measurement of turgor pressure in cells of plants under physiologically relevant conditions. [14, 248, 392] The first such pressure probe was suitable only for giant algal cells. [393] However the group under Zimmermann & Steudle developed from it a practical instrument based on an electronic pressure transducer. Initially also used with giant algal cells the understanding of higher plant cell water relations was given a powerful new tool by its modification for use with higher plants. [394] This probe is a miniaturised manometer that allows not only the continuous measurement of turgor pressure in individual higher plant cells but also the transient manipulation of that pressure to allow L_p, ε and σ to be measured *in situ* under physiological conditions. (A thorough evaluation of errors due to puncturing the wall has been reported. [274])

The elastic modulus is determined directly from observed values of dP, dV and cell volume according to equation (3.33). [23] Various approaches have allowed L_p to be measured from pressure relaxations following both hydrostatic and osmotic perturbation [23, 149, 283, 315] using the equation

$$T_{1/2} = \frac{(\ln 2)V}{AL_p(\varepsilon + \pi_i)} \tag{3.52}$$

(where $T_{1/2}$ is the half time of turgor relaxation following perturbation of water potential of the system. The other parameters are defined in the text.)

Following osmotic perturbation the relative values of initial and equilibrium pressures allow the measurement of the reflection coefficient for osmotica in the bathing medium. [315] This has been extended by the 'turgor minimum method' [139, 154, 289] to allow measurement of solute permeability. Although in this treatment unstirred layers will influence values obtained, it is argued that such layers are physiologically relevant within plants and should not be considered purely as experimental artefacts.

Finally the modification of the procedure to allow measurement of volume changes at constant pressure ('pressure clamp') permits relaxation parameters to be calculated independently of ε. [276, 283, 395] In practice, this also allows determination of cell volume free from the errors inherent in visual estimation and also of cells buried within tissues. The relaxation times in this mode are longer than those used otherwise. This may be of advantage when this approach is applied to higher plant cells that have very rapid relaxation rates. It has currently only been used for giant algal cells. An electronic microelectrode method for determining cell volumes of giant algal cells independently of visual inspection has been described by Zimmermann, Benz & Koch. [396]

Recently progress has been made to automate the retrieval of information from the pressure probe. For giant algal cells, where the elasticity of the equipment is of relatively little importance the pressure transducer output may be fed directly to a computer (e.g. [395, 278]). For higher plant cells a resistance feedback system was proposed [394], however, natural variation of potential difference from vacuole to bathing medium rendered this method unreliable. Currently methods based on the monitoring of the cell sap/silicone oil boundary using electronic photosensors [397] or video [398] are being developed. A detailed description of a video data retrieval system linked to a computer has recently been published. [398]

Illumination of the relevant cells, essential for the operation of the probe, has presented some particular problems. Not only do conventional fibre-optic illuminators tend to warm the cells, but they rule out turgor pressure measurements in the dark or under subdued light. Recently Shackel & Brinckmann [14] have refined a technique by which cell and probe tip are illuminated by a single optical fibre threaded into the probe capillary allowing low light intensity to be used and yet maintaining meniscus visibility. Manipulation of this same fibre allows temporary sealing of the tip as it is inserted into cells, allowing the probe to be 'prepressurised'. This prevents the transient loss of turgor pressure usually associated with insertion.

A3.7 Shrinking and swelling

Hydraulic conductivity of cell membranes of *Nitella* have been measured by monitoring the time constant of shrinking and swelling of the cells following transfer from one osmoticum to another. [61, 399]

Acknowledgements

I would like to thank Prof. U. Zimmermann, Dr. E. Steudle and Prof. R. G. Wyn Jones for introducing me to the field of plant water relations. If I still do not understand them it is no fault of theirs!

References

1. D. Kramer. *Water Relations of Plants*. Academic Press: New York, 1983.
2. J. D. Bewley & M. Black. *Physiology and Biochemistry of Seeds*, Vol. 2. Springer-Verlag: Berlin, 1982.
3. J. D. Bewley & J. E. Krochko, in *Encyclopedia of Plant Physiology*, Vol. 12B (eds. O. L. Lange, P. S. Nobel, C. B. Osmond & H. Ziegler). Springer-Verlag: Berlin, 1982, pp. 325–78.
4. H. Meidner & D. W. Sheriff. *Water and Plants*. Blackie: Glasgow, 1976.
5. R. P. Ambroggi. *Scientific American* **243** (September 1980), 90–104.
6. F. Franks (ed.). *Water – A Comprehensive Treatise*, Vols. 1–5. Plenum: New York, 1975.
7. M. H. Zimmermann. *Xylem and the Ascent of Sap*. Springer-Verlag: Berlin, 1983.
8. C. M. Willmer. *Stomata*. Longman: London, 1983.
9. A. D. Tomos, E. Steudle, U. Zimmermann & E.-D. Schulze. *Plant Physiol.* **68** (1981), 1135–43.
10. H. Jones, A. D. Tomos, R. A. Leigh & R. G. Wyn Jones. *Planta* **158** (1983), 230–6.
11. E. Steudle & W. D. Jeschke. *Planta* **158** (1983), 237–48.
12. E. Steudle & J. S. Boyer. *Planta* **164** (1985), 189–200.
13. M. E. Westgate & E. Steudle. *Plant Physiol.* **78** (1985), 183–91.
14. K. A. Shackel & E. Brinckmann. *Plant Physiol.* **78** (1985), 66–70.
15. U. Lüttge. *Planta* **168** (1986), 287–9.
16. J. Pritchard, A. D. Tomos & R. G. Wyn Jones. *J. Exp. Bot.* **38** (1987), 948–59.
17. R. Hook. *Micrographia*. London, 1667. (Recent reprint Weinheim: New York, 1961.)
18. A. B. Hope & N. A. Walker. *Physiology of Giant Algal Cells*. Cambridge University Press: Cambridge, 1975.
19. J. Gutknecht, D. F. Hastings & M. A. Bisson, in *Membrane Transport in Biology*, Vol. 3. (eds. G. Giebisch, D. Tosteson & H. H. Ussing). Springer-Verlag: Berlin, 1978, pp. 125–74.
20. J. Dainty. *Advan. Bot. Res.* **1** (1963), 279–326.
21. J. Dainty. *Encyclopedia of Plant Physiology*, Vol. 2A. (eds. U. Lüttge & M. Pitman). Springer-Verlag: Berlin, 1976, pp. 12–35.
22. U. Zimmermann, in *Integration of Activity in the Higher Plant*. (ed. D. Jennings). Cambridge University Press: Cambridge, 1977, pp. 117–54.
23. U. Zimmermann & E. Steudle. *Advanc. Bot. Res.* **6** (1978), 45–117.
24. U. Zimmermann. *Ann. Rev. Plant Physiol.* **29** (1978), 121–48.
25. M. T. Tyree & P. G. Jarvis. *Encyclopedia of Plant Physiology*, Vol. 12B (eds. O. L. Lange, P. S. Nobel, C. B. Osmond & H. Ziegler). Springer-Verlag: Berlin, 1982, pp. 35–77.
26. J. B. Passioura. *Encyclopedia of Plant Physiology*, Vol. 12B (eds. O. L. Lange, P. S. Nobel, C. B. Osmond & H. Ziegler). Springer-Verlag: Berlin, 1982, pp. 1–33.
27. C. R. House. *Water Transport in Cells and Tissues*. Arnold: London, 1974.
28. P. E. Weatherley. *Advanc. Bot. Res.* **3** (1970), 171–206.
29. T. C. Hsiao. *Ann. Rev. Plant. Physiol.* **24** (1973), 519–70.

30. T. C. Hsiao, E. Acevedo, E. Fereres & D. W. Henderson. *Phil. Trans. Roy. Soc. Lond. B.* **273** (1976), 479–500.
31. P. E. Weatherley. *Encyclopedia of Plant Physiology*, Vol. 12B (eds. O. L. Lange, P. S. Nobel, C. B. Osmond & H. Ziegler). Springer-Verlag: Berlin, 1982, pp. 79–109.
32. K. J. Bradford & T. C. Hsiao. *Encyclopedia of Plant Physiology*, Vol. 12B (eds. O. L. Lange, P. S. Nobel, C. B. Osmond & H. Ziegler). Springer-Verlag: Berlin, 1982, pp. 263–324.
33. R. G. Wyn Jones & J. Gorham. *Encyclopedia of Plant Physiology*, Vol. 12C (eds. O. L. Lange, P. S. Nobel, C. B. Osmond & H. Ziegler). Springer-Verlag: Berlin, 1983, pp. 35–58.
34. J. M. Morgan. *Ann. Rev. Plant. Physiol.* **35** (1984), 299–319.
35. J. S. Boyer. *Ann. Rev. Plant Physiol.* **36** (1985), 473–516.
36. D. J. Cosgrove. *Ann. Rev. Plant Physiol.* **37** (1986), 377–405.
37. R. O. Slatyer. *Plant–Water Relationships*. Academic Press: New York, 1967.
38. P. S. Nobel. *Biophysical Plant Physiology and Ecology*. Freeman: San Francisco, 1983.
39. D. A. Baker, in *Advanced Plant Physiology* (ed. M. B. Wilkins). Pitman: London, 1984, pp. 297–318.
40. W. S. Cram. *Encyclopedia of Plant Physiology*, Vol. 2A (eds. U. Lüttge & M. Pitman). Springer-Verlag: Berlin, 1976, pp. 284–361.
41. J. A. Raven. *Encyclopedia of Plant Physiology*, Vol. 2A (eds. U. Lüttge & M. Pitman). Springer-Verlag: Berlin, 1976, pp. 129–188.
42. A. Läuchli. *Encyclopedia of Plant Physiology*, Vol. 2B (eds. U. Lüttge & M. Pitman). Springer-Verlag: Berlin, 1976, pp. 3–34.
43. R. M. Spanswick. *Encyclopedia of Plant Physiology*, Vol. 2B (eds. U. Lüttge & M. Pitman). Springer-Verlag: Berlin, 1976, pp. 35–53.
44. R. O. Slatyer & S. A. Taylor. *Nature* **187** (1960), 922–4.
45. J. F. Thain. *Principles of Osmotic Phenomena*. Roy. Inst. Chem.: London, 1967.
46. B. Lucke & M. McCutcheon. *Physiol. Rev.* **12** (1932), 68–139.
47. E. Ponder. *Hemolysis and Related Phenomena*. Grune & Stratton: New York, 1948.
48. A. Mauro. *Science* **126** (1957), 252–3.
49. E. Robbins & A. Mauro. *J. Gen. Physiol.* **43** (1960), 523–32.
50. P. M. Ray. *Plant Physiol.* **35** (1960), 783–95.
51. J. W. Gibbs. *The Collected Works of J. Willard Gibbs*. Yale University Press: New Haven, 1948.
52. A. Katchalsky & P. F. Curran. *Nonequilibrium Thermodynamics in Biophysics*. Havard University Press: Cambridge, 1965.
53. J. J. Oertli. *Z. Pflazenphysiol.* **61** (1969), 264–5.
54. M. T. Tyree, in M. H. Zimmermann, *Trees: Structure and Function*. Springer-Verlag: Berlin, 1971, pp. 281–305.
55. A. Hill. *Q. Rev. Biophys.* **12** (1979), 67–99.
56. B. L. Silver. *The Physical Chemistry of Membranes*. Allen & Unwin: Boston, 1985.
57. L. Onsager. *Phys. Rev.* **37** (1931a), 405–26.
58. L. Onsager. *Phys. Rev.* **38** (1931b), 2265–79.
59. A. J. Staverman. *Rec. Trav. Chim.* **70** (1951), 344–52.

60. E. Steudle, in *Control of Leaf Growth* (eds. N. R. Baker, W. J. Davies & C. K. Ong). Cambridge University Press: Cambridge, 1985, pp. 35–55.

61. R. Fettiplace & D. A. Haydon. *Physiological Rev.* **60** (1980), 510–50.

62. T. Conlon & R. Outhred. *Biochim. Biophys. Acta* **288** (1972), 354–61.

63. R. I. Macey, in *Membrane Transport in Biology*, Vol. 2 (eds. G. Giebisch, D. Toteson & H. H. Ussing). Springer-Verlag: Berlin, 1979, pp. 1–58.

64. J. R. Philip. *Plant Physiol.* **33** (1958), 264–71.

65. J. D. Ferry. *Viscoelastic Properties of Polymers*, 2nd edn. Wiley: Chichester, 1970.

66. Y. Masuda. *Bot. Mag. Tokyo, Special Issue* **1** (1978), 103–23.

67. R. Yamamoto, K. Makai & Y. Masuda. *Plant & Cell Physiol.* **15** (1974), 1027–38.

68. S. A. Wainwright, W. D. Briggs, J. D. Currey & J. M. Gosline. *Mechanical Design in Organisms*. Edward Arnold: London, 1976.

69. J. F. Vincent. *Structural Biomaterials*. MacMillan: London, 1982.

70. K. L. Dorrington. *Soc. Exp. Biol. Symp.* **34** (1980), 289–314.

71. U. Zimmermann & D. Hüsken. *J. Membrane Biol.* **56** (1980), 55–64.

72. R. D. Preston. *The Physical Biology of Plant Cell Walls*. Chapman & Hall: London, 1974.

73. R. S. Pearce & P. Beckett. *Planta* **166** (1985), 335–40.

74. E. D. T. Atkins & K. D. Parker. *J. Polym. Sci. C.* **28** (1969), 69–81.

75. A. Suggett, in *Water – A Comprehensive Treatise*, Vol. 4 (ed. F. Franks). Plenum: New York, 1975, pp. 519–67.

76. R. D. Preston & A. B. Wardrop. *Biochim. Biophys. Acta.* **3** (1949), 549–59.

77. G. N. Christensen & K. Kelsey. *Australian J. Applied Sci.* **9** (1958), 265–82.

78. C. Skaar. *Water in Wood*. Syracuse University Press: Syracuse, 1972.

79. D. F. Gaff & D. J. Carr. *Australian J. Biol. Sci.* **14** (1961), 299–311.

80. G. E. Briggs, A. B. Hope & R. N. Robertson. *Electrolytes and Plant Cells*. Blackwell: London, 1961.

81. R. D. Preston, E. Nicolai, R. Reed & A. Millard. *Nature* **162** (1948), 665.

82. A. Frey-Wyssling, R. W. G. Wyckoff & K. Mühlethaler. *Experientia* **4** (1948), 475–6.

83. J. Sugiyama, H. Harada, Y. Fujiyoshi & N. Uyeda. *Planta*. **166** (1985), 161–68.

84. N. Carpita, D. Sabularse, D. Montezinos & D. P. Delmer. *Science* **205** (1979), 1144–7.

85. J. H. Miller. *Am. Fern J.* **70** (1980), 119–23.

86. M. Tepfer & I. E. P. Taylor. *Science* **213** (1981), 761–3.

87. J. Clifford, in *Water – A Comprehensive Treatise*, Vol. 5 (ed. F. Franks). Plenum: New York, 1975, pp. 75–132.

88. B. W. Ninham, in *Biophysics of Water* (eds. F. Franks & S. F. Mathias). Wiley: Chichester, 1982, pp. 105–19.

89. W. Drost-Hansen, in *Chemistry of the Cell Interface*, Vol. B (ed. H. D. Brown). Academic Press: New York, 1971, pp. 1–84.

90. W. Drost-Hansen, in *Biophysics of Water* (eds. F. Franks & S. F. Mathias). Wiley: Chichester, 1982, pp. 163–9.

91. J. S. Clegg, in *Biophysics of Water* (eds. F. Franks & S. F. Mathias). Wiley: Chichester, 1982, pp. 365–83.

92. W. H. Wade & N. Hackerman. *J. Phys. Chem.* **65** (1961), 1681–3.

93. J. L. Anderson & J. A. Quinn. *J. Chem. Soc. Faraday I* **68** (1972), 744–8.
94. F. Franks & S. F. Mathias (eds.). *Biophysics of Water*. Wiley: Chichester, 1982.
95. B.-Z. Ginzburg & A. Katchalsky. *J. Gen. Physiol.* **47** (1963), 401–18.
96. L. Layton. *Fluid Behaviour in Biological Systems*. Clarendon Press: Oxford, 1975.
97. E. Richter & R. Ehwald. *Plant Sci. Letts.* **32** (1983), 177–81.
98. R. Brdicka. *Grundlagen der physikalischen Chemie*. Deutscher Verlag der Wissenschaften: Berlin, 1982, p. 362.
99. D. P. Aikman, R. Harmer & T. S. O. Rust. *Physiol. Plant.* **48** (1980), 395–402.
100. U. Zimmermann & E. Steudle. *Australian J. Plant Physiol.* **2** (1975), 1–12.
101. M. T. Tyree. *J. Exp. Bot.* **24** (1973), 33–7.
102. M. T. Tyree. *Can. J. Bot.* **46** (1968), 317–27.
103. J. Schönherr. *Ber. Deut. Botan. Gez.* **87** (1974), 389–402.
104. J. Schönherr & M. J. Bukow. *Planta* **109** (1973), 73–93.
105. R. H. Atalla & U. P. Agarwal. *Science* **227** (1985), 636–8.
106. N. A. Walker & M. G. Pitman. *Encyclopedia of Plant Physiology*, Vol. 2A (eds. U. Lüttge & M. Pitman). Springer-Verlag: Berlin, 1976, pp. 93–126.
107. J. Dainty & A. B. Hope. *Australian J. Biol. Sci.* **12** (1959), 395–411.
108. P. J. Moore, A. G. Darvill, P. Albersheim & L. A. Staehelin. *Plant Physiol.* **82** (1986), 787–94.
109. J. Dainty, A. B. Hope & C. Denby. *Australian J. Biol. Sci.* **13** (1960), 267–76.
110. S. Petterson. *Physiol. Plant.* **19** (1966), 459–92.
111. J. B. Passioura. *J. Exp. Bot.* **31** (1980), 1161–9.
112. B.-Z. Ginzburg & D. Cohen. *Trans. Faraday Soc.* **60** (1964), 185–9.
113. R. Hayashi, H. Morikawa, N. Nakajima, Y. Ichikawa & M. Senda. *Plant & Cell Physiol.* **21** (1980), 999–1005.
114. J. T. Ogilvie, J. R. McIntosh & P. F. Curran. *Biochim. Biophys. Acta.* **66** (1963), 441–4.
115. M. Tazawa & N. Kamiya. *Ann. Rep. Sci. Works Fac. Sci. Osaka University* **13** (1965), 123–57.
116. M. Tazawa & N. Kamiya. *Australian J. Biol. Sci.* **19** (1966), 399–419.
117. P. H. Barry & A. B. Hope. *Biochim. Biophys. Acta.* **193** (1969), 124–8.
118. A. Katchalsky & O. Kedem. *Biophys. J.* **2** (1962), 53–78.
119. R. M. Hurtado & W. Drost-Hansen, in *Cell-Associated Water* (eds. W. Drost-Hansen & J. Clegg). Academic Press: New York, 1979, pp. 115–23.
120. P. M. Wiggins. *Biophys. J.* **13** (1973), 385–98.
121. H. G. Hempling, in *Biophysics of Water* (eds. F. Franks & S. F. Mathias). Wiley: Chichester, 1982, pp. 205–14.
122. G. N. Ling, in *The Aqueous Cytoplasm* (ed. A. D. Keith). Marcel Dekker: New York, 1979, pp. 23–60.
123. B.-Z. Ginzburg & M. Ginzburg, in *Biophysics of Water* (eds. F. Franks & S. F. Mathias). Wiley: Chichester, 1982, pp. 340–2.
124. P. Matile. *Ann. Rev. Plant Physiol.* **29** (1978), 193–213.
125. R. A. Leigh & R. G. Wyn Jones. *Advanc. Plant Nutrition*, Vol. 2 (eds. B. Tinker & A. Lauchli). Praeger: New York, 1986, pp. 249–79.
126. J. Dainty. *Ann. Rev. Plant Physiol.* **13** (1962), 379–402.
127. D. C. Tosteson. *Fed. Proc.* **22** (1963), 19–26.
128. J. Gray. *J. Exp. Biol.* **9** (1932), 277–99.

129. W. T. W. Potts & P. P. Rudy. *J. Exp. Biol.* **50** (1969), 223–37.
130. J. Dainty, in *Plant Cell Organelles* (ed. J. B. Pridham). Academic Press: London, 1968, pp. 40–6.
131. A. D. Tomos & R. G. Wyn Jones, in *Transport in Plant Cells and Tissues* (eds. D. A. Baker & J. L. Hall). Pitman: London, 1988, pp. 222–50.
132. R. E. Cleland. *Ann. Rev. Plant Physiol.* **22** (1971), 197–222.
133. K. Raschke, in *Encyclopedia of Plant Physiology*, Vol. 7 (eds. W. Haupt & M. E. Feisleib). Springer-Verlag: Berlin, 1979, pp. 383–441.
134. B. S. Hill & G. P. Findlay. *Q. Rev. Biophys.* **14** (1981), 173–222.
135. Ch. Coster. *Rec. Trav. Bot. Neer.* **24** (1927), 257–305.
136. T. C. Hsiao. *Encyclopedia of Plant Physiology*, Vol. 2B (eds. U. Lüttge & M. Pitman). Springer-Verlag: Berlin, 1976, pp. 195–221.
137. D. J. Patterson. *Biol. Rev.* **55** (1980), 1–46.
138. N. C. Turner & M. M. Jones, in *Adaptation of Plants to Water and High Temperature Stress* (eds. N. C. Turner & P. J. Kramer). Wiley: New York, 1980, pp. 87–103.
139. S. D. Tyerman & E. Steudle. *Plant Physiol.* **74** (1984), 464–8.
140. J. S. Graves & J. Gutknecht. *J. Gen. Physiol.* **67** (1976), 579–97.
141. U. Zimmermann & E. Steudle, in *Membrane Transport in Plants* (eds. U. Zimmermann & J. Dainty). Springer-Verlag: Berlin, 1974, pp. 64–71.
142. U. Zimmermann, E. Steudle & P. I. Lelkes. *Plant Physiol.* **58** (1976), 608–13.
143. R. H. Robichaux, K. E. Holsinger & S. R. Morse, in *On the Economy of Plant Form and Function* (ed. T. J. Givnish). Cambridge University Press: Cambridge, 1986, pp. 353–80.
144. J. Levitt. *Plant Physiol.* **82** (1986), 147–53.
145. J. Rygol & U. Lüttge. *Plant Cell & Env.* **6** (1983), 545–53.
146. J. C. Rankin & J. A. Davenport. *Animal Osmoregulation.* Blackie: Glasgow, 1981.
147. K.-H. Büchner, U. Zimmermann & F. W. Bentrup. *Planta* **151** (1981), 95–102.
148. A. D. Tomos & R. G. Wyn Jones, in *Biophysics of Water* (eds. F. Franks & S. F. Mathias). Wiley: Chichester, 1982, pp. 327–31.
149. A. D. Tomos & U. Zimmermann, in *Biophysics of Water* (eds. F. Franks & S. F. Mathias). Wiley: Chichester, 1982, pp. 256–61.
150. E. Steudle, U. Zimmermann & J. Zillikens. *Planta* **154** (1982), 371–80.
151. E. Brinckmann, S. D. Tyerman, E. Steudle & E.-D. Schulze. *Oecologia* **62** (1984), 110–17.
152. E. Steudle & J. Wieneke. *J. Amer. Soc. Hort. Sci.* **110** (1985), 824–9.
153. W. Wenkert, E. R. Lemon & T. R. Sinclair. *Ann. Bot.* **42** (1978), 295–307.
154. S. D. Tyerman & E. Steudle. *Australian J. Plant Physiol.* **9** (1982), 461–79.
155. R. A. Leigh & A. D. Tomos. In preparation.
156. E. Steudle, J. A. C. Smith & U. Lüttge. *Plant Physiol.* **66** (1980), 1155–63.
157. C. Wienecke, J. Gorham, A. D. Tomos & J. Davenport. In preparation.
158. S. Nakagawa, H. Kataoka & M. Tazawa. *Plant & Cell Physiol.* **15** (1974), 457–68.
159. M. A. Bisson & D. Bartholomew. *Plant Physiol.* **74** (1984), 252–5.
160. H. Kauss. *Ber. Deutsch. Bot. Ges.* **82** (1969), 115–25.
161. H. Kauss. *Plant Physiol.* **52** (1973), 613–15.
162. R. G. Hiller & H. Greenway. *Planta* **78** (1968), 49–59.
163. H. Kesseler. *Helgol. Wiss. Meeresunters.* **10** (1964), 73–90.

164. H. Kesseler. *Bot. Gothob.* **3** (1965), 103–11.
165. D. F. Hastings & J. Gutknecht. *J. Membrane Biol.* **28** (1976), 263–75.
166. M. A. Bisson & J. Gutknecht. *J. Membrane Biol.* **24** (1975), 183–200.
167. M. A. Bisson & G. O. Kirst. *Australian J. Plant Physiol.* **6** (1979), 523–38.
168. D. M. Dickson, R. G. Wyn Jones & J. Davenport. *Planta* **155** (1982), 409–15.
169. Y. Okazaki, T. Shimmen & M. Tazawa. *Plant & Cell Physiol.* **25** (1984), 565–71.
170. Y. Okazaki, T. Shimmen & M. Tazawa. *Plant & Cell Physiol.* **25** (1984), 573–81.
171. R. J. Reid, R. L. Jefferies & M. G. Pitman. *J. Exp. Bot.* **35** (1984), 925–37.
172. R. H. Reed & W. D. P. Stewart. *Biochim. Biophys. Acta* **812** (1985), 155–62.
173. R. Hoffmann & M. A. Bisson. *Can. J. Bot.* **64** (1986), 1599–605.
174. N. J. W. Clipson, A. D. Tomos, T. J. Flowers & R. G. Wyn Jones. *Planta* **165** (1985), 392–6.
175. E. Acevedo, E. Fereres, T. C. Hsiao & D. W. Henderson. *Plant Physiol.* **64** (1979), 476–80.
176. R. F. Meyer & J. S. Boyer. *Planta* **108** (1972), 77–87.
177. M. M. Jones & N. C. Turner. *Plant Physiol.* **61** (1978), 122–6.
178. E. Fereres, E. Acevedo, D. Henderson & T. C. Hsiao. *Physiol. Plant.* **44** (1978), 261–7.
179. S. A. Gavande & S. A. Taylor. *Agron. J.* **59** (1967), 4–7.
180. E. L. Gracean & J. S. Oh. *Nature* **235** (1972), 24–5.
181. P. V. Biscoe. *J. Exp. Bot.* **23** (1972), 930–40.
182. J. E. Goode & K. H. Higgs. *J. Hortic. Sci.* **48** (1973), 203–15.
183. L. Kappen & M. Mauer. *Oecologia* **12** (1975), 241–50.
184. A. P. Hartgerink & J. M. Mayo. *Can. J. Bot.* **54** (1976), 1884–95.
185. J. M. Morgan. *Nature* **270** (1977), 235.
186. J. M. Morgan. *J. Exp. Bot.* **31** (1980), 655–65.
187. R. Munns, C. J. Brady & E. W. R. Barlow. *Australian J. Plant Physiol.* **6** (1979), 379–89.
188. O. Osonubi & W. J. Davies. *Oecologia* **32** (1978), 323–32.
189. W. H. Bennert & H. A. Mooney. *Flora* **168** (1979), 405–27.
190. E. Karami, D. R. Kreig & J. E. Quisenberry. *Crop Sci.* **20** (1980), 421–6.
191. R. C. Ackerson. *Plant Physiol.* **67** (1981), 489–93.
192. S. Takami, H. M. Rawson & N. C. Turner. *Plant Cell & Environ.* **5** (1982), 279–86.
193. I. E. Henson, V. Mahalakshmi, F. R. Bidinger & G. Alagarswamy. *Plant Cell & Environ.* **5** (1982), 147–54.
194. T. C. Hsiao, J. C. O'Toole, E. B. Yambao & N. C. Turner. (Unpublished, quoted in [34]).
195. J. Gutknecht & M. A. Bisson, in *Water Relations in Membrane Transport in Plants and Animals* (eds. A. M. Jungreis, T. K. Hodges, A. Kleinzeller & S. G. Schultz). Academic Press: New York, 1977, pp. 3–14.
196. D. F. Hastings & J. Gutknecht, in *Membrane Transport in Plants* (eds. U. Zimmermann & J. Dainty). Springer-Verlag: Berlin, 1974, pp. 79–83.
197. H. Kauss & U. Rausch, in *Compartmentation in Algal Cells and their Interactions* (eds. W. Wiessner, D. G. Robinson & R. C. Carr). Springer-Verlag: Berlin, 1984, pp. 147–56.
198. G. O. Kirst & D. Kramer. *Plant Cell & Environ.* **4** (1981), 455–62.
199. D. L. McNeil. *Australian J. Plant Physiol.* **3** (1976), 311–24.

200. L. de Meis. *Biochem. Soc. Symp.* **50** (1985), 97–125.
201. A. D. Brown & J. R. Simpson. *J. Gen. Microbiol.* **72** (1972), 589–91.
202. L. J. Borowitzka & A. D. Brown. *Arch. Microbiol.* **96** (1974), 37–52.
203. R. G. Wyn Jones & A. Pollard. *Encyclopedia of Plant Physiology*, Vol. 15B (eds. A. Läuchli & R. L. Bieleski). Springer-Verlag: Berlin, 1982, pp. 528–63.
204. T. C. Hsaio & A. Läuchli. *Advanc. Plant Nutrition*, Vol. 2 (eds. B. Tinker & A. Läuchli). Praeger: New York, 1986, pp. 281–312.
205. S. N. Timasheff, in *Biophysics of Water* (eds. F. Franks & S. F. Mathias). Wiley: Chichester, 1982, pp. 70–2.
206. P. F. Scholander, H. T. Hammel, E. A. Hemmingsen & E. D. Bradstreet. *Proc. Nat. Acad. Sci. USA* **52** (1964), 119–25.
207. S. A. Sovonick, B. R. Geiger & R. J. Fellows. *Plant Phys.* **54** (1974), 886–91.
208. J. J. Oertli. *Z. Pflanzenernaehr. Bodenkd.* **147** (1984), 187–97.
209. M. T. Tyree & H. T. Hammel. *J. Exp. Bot.* **23** (1972), 267–82.
210. R. Storey, M. G. Pitman, R. Stelzer & C. Carter. *J. Exp. Bot.* **34** (1983), 778–94.
211. R. Storey, M. G. Pitman, R. Stelzer & C. Carter. *J. Exp. Bot.* **34** (1983), 1196–206.
212. L. Bernstein. *Plant Physiol.* **47** (1971), 361–5.
213. S. L. Jacobson. *Can. J. Bot.* **49** (1971), 121–7.
214. D. J. Cosgrove & R. E. Cleland. *Plant Physiol.* **72** (1983), 326–31.
215. R. A. Leigh & A. D. Tomos. *Planta* **159** (1983), 469–75.
216. N. J. W. Clipson & T. J. Flowers. *New Phytol.* **105** (1987), 359–66.
217. J. Rozema, H. Gude, F. Bijl & H. Wesselman. *Acta. Bot. Neer.* **30** (1981), 309–11.
218. R. S. Ownbey & B. E. Mahall. *Physiol. Plant.* **57** (1983), 189–95.
219. A. D. Tomos. Unpublished.
220. D. J. Cosgrove, J. K. E. Ortega & W. Shropshire. *Biophysical J.* **51** (1987), 413–24.
221. G. W. Gould & G. J. Dring. *Nature* **258** (1975), 402–5.
222. D. J. Ellar, in *Relations between Structure and Function in the Prokaryotic Cell* (eds. R. Y. Stanier, H. J. Rogers & J. B. Ward). Cambridge University Press: Cambridge, 1979, pp. 295–325.
223. J. Gutknecht. *Science* **160** (1968), 68–70.
224. E. Steudle, U. Zimmermann & P. I. Lelkes, in *Transmembrane Ion Exchanges in Plants* (eds. G. Ducet, R. Heller & M. Thellier). CNRS: Paris, 1977, pp. 123–32.
225. U. Zimmermann & E. Steudle, in *Regulation of Cell Membrane Activities in Plants* (eds. E. Marre & O. Ciferri). Elsevier/North-Holland: Amsterdam, 1977, pp. 231–42.
226. E. Steudle & P. I. Lelkes. Unpublished, quoted in [24].
227. U. Zimmermann & F. Beckers. *Planta* **138** (1978), 173–9.
228. S. Wendler, U. Zimmermann & F.-W. Bentrup. *J. Membrane Biol.* **72** (1983), 75–84.
229. H. Mummert. *Transportmechanismen für K^+, Na^+ und Cl^- in stationären und dynamischen Zuständen bei Acetabularia.* Ph.D. Thesis, Universität Tübingen, 1979.
230. E. Marre, P. Lado, F. Rasi Caldogno & R. Colombo. *Plant Sci. Letts.* **1** (1973), 179–84.
231. R. E. Cleland. *Planta* **127** (1975), 233–43.
232. E. Van Volkenburgh & J. S. Boyer. *Plant Physiol.* **77** (1985), 190–4.

233. S. Delrot & J. L. Bonnemain. *Comptes Rendus Acad. Sci. Paris Ser. D.* **287** (1978), 125–30.
234. I. Hagege, D. Hagege, S. Delrot, J.-P. Despegnel and J.-L. Bonnemain. *Compt. Rend. Acad. Sci. Paris Ser. III* **299** (1984), 435–40.
235. J. A. C. Smith & J. A. Milburn. *Planta* **148** (1980), 42–8.
236. S. Enoch & Z. Glinka. *Physiol. Plant.* **53** (1981), 548–52.
237. S. Enoch & Z. Glinka. *Physiol. Plant.* **59** (1983), 203–7.
238. J. W. Patrick. *Z. Pflanzenphysiol.* **111** (1983), 9–18.
239. J. W. Patrick. *J. Plant Physiol.* **115** (1984), 297–310.
240. P. Wolswinkel & A. Ammerlaan. *Physiol. Plant.* **61** (1984), 172–82.
241. P. Wolswinkel & A. Ammerlaan. *Plant Cell & Environ.* **9** (1986), 133–40.
242. J. H. Thorne. *Ann. Rev. Plant. Physiol.* **36** (1985), 317–43.
243. P. Wolswinkel. *Physiol. Plant.* **65** (1985), 331–9.
244. P. Wolswinkel, E. Kraus & A. Ammerlaan. *J. Exp. Bot.* **37** (1986), 1462–71.
245. L. Reinhold, A. Seiden & M. Volokita. *Plant. Physiol.* **75** (1984), 846–9.
246. A. H. G. C. Rijven & R. M. Gifford. *Plant, Cell & Environ.* **6** (1983), 417–25.
247. R. E. Wyse, E. Zamski & A. D. Tomos. *Plant. Physiol.* **81** (1986), 478–81.
248. J. Palta, R. G. Wyn Jones & A. D. Tomos. *Plant Cell & Environ.* **10** (1987), 735–40.
249. D. C. Chalmers, J. O. D. Coleman & N. J. Walton. *Biochem. Soc. Trans.* **14** (1986), 108–9.
250. F. W. Bentrup, in *Encyclopedia of Plant Physiology*, Vol. 7 (eds. W. Haupt & M. E. Feinleib). Springer-Verlag: Berlin, 1979, pp. 42–70.
251. U. Zimmermann, in *Animals and Environmental Fitness* (ed. R. Gilles). Pergamon: Oxford, 1980, pp. 441–59.
252. H. G. L. Coster, E. Steudle & U. Zimmermann. *Plant. Physiol.* **58** (1976), 636–43.
253. U. Zimmermann, K.-H. Büchner & R. Benz. *J. Membrane Biol.* **67** (1982), 183–97.
254. R. Pethig. *Dielectric and electronic properties of biological materials.* Wiley: Chichester, 1979.
255. N. Unwin & R. Henderson. *Scientific American* **250** (February 1984), 56–66.
256. R. Benz & U. Zimmermann. *Biophys. J.* **43** (1983), 13–26.
257. K.-H. Büchner, K. Rosenheck & U. Zimmermann. *J. Membrane Biol.* **88** (1985), 131–7.
258. W. J. Lucas & J. M. Alexander. *Plant Physiol.* **68** (1981), 553–9.
259. A. D. Tomos, R. A. Leigh, C. A. Shaw & R. G. Wyn Jones. *J. Exp. Bot.* **35** (1984), 1675–83.
260. M. T. Tyree. *Can. J. Bot.* **54** (1976), 2738–46.
261. M. T. Tyree & H. Richter. *Can. J. Bot.* **60** (1981), 911–16.
262. F. J. Molz & J. M. Ferrier. *Plant Cell & Environ.* **5** (1982), 191–206.
263. H. Jones, R. A. Leigh, A. D. Tomos & R. G. Wyn Jones. *Planta* **170** (1987), 257–62.
264. E. Münch. *Die Stoffbewegungen in der Pflanze.* Fischer: Jena, 1930.
265. M. T. Tyree. *J. Theor. Biol.* **26** (1970), 181–214.
266. U. Lüttge & N. Higinbotham. *Transport in Plants.* Springer-Verlag: Berlin, 1979.
267. H. Ginsburg & B.-Z. Ginzburg. *J. Exp. Bot.* **21** (1970), 580–92.
268. D. J. Cosgrove & E. Steudle. *Planta* **153** (1981), 343–50.

269. B. E. S. Gunning & R. L. Overall. *Bio Science* **93** (1983), 260–5.
270. M. G. Erwee & P. B. Goodwin. *Planta* **163** (1985), 9–19.
271. A. W. Robards & D. T. Clarkson, in *Intercellular Communication in Plants: Studies in Plasmodesmata*. Springer-Verlag: Berlin, 1976, pp. 181–203.
272. P. B. Goodwin. *Planta* **157** (1983), 124–30.
273. J. Graham, D. T. Clarkson & J. Sanderson. *Annual Report: Agricultural Research Council*, Letcombe Laboratory, Wantage, Oxon., UK, 1974.
274. U. Zimmermann & D. Hüsken. *Plant Physiol.* **64** (1980), 18–24.
275. D. Eamus & A. D. Tomos. *Plant Sci. Letts.* **31** (1983), 253–9.
276. S. Wendler & U. Zimmermann. *Planta* **164** (1985), 241–5.
277. S. Wendler & U. Zimmermann. *J. Membrane Biol.* **85** (1985), 121–32.
278. S. Wendler & U. Zimmermann. *J. Membrane Biol.* **85** (1985), 133–42.
279. K. Kiyosawa & M. Tazawa. *J. Membrane Biol.* **37** (1977), 157–66.
280. R. Collander. *Physiol. Plant.* **2** (1949), 300–11.
281. R. Collander. *Physiol. Plant.* **7** (1954), 420–45.
282. R. J. Lannoye, S. E. Tarr & J. Dainty. *J. Exp. Bot.* **21** (1970), 543–51.
283. E. Steudle, S. D. Tyerman & S. Wendler, in *Effects of Stress on Photosynthesis* (eds. R. Marcelle, H. Clijsters & M. van Poucke). Martinus Nijhoff/Junk: The Hague, 1983, pp. 95–109.
284. U. Zimmermann & E. Steudle. *J. Membrane Biol.* **16** (1974), 331–52.
285. E. L. Fiscus. *Plant Physiol.* **55** (1975), 917–22.
286. C. S. Spyropoulos. *J. Membrane Biol.* **76** (1983), 17–26.
287. J. Dainty & B.-Z. Ginzburg. *Biochim. Biophys. Acta.* **79** (1964), 102–11.
288. J. Dainty & B.-Z. Ginzburg. *Biochim. Biophys. Acta.* **79** (1964), 129–37.
289. E. Steudle & S. D. Tyerman. *J. Membrane Biol.* **75** (1983), 85–96.
290. D. C. Walton. *Ann. Rev. Plant. Physiol.* **31** (1980), 453–89.
291. Z. Glinka & L. Reinhold. *Plant. Physiol.* **48** (1971), 103–5.
292. Z. Glinka & L. Reinhold. *Plant. Physiol.* **49** (1972), 602–6.
293. G. N. St. G. Ord, I. F. Cameron & D. S. Fensom. *Can. J. Bot.* **55** (1977), 1–4.
294. H. Jones. *The Water Relations of Cereal Roots as Studied at the Cellular Level.* Ph.D. Thesis, University of Wales, 1985.
295. A. H. Markhart. *Plant. Physiol.* **69** (1982), Suppl. p. 38.
296. J. S. Boyer & G. Wu. *Planta* **139** (1978), 227–37.
297. M. J. Dowler, D. L. Rayle, W. T. Cande & P. M. Ray. *Plant. Physiol.* **53** (1974), 229–32.
298. M. Cailloux. *Can. J. Bot.* **50** (1972), 557–73.
299. J. Rygol & U. Lüttge. *Physiol. Veg.* **22** (1984), 783–92.
300. M. Wiesenseel & E. Schmeibidl. *Z. Pflanzenphysiol.* **70** (1973), 420–31.
301. J. Gutknecht. *Science* **158** (1967), 787–8.
302. G. Hevesy, E. Höfer & A. Krogh. *Skandinavisches Archiv für Physiol.* **72** (1935), 199–214.
303. K.-H. Büchner & U. Zimmermann. *Planta* **154** (1982), 318–25.
304. D. G. Stout, P. L. Steponkus & R. M. Cotts. *Can. J. Bot.* **55** (1977), 1623–31.
305. D. G. Stout, P. L. Steponkus & R. M. Cotts. *Plant. Physiol.* **62** (1978), 146–51.
306. D. G. Stout, P. L. Steponkus & R. M. Cotts. *Plant. Physiol.* **62** (1978), 636–41.
307. R. I. Macey, D. M. Karan & R. E. L. Farmer, in *Biomembranes*, Vol. 3 (eds. F. Kreuzer & J. F. G. Slegers). Plenum: New York, 1972, pp. 331–40.
308. J. Weigl. *Z. Naturforschung.* **22b** (1967), 885–90.

309. M. T. Tyree & Y. N. S. Cheung. *Can. J. Bot.* **55** (1977), 2591–9.
310. H. D. Price & T. E. Thompson. *J. Mol. Biol.* **41** (1969), 443–57.
311. A. H. Markhart, E. L. Fiscus, A. W. Naylor & P. J. Kramer. *Plant Physiol.* **64** (1979), 83–7.
312. K. Höfler. *J. Wiss. Botan.* **73** (1930), 300–50.
313. J. Levitt, G. W. Scarth & R. D. Gibbs. *Protoplasma* **26** (1936), 237–48.
314. E. Kiyosawa & M. Tazawa. *Protoplasma* **78** (1973), 203–14.
315. E. Steudle & U. Zimmermann. *Biochim. Biophys. Acta* **332** (1974), 399–412.
316. S. Rabinowitch, N. B. Grover & B.-Z. Ginzburg. *J. Membrane Biol.* **22** (1975), 211–30.
317. H. Ginsburg & B.-Z. Ginzburg. *J. Memb. Biol.* **4** (1971), 29–41.
318. H. Ginsburg. *J. Theor. Biol.* **32** (1971), 147–58.
319. J. Meyer, F. Sauer & D. Woermann, in *Membrane Transport in Plants* eds. U. Zimmermann & J. Dainty). Springer-Verlag: Berlin, 1974, pp. 28–35.
320. K. Katou & M. Furumoto. *Protoplasma* **130** (1986), 80–2.
321. K. Katou & M. Furumoto. *Protoplasma* **133** (1986), 174–85.
322. D. F. Hastings & J. Gutknecht. *J. Theor. Biol.* **73** (1978), 363–6.
323. H. Tedeschi & D. L. Harris. *Arch. Biochem. Biophys.* **58** (1955), 52–67.
324. C. J. Bentzel & A. K. Solomon. *J. Gen. Physiol.* **50** (1967), 1547–63.
325. H. Gimmler, R. Schirling & U. Tobler. *Z. Pflanzenphysiol.* **83** (1977), 145–58.
326. P. S. Nobel. *J. Membrane Biol.* **12** (1973), 287–99.
327. A. B. Tolberg & R. I. Macey. *Biochim. Biophys. Acta.* **109** (1965), 424–30.
328. P. S. Nobel. *J. Theor. Biol.* **23** (1969), 375–9.
329. E. L. Gross & L. Packer. *Arch. Biochem. Biophys.* **121** (1967), 779–89.
330. P. S. Nobel, in *Ion Transport in Plant Cells and Tissues* (eds. D. A. Baker & J. L. Hall). Elsevier/North Holland: Amsterdam, 1975, pp. 101–24.
331. C.-T. Wang & P. S. Nobel. *Biochim. Biophys. Acta.* **241** (1971), 200–12.
332. P. S. Nobel & Y.-N. S. Cheung. *Nature* **237** (1972), 207–8.
333. P. S. Nobel, in *Membrane Transport in Plants* (eds. U. Zimmermann & J. Dainty). Springer-Verlag: Berlin, 1974, pp. 289–95.
334. P. S. Nobel & C.-T. Wang. *Arch. Biochem. Biophys.* **157** (1973), 388–94.
335. A. D. Tomos, in *Control of Leaf Growth* (eds. N. R. Baker, W. J. Davies & C. K. Ong). Cambridge University Press: Cambridge, 1985, pp. 1–33.
336. D. J. Cosgrove. *Plant. Physiol.* **78** (1985), 347–56.
337. J. A. Lockhart. *J. Theor. Biol.* **8** (1965), 264–75.
338. P. M. Ray, P. B. Green & R. E. Cleland. *Nature* **239** (1972), 163–4.
339. D. J. Cosgrove. *Plant Physiol.* **68** (1981), 1439–46.
340. D. J. Cosgrove. *Phil. Trans. Roy. Soc. Lond., B* **303** (1983), 453–65.
341. P. B. Green, R. O. Erickson & J. Buggy. *Plant Physiol.* **47** (1971), 423–30.
342. E. van Volkenburgh & R. E. Cleland. *Planta* **148** (1980), 273–8.
343. G. Taylor & W. J. Davies. *New Phytol.* **101** (1985), 259–68.
344. G. Taylor & W. J. Davies. *New Phytol.* **104** (1986), 347–53.
345. R. E. Cleland. *Encyclopedia of Plant Physiology*, Vol. 13B (eds. W. Tanner & F. A. Loewus). Springer-Verlag: Berlin, 1981, pp. 225–76.
346. L. Taiz. *Ann. Rev. Plant Physiol.* **35** (1984), 585–657.
347. H. Greenway & R. Munns. *Plant Cell & Environ.* **6** (1983), 575–89.
348. K. Matsuda & A. Riazi. *Plant Physiol.* **68** (1981), 571–6.
349. V. A. Michelena & J. S. Boyer. *Plant Physiol.* **69** (1982), 1145–9.

350. M. A. Matthews, E. van Volkenburgh & J. S. Boyer. *Plant Cell & Environ.* **7** (1984), 199–206.

351. A. Thomas, A. D. Tomos, J. L. Stoddart & R. G. Wyn Jones. *5th Fed. Eur. Soc. Plant Physiol.*, Hamburg, 1986, Abs. 11.90.

352. T. Rich & A. D. Tomos. *J. Exp. Bot.* **39** (1988), in press.

353. E. van Volkenburgh & R. E. Cleland. *Planta* **167** (1986), 37–43.

354. E. van Volkenburgh & R. E. Cleland. *What's New in Plant Physiol.* **15** (1984), 25–8.

355. E. Acevedo, T. C. Hsiao & D. W. Henderson. *Plant Physiol.* **48** (1971), 631–6.

356. J. S. Boyer. *Ann. Rev. Plant Physiol.* **20** (1968), 351–64.

357. J. A. Bunce. *J. Exp. Bot.* **28** (1977), 156–61.

358. R. Goldenberg & R. Pratt. *Physiol. Veg.* **19** (1981), 523–32.

359. W. J. Davies & E. van Volkenburgh. *J. Exp. Bot.* **34** (1983), 987–99.

360. T. C. Hsaio, W. K. Silk & J. Jing, in *Control of Leaf Growth* (eds. N. R. Baker, W. J. Davies & C. K. Ong). Cambridge University Press: Cambridge, 1985, pp. 239–66.

361. J. W. Radin & J. S. Boyer. *Plant Physiol.* **69** (1982), 771–5.

362. G. Taylor & W. J. Davies. *Oecologia* **69** (1986), 589–93.

363. H. D. Hooker. *Bull. Torrey Bot. Club.* **43** (1916), 1–27.

364. H. D. Hooker. *Bull. Torrey Bot. Club.* **44** (1917), 389–403.

365. H. Ziegenspeck. *Bot. Arch.* **39** (1938), 268–309, 332–372.

366. D. E. Aylor, J. Y. Parlange & A. D. Krikorian. *Amer. J. Bot.* **60** (1973), 163–71.

367. H. Meidner & M. Edwards. *J. Exp. Bot.* **26** (1975), 319–30.

368. G. P. Findlay & C. K. Pallaghy. *Australian J. Plant Physiol.* **5** (1978), 219–29.

369. G. P. Findlay & N. Findlay. *Australian J. Plant Physiol.* **2** (1975), 597–621.

370. R. D. Keynes & H. Martins-Ferreira. *J. Physiol.* **119** (1953), 315–17.

371. A. Lang. *Plant Cell & Environ.* **6** (1983), 683–9.

372. M. J. Canny, in *Advanced Plant Physiology* (ed. M. B. Wilkins). Pitman: London, 1984, pp. 277–96.

373. J. P. Wright & D. B. Fisher. *Plant Physiol.* **65** (1980), 1133–5.

374. H. T. Hammel. *Plant Physiol.* **43** (1968), 1042–8.

375. S. N. Sheikholeslam & H. B. Currier. *Plant Physiol.* **59** (1975), 376–80.

376. E. J. Stadelmann. *Methods in Cell Physiology*, Vol. 2 (ed. D. M. Prescott). Academic Press: New York, 1966, pp. 143–216.

377. J. P. Palta & O. Y. Lee-Stadelmann. *Plant Cell & Environ.* **6** (1983), 601–10.

378. K. Höfler. *Ber. Deut. Botan. Ges.* **35** (1917), 706–26.

379. G. A. Ritchie. *Advanc. Ecol. Res.* **9** (1975), 165–254.

380. N. C. Turner. *Plant & Soil* **58** (1981), 339–66.

381. A. J. B. Talbot, M. T. Tyree & J. Dainty. *Can. J. Bot.* **53** (1975), 784–8.

382. W. J. V. Osterhout. *J. Gen. Physiol.* **32** (1949), 559–66.

383. N. Kamiya & M. Tazawa. *Protoplasma* **46** (1956), 394–422.

384. J. Dainty, H. Vinters & M. T. Tyree, in *Membrane Transport in Plants* (eds. U. Zimmermann & J. Dainty). Springer-Verlag: Berlin, 1974, 59–63.

385. R. W. Brown & B. P. van Haveren. *Psychrometry in Water Relations Research*. Utah State University: Logan, 1972.

386. D. C. Spanner. *J. Exp. Bot.* **2** (1951), 145–68.

387. L. A. Richards & G. Ogata. *Science* **128** (1958), 1089–90.

388. S. L. Rawlins, in *Psychrometry in Water Relations Research* (eds. R. W. Brown & B. P. van Haveren). Utah State University: Logan, 1972, pp. 43–50.

389. E. C. Campbell, G. S. Campbell & W. K. Barlow. *Agric. Meteorol.* **12** (1973), 113–21.

390. K. A. Shackel. *Plant Physiol.* **75** (1984), 766–72.

391. J. S. Boyer & E. B. Knipling. *Proc. Nat. Acad. Sci. USA* **54** (1965), 1044–51.

392. J. Rygol, K.-H. Büchner, K. Winter & U. Zimmermann. *Oecologia* **69** (1986), 171–5.

393. P. B. Green & F. W. Stanton. *Science* **155** (1967), 1675.

394. D. Hüsken, E. Steudle & U. Zimmermann. *Plant Physiol.* **61** (1978), 158–63.

395. S. Wendler & U. Zimmermann. *Plant Physiol.* **69** (1982), 998–1003.

396. U. Zimmermann, R. Benz & H. Koch. *Planta* **152** (1981), 352–5.

397. K.-H. Büchner. Personal communication.

398. D. J. Cosgrove & D. M. Durachko. *Rev. Sci. Instrum.* **57** (1986), 2614–19.

399. R. B. Kelly, P. G. Kohn & J. Dainty. *Trans. Bot. Soc. Edinburgh* **39** (1963), 373–91.

400. P. H. Barry & A. B. Hope. *Biophys. J.* **9** (1969), 700–28.

401. N. Kamiya, M. Tazawa & T. Takata. *Plant Cell Physiol.* **3** (1962), 285–92.

402. E. Brinckmann, M. Wartinger & D. J. von Willert. *Ber. Deutsch. Bot. Ges.* **98** (1985), 447–54.

403. E. Steudle, H. Ziegler & U. Zimmermann. *Planta* **159** (1983), 85–96.

404. U. Zimmermann, D. Hüsken & E.-D. Schulze. *Planta* **149** (1980), 445–53.

405. H. Arif & A. D. Tomos. Unpublished.

406. D. J. Cosgrove, E. van Volkenburgh & R. E. Cleland. *Planta* **162** (1984), 46–54.

407. V. Wartinovaara. *Acta. Bot. Fennica.* **34** (1944), 1–22.

408. J. P. Palta & E. J. Stadelmann. *J. Membrane Biol.* **33** (1977), 231–47.

409. J. S. Boyer, A. J. Cavalieri & E.-D. Schulze. *Planta* **163** (1985), 527–43.

410. C. A. Perry, R. A. Leigh, A. D. Tomos, R. E. Wyse & J. L. Hall. *Planta* **170** (1987), 353–61.

411. A. D. Tomos, R. A. Leigh, R. E. Wyse, R. G. Wyn Jones & J. Palta. Unpublished.

412. R. H. Reed. *Plant, Cell and Environ.* **7** (1984), 165–70.

413. J. Palta, R. E. Wyse, R. G. Wyn Jones and A. D. Tomos. In preparation.

414. T. B. Kinrade & R. E. Wyse. *Plant Physiol.* **82** (1986), 1148–50.

4

Transport of water across synthetic membranes

PROFESSOR W. PUSCH

Max-Planck-Institut für Biophysik, 6000 Frankfurt am Main 70, Kennedy-Allee 70

4.1 Introduction

In biological systems the transport of water and solutes from one compartment of the system (e.g. cell, organ, lumen, extra- and intracellular space) to another is generally controlled by membranes. The common biological membrane consists of a closely packed lipid bilayer (figure 4.1) where the lipid molecules are comparatively regularly arranged. The hydrophilic head groups of the lipid molecules are in contact with the adjacent aqueous phases whereas their hydrophobic head groups are facing each other within the membrane. The long hydrophobic chains of the lipid molecules are completely located within the membrane. The membrane thickness is about 75 Å. Proteins are embedded in the lipid bilayer which can move more or less freely in lateral directions inside the lipid phase. Proteins immersed only partially in the bilayer and thus belonging only to one side of the membrane can be differentiated from proteins completely penetrating the bilayer (the so-called integral proteins). Due to the comparatively low dielectric constant of the hydrophobic interior of the lipid bilayer, hydrophilic compounds (e.g. electrolytes and nonelectrolytes) are nearly insoluble within the lipid bilayer.

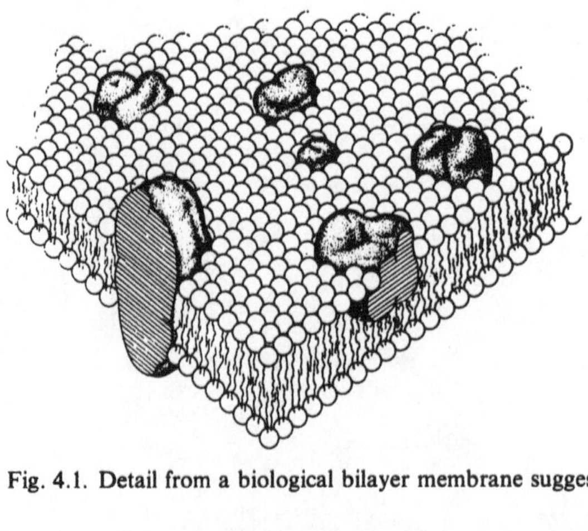

Fig. 4.1. Detail from a biological bilayer membrane suggested by Singer & Nicolson. [1]

Specific transport mechanisms (e.g. carrier-mediated transport, pore-formation, channels, active transport) are thus required to transport hydrophilic substances across the lipid bilayer. The integral proteins are of great importance in relation to the active transport of ions across lipid bilayers. In the presence of Ca^{2+} and Mg^{2+} ions, several integral proteins might combine to form a channel which is permeable only for distinct ions (e.g. Na^+, K^+, Li^+). Biological membranes are not only asymmetrically organized but also function asymmetrically due to the attachment of different proteins on each side of the membrane. This means, for example, that different processes might take their course at each membrane surface (e.g. different chemical reactions). The asymmetric function of biological membranes is regularly a precondition for so-called active transport mechanisms.

Transport of matter across thin barrier layers (membranes) is very important in synthetic membrane systems, for instance. In general, transport of matter across synthetic membranes in the presence of aqueous phases on both sides of the membrane requires the presence of water within the membrane (e.g. at least water of hydration for electrolytes and nonelectrolytes). The transport of water as well as that of solutes may be both convective and diffusive in nature. In the case of hydrodynamic permeabilities of the membrane which are of the same order of magnitude as diffuse permeabilities, a distinction between convective and diffusive fluxes is made difficult by the fact that both fluxes depend in the same way on the viscosity of the solution, η, and thus on temperature. Furthermore, diffusive transport can be free diffusion and so-called hindered diffusion. The mode of transport is closely correlated with the structure of water within the membrane. In addition to knowledge of the water structure within the membrane, the elaboration of the mechanisms of water and solute transport across biological and synthetic membranes requires a clear formulation of the transport relationships applied for the evaluation of corresponding transport experiments. Thus, firstly the transport relationships will be critically summarized. Secondly, the structure of water in biological and synthetic membranes will be discussed. Finally, an attempt will be made to correlate the water structure with the transport mechanisms of water and solute.

4.2 Transport relationships

Transport of matter across membranes can be treated by membrane model-independent transport relationships and membrane model-dependent ones. In the case of membrane model-independent relationships, the phenomenological relationships of the thermodynamics of irreversible processes offer themselves. Consider systems of the type solution-(′)/membrane/solution-(″) (figure 4.2) where the two external phases (solution-(′) and -(″)) are well stirred and are thus uniform, then no concentration, pressure, electrical potential, and temperature gradients will

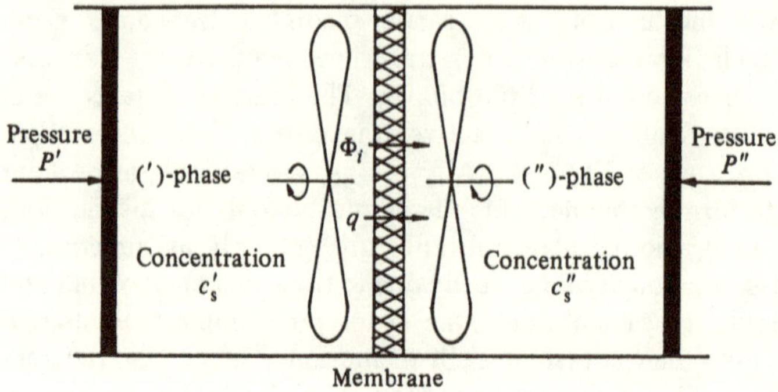

Fig. 4.2. Diagrammatic presentation of an isothermal membrane/solution system.

exist in the external phases. The gradients are restricted to the membrane phase. Treating the system as a discontinuous one, changes of the thermodynamic quantities of the two external phases only are considered. Using the laws of conservation of mass and energy as well as the laws of thermodynamics, linear relations between generalized fluxes and conjugated generalized forces can be set up for systems near equilibrium. [2, 3] Material fluxes are defined as follows:

$$\Phi_i = -(1/A)(\mathrm{d}n_i'/\mathrm{d}t) = (1/A)(\mathrm{d}n_i''/\mathrm{d}t) \tag{4.1}$$

where n_i is the mol number of component i and $\mathrm{d}n_i/\mathrm{d}t$ is the variation of the mol number with time in the corresponding phase. No distinction between diffusive and convective fluxes exists with this definition of Φ_i. Depending on the type of membrane (e.g. coarse porous, fine porous, dense), the fluxes may be either mainly convective or purely diffusive or both convective and diffusive. In the last case, the hydrodynamic permeability, l_p, will be of the same order of magnitude as the osmotic permeability, l_π. Since water transport across membranes separating electrolyte solutions can be generated by pressure, osmotic, and electrical potential differences, the transport relationships for a binary electrolyte solution will be discussed first.

4.2.1 *Phenomenological relationships*

Consider an isothermal membrane/solution system with a membrane separating two solutions of different solute concentrations, c_s' and c_s'' (figure 4.2), which are kept under different hydrostatic pressures, P' and P'', and different electrical potentials, E' and E''. Then a water flux, Φ_w (mol/cm^{2}M s), ion fluxes, Φ_i (eq/cm^{2}M s), or a solute flux, Φ_s (mol/cm^{2}M s), and eventually an electric current, characterized by its density j (mA/cm^{2}M s), across the membrane will result in a steady or quasi-steady state of the system. Since the individual fluxes Φ_w and Φ_s are not directly measurable, it is convenient to introduce other fluxes. As demonstrated by Schlögl [3], experimentally accessible fluxes, other than the electric current, are the volume flux, q

$(\mathrm{cm^3/cm^2M\ s})$, and the so-called chemical flux of the neutral species or salt, χ $(\mathrm{cm^3/cm^2M\ s})$. For a binary salt solution the following linear relationships exist between the original fluxes, Φ_w, Φ_i, Φ_s, and the experimentally accessible ones, j, q, and χ:

$$j = F(z_+\Phi_+ + z_-\Phi_-) \tag{4.2}$$

$$q = \bar{V}_s\Phi_s + \bar{V}_w\Phi_w \tag{4.3}$$

$$\chi = \Phi_s/\tilde{c}_s - \Phi_w/\tilde{c}_w \tag{4.4}$$

where F = Faraday number (C/eq), $\Phi_s = \Phi_+/\nu_+$ for reversible electrodes which are impermeable to the cations (Ag/AgCl electrodes for chlorides), ν_+ is the stoichiometric number of the cation within the corresponding salt; $\bar{V}_w$ and $\bar{V}_s$ are the partial molar volumes of water and solute, respectively, $(\mathrm{cm^3/mol})$; $\tilde{c}_s$ and $\tilde{c}_w$ are mean molar concentrations of water and solute, respectively, $(\mathrm{mol/cm^3})$; and z_+ and z_- are the valencies of the cation and the anion. As can be seen from equation (4.4), the chemical flux is equal to the difference in the 'superficial' velocity of solute and solvent. It should be pointed out in this connection that the chemical flux varies with the distance, x, normal to the membrane surfaces, even under steady-state conditions; i.e. the chemical flux is not constant across the membrane in a steady state. It is a measure of the variation of solute or salt concentration with time within the (′)- and (″)-phases and is related to these variations as follows:

$$\chi = -(V'/A\tilde{c}_s)(\mathrm{d}c_s'/\mathrm{d}t) = (V''/A\tilde{c}_s)(\mathrm{d}c_s''/\mathrm{d}t) \tag{4.4a}$$

where V' and V'' are the volumes of the (′)- and (″)-phases, respectively $(\mathrm{cm^3})$; $\tilde{c}_s = \frac{1}{2}(c_s' + c_s'')$; A is the effective membrane area ($\mathrm{cm^2}$ wet membrane $= \mathrm{cm^2M}$); and $\mathrm{d}c_s/\mathrm{d}t$ is the variation of solute or salt concentration with time t in the corresponding compartment (phase; $\mathrm{mol/cm^3\,s}$). The following linear relationships are usually set up between general fluxes and conjugated general forces for a system near equilibrium:

$$j = L_E\Delta E + L_{EP}\Delta P + L_{E\Pi}\Delta\Pi \tag{4.5a}$$

$$q = L_{PE}\Delta E + L_P\Delta P + L_{P\Pi}\Delta\Pi \tag{4.5b}$$

$$\chi = L_{\Pi E}\Delta E + L_{\Pi P}\Delta P + L_{\Pi}\Delta\Pi \tag{4.5c}$$

where $\Delta P = P' - P''$ (atm), $\Delta E = E' - E''$ (mV), and $\Delta\Pi = \Pi' - \Pi''$ (atm) are the pressure, electrical potential, and osmotic pressure difference, respectively, across the membrane. The electrical potential difference must be measured by use of reversible electrodes such as Ag/AgCl electrodes in the presence of $\mathrm{Cl^-}$ ions in the external solutions. As a consequence of the Onsager reciprocal relations, which require $L_{ik} = L_{ki}$, the number of independent transport coefficients, L_{ik}, reduces to six.

From an experimental point of view, a constant electric current $I = Aj$ is more easily achieved than a constant electrical potential difference, ΔE, across the membrane. Therefore, it is more convenient to use j instead of ΔE as the independent variable. Thus, rearranging equations (4.5a)–(4.5c) yields the following equivalent relationships:

$$-\Delta E = -r_e j + l_{ep}\Delta P + l_{e\pi}\Delta\Pi \tag{4.6a}$$

$$q = l_{pe} j + l_p \Delta P + l_{p\pi}\Delta\Pi \tag{4.6b}$$

$$\chi = l_{\pi e} j + l_{\pi p}\Delta P + l_\pi \Delta\Pi \tag{4.6c}$$

where

$l_{pe} = l_{ep} = L_{PE}/L_E =$ electro-osmotic permeability (cm^3/A s)
$l_{ep} = l_{pe} = L_{EP}/L_E =$ streaming potential coefficient (mV/atm)
$r_e = 1/L_E =$ ohmic area resistance of the membrane (Ω cm^{2}M)
$L_E = 1/r_e =$ electrical conductivity of the membrane (S/cm^{2}M).

If the resistance, r_e, is divided by the membrane thickness, d, the specific membrane resistance, ρ_m (Ω cmM), results. All further transport coefficients will be specified in the next subsection.

Under the usual boundary condition $j = 0$, the linear relationships (4.6a)–(4.6c) reduce to the following transport relationships:

$$\Delta E = -l_{ep}\Delta P - l_{e\pi}\Delta\Pi \tag{4.7a}$$

$$q = l_p \Delta P + l_{\pi p}\Delta\Pi \tag{4.7b}$$

$$\chi = l_{\pi p}\Delta P + l_\pi \Delta\Pi \tag{4.7c}$$

where

$l_p = L_P - (L_{EP}^2/L_E) =$ hydrodynamic permeability of the membrane at $j = 0$ (cm^3/cm^{2}M atm s)
$l_\pi = L_\Pi - (L_{E\Pi}^2/L_E) =$ osmotic permeability of the membrane at $j = 0$ (cm^3/cm^{2}M atm s)
$l_{\pi p} = L_{\Pi P} - (L_{EP}L_{E\Pi}/L_E) =$ coupling coefficient at $j = 0$ (cm^3/cm^{2}M s atm).

As is obvious from these correlations, the hydrodynamic and osmotic permeabilities, measured under the boundary condition $j = 0$, are always smaller than those measured under the boundary condition $\Delta E = 0$, since $L_{EP}^2 \geqslant 0$, $L_{E\Pi}^2 \geqslant 0$, and $L_E \geqslant 0$. This effect, although mainly small for l_p, is especially pronounced with ion-exchange membranes which possess a high fixed charge capacity. Now, with the definition of the reflection coefficient, $\sigma \equiv -l_{\pi p}/l_p$, given by Staverman [4], equations (4.7b) and (4.7c) finally yield:

$$q = l_p(\Delta P - \sigma\Delta\Pi) \tag{4.8a}$$

$$\chi = -\sigma l_p \Delta P + l_\pi \Delta\Pi \tag{4.8b}$$

A volume flux, q, is thus generated by both a pressure and an osmotic difference across the membrane. The pressure driven part of the water flux might be both convective and diffusive flow, where the diffusive component is then due to pressure diffusion. When convective water transport predominates, a substantial part of the solute flux will be convective flow too, provided $\sigma < 1$. This is obvious from a relationship which results by solving equations (4.8a) and (4.8b) for Φ_s using in addition equation (4.3) and the approximations $c_w \bar{V}_w \approx 1$ and $q \approx \bar{V}_w \Phi_w$:

$$\Phi_s = (1 - \sigma)\tilde{c}_s q + \omega \Delta \Pi \tag{4.9}$$

where

$$\omega = (l_\pi / l_p - \sigma^2)\tilde{c}_s l_p \tag{4.10}$$

As can be seen from relationship (4.9), the solute flux is composed of two contributions, $(1 - \sigma)\tilde{c}_s q$ and $\omega \Delta \Pi$, where the first contribution results from a coupling of solute transport with volume flow. Assuming that this coupling is mainly due to convective solute transport, a determination of the full set of transport coefficients l_p, l_π, and σ makes a formal distinction between convective and diffusive solute transport possible. However, it is still impossible to differentiate experimentally between convective and diffusive transport by checking, for instance, whether the same value of l_p is obtained when q is measured firstly as a function of ΔP at $\Delta \Pi = 0$ and secondly as a function of $\Delta \Pi$ at $\Delta P = 0$ since the second experiment yields σl_p and not l_p alone.

The transport behaviour of the synthetic membrane/binary solution system under the boundary condition $j = 0$ is thus characterized by the following three independent transport coefficients: l_π, l_p, and σ. The reflection coefficient σ is a measure of the permeability of the membrane with respect to the solute. If $\sigma = 1$, no solute passes the membrane (i.e. it is a semipermeable membrane); on the other hand, if $\sigma = 0$, solute and solvent are transported across the membrane with a concentration ratio equal to that present in one of the adjacent bulk solutions. It should be mentioned that σ can also assume negative values.

Adjusting the boundary condition to $\Delta \Pi = 0$, equation (4.6b) yields:

$$q = l_p \Delta P + l_{ep} j \tag{4.8c}$$

This equation correlates a volume flux, q, to a pressure difference and an electric current across the membrane (electroosmotic flow) provided that the term $l_{ep}j$ is not counterbalanced by the term $l_p \Delta P$. If the hydrodynamic volume flux, $l_p \Delta P$, is counterbalanced by the electroosmotic flux, $l_{ep}j$, a pressure difference, ΔP, causes an electric current, $I = Aj$, across the membrane at $q = 0$. As is obvious from equations (4.7a) and (4.8c), this situation requires a specific value of ΔE.

4.2.2 *Solution–diffusion model relationships*

Starting with the relationships of the thermodynamics of irreversible processes, Merten [5] developed a theoretical description of solute and solvent transport across synthetic membranes assuming the solute and solvent transport to be uncoupled and occurring by diffusion only. In addition, the effect of pressure upon the solute transport was neglected. The conventional solution–diffusion model characterizes pressure diffusion of water by the $l_p \Delta P$ term as well as Fickian diffusion of water and the solute by the terms $l_p \Delta \Pi$ and $(P_s/d)\Delta c_s$ respectively. Therefore, this transport model cannot characterize pressure diffusion of the solute. When dealing, for instance, with aqueous solutions of organic solutes such as phenol, pressure diffusion of the solute may also exist. Therefore, an extended solution–diffusion model was recently developed. [6] As this extended solution–diffusion model can be applied generally, it is presented first. Thereafter, the conventional solution–diffusion model is reviewed.

According to the assumptions underlying the solution–diffusion model, no coupling exists between solute and solvent flux: i.e. the solute and water fluxes are independent of each other and are related only to their respective chemical potential differences across the membrane, $\Delta\mu_s$ and $\Delta\mu_w$. The corresponding phenomenological relationships, relating generalized fluxes to conjugated generalized forces, are as follows:

$$\Phi_w = (L_w/d)\Delta\mu_w = (L_w \bar{V}_w/d)(\Delta P - \Delta\Pi) \tag{4.11}$$

$$\Phi_s = (L_s/d)\Delta\mu_s = (L_s/d)(\Delta\Pi/\tilde{c}_s + \bar{V}_s \Delta P) \tag{4.12}$$

where L_w and L_s (mol^2/J s cmM) are the corresponding original transport parameters. Now, let $q \simeq \bar{V}_w \Phi_w$ and $L_s RT/\tilde{c}_s = P_s = \varepsilon K_s D_{sm}$ where $K_s = C_s/c_s$ is the solute partition coefficient in ((mol/l pore vol.):(mol/l)), D_{sm} (cm^{2}M/s) is the solute diffusion coefficient within the membrane relative to the membrane-fixed reference system, and ε is the membrane porosity (cm^3 pore vol./cm^{3}M). Then, by introducing an additional transport parameter $l_{sp} = \bar{V}_s L_s/d$ (mol/cm^{2}M s atm), relationships (4.11) and (4.12) can be rewritten as follows where l_p is again the hydrodynamic permeability of the membrane, equal to $L_w \bar{V}_w^2/d$:

$$q = l_p(\Delta P - \Delta\Pi) \tag{4.13a}$$

$$\Phi_s = (\varepsilon K_s D_{sm}/d)\Delta c_s + l_{sp}\Delta P \tag{4.13b}$$

For completeness, a diffusion coefficient for water within the membrane is defined [3, p. 64]:

$$\varepsilon \bar{D}_w = l_p RTd/\bar{V}_w = L_w \bar{V}_w RT \tag{4.14}$$

This water diffusion coefficient is related to the membrane-fixed reference system. As Prigogine pointed out [7], this diffusion coefficient does not characterize only the thermal motion of the water molecules but may also include convective motion. In order to arrive at a purely thermal diffusion coefficient, characterizing only the Brownian motion of water molecules, the local centre of the mass-fixed reference system must be considered. Considering the membrane as a homogeneous phase composed of polymer, water and solutes, the following relation holds between the effective water diffusion coefficient, $\bar{D}_w$, and the one related to the barycentric motion, $\bar{D}_w^*$ [8]:

$$\bar{D}_w^* = \bar{D}_w(1 - w)^2 \tag{4.15}$$

where w is the water content of the membrane in g H_2O/g wet membrane. At very low water contents, $w \ll 1$, equation (4.15) yields $\bar{D}_w^* \approx \bar{D}_w$. On the other hand, at higher water contents, $w \to 1$, $\bar{D}_w$ will itself approach infinity, $\bar{D}_w \to \infty$ at $w \to 1$ [6, p. 329].

The transport relationships of the extended solution–diffusion model can be used for aqueous solutions of organic solutes if the organic solute is preferentially sorbed by the membrane indicating strong solute–membrane interactions. In the absence of such strong solute–membrane interactions it is sufficient to apply the standard solution–diffusion model relationships which result from equations (4.11) and (4.12) if the pressure-dependent part of the chemical potential of the solute is neglected. In that case, equation (4.11) remains valid whereas equation (4.12) reduces to:

$$\begin{aligned}\Phi_s &= (L_s/d)(\Delta\Pi/\tilde{c}_s) = (P_s/d)\Delta c_s \\ &= (\varepsilon K_s D_{sm}/d)\Delta c_s\end{aligned} \tag{4.12a}$$

4.2.3 *Fine porous membrane models*

Fine porous membrane models consider the membrane to be made up of a system of interconnected capillaries penetrating the rigid membrane material. The diameters, lengths, and orientations of the capillaries are subject to statistical variations. The statistical variation of the pore diameters, for instance, can be characterized by the pore size distribution curve. Applying the formalism of the thermodynamics of irreversible processes to the solute and solvent transport across a fine porous membrane, either the gradients of the generalized forces are linearly related to the conjugated generalized fluxes or the generalized fluxes are linearly related to the gradients of the conjugated generalized forces. In addition, local thermodynamic equilibrium is assumed to exist within the membrane system. When dealing with electrolyte solutions, further model assumptions regarding the electrical interaction of the ions with the charged or uncharged membrane matrix have to be added. The two models offering themselves for this purpose are the fine porous membrane model, introduced by Schmid [9], and the capillary model [10] based on the electrochemical double layer [11, 12] and treated by von Helmholtz and von Smoluchowski [13, 14]. Applying the fixed-charge model, it is assumed that the fixed charges are homogeneously distributed over the membrane phase yielding a constant fixed-charge concentration, independent of the local position x within the membrane. Schlögl [3, 15] used this model as the basis for his theoretical work on membrane transport. He integrated the Nernst–Planck equations across the membrane under several distinct boundary conditions as summarized in his book. [3] After a short discussion of the general Nernst–Planck equations, only those solutions which are relevant to water transport will be reported in this subsection.

Using the fine porous and fixed-charge membrane model, the following assumptions are made:

(*a*) The concentrations C_i of all species in the membrane phase depend only on the *x*-coordinate, implying that no concentration profile across the membrane pores parallel to the membrane surfaces exists.

(*b*) If the membrane contains fixed charges, there is excess of counterions in the pore liquid and thus the pore solution possesses a charge density, ρ_{el}, which also depends only on the *x*-coordinate.

(*c*) The solvent is assumed to have a mol fraction close to unity; therefore, the pore solution and the external solutions can be regarded as dilute solutions.

(*d*) Furthermore, no coupling of transport between solutes and/or ions is taken into account except the coupling between co- and counterions which is caused by the electrostatic interaction and leads, for example, to diffusion potentials. Only coupling of transport between solvent and solute or ions is taken into account. The equivalent concentration of the fixed charges of the membrane, C_X, referred to the pore volume of the membrane (eq/pore vol.) is given by:

$$\rho_{el} = -\omega F C_X \tag{4.16}$$

where F is the Faraday number (C/eq) and $\omega = +1$ or -1 for an anion or cation exchange membrane, respectively. In principle, the Poisson equation, $\Delta\varphi_m = -\rho_m/\varepsilon$, must be solved in order to obtain the electrical potential in the membrane phase, φ_m (here ρ_m is the total charge density, made up of the fixed charges and ρ_{el}, and ε is the dielectric constant of the membrane phase). Since no analytical solution of the corresponding Poisson equation is available, Schlögl assumed that electroneutrality exists in the membrane $\rho_m \approx 0$, and thus the co- and counterion concentrations, C_{co} and C_g, in the membrane phase are subject to the following restriction:

$$z_{co}C_{co} + z_g C_g + \omega C_X = 0 \tag{4.17}$$

The valencies, z_{co} and z_g, are counted positive for cations and negative for anions. As shown recently [16], this electroneutrality condition is obtained as a solution of the corresponding Poisson equation in a first approximation.

The ion or solute fluxes across the membrane are linearly related to the gradient of their electrochemical potentials, η_i. Neglecting activity coefficients, $\gamma_i \approx 1$, as well as the pressure-dependent term of the respective chemical potential, $\bar{V}_i \Delta P$, where $\bar{V}_i$ is the partial molar volume of the ions or solute (cm^3/mol), the following relationships result for the ion and solute fluxes, respectively (i = co or g):

$$\Phi_i = C_i q - \varepsilon D^{v}_{im}\{(\mathrm{d}C_i/\mathrm{d}x) + (z_i F C_i/RT)(\mathrm{d}\bar{\varphi}/\mathrm{d}x)\} \tag{4.18a}$$

and for neutral solutes ($z_i = 0$, s = solute)

$$\Phi_s = C_s q - \varepsilon D^{v}_{sm}(dC_s/dx) \tag{4.18b}$$

where D^{v}_{im} and D^{v}_{sm} are the diffusion coefficients of the ion or solute, respectively, relative to a volume-fixed reference system, and $\bar{\varphi}$ is the internal electrical potential at the position x within the membrane. The term $C_i q$ characterizes convective transport of the ith component and the terms $D^{v}_{im}(dC_i/dx)$ and $D^{v}_{im}(z_i/FC_i/RT)(d\bar{\varphi}/dx)$ diffusive transport due to a concentration gradient (dC_i/dx) and an electrical potential gradient ($d\bar{\varphi}/dx$). The internal electric field acts only on ions ($z_i \neq 0$) and thus causes electrodiffusion of ions.

The volume flux, q, is related to the corresponding driving forces by a hydrodynamic relationship:

$$q = d_h[-(d\bar{P}/dx) - \rho_{el}(d\bar{\varphi}/dx)] \tag{4.19}$$

where d_h is the hydrodynamic permeability coefficient of the membrane (cm^3/cmM s atm). The volume flux across the membrane is thus governed by the pressure gradient, $-(d\bar{P}/dx)$, within the membrane pores and the electric field, $\mathbf{E} = -d\bar{\varphi}/dx$, acting on the charged pore solution. It should be noted that the relationships (4.18) and (4.19), governing the transport of ions and solute across the membrane, do not directly comprise a term responsible for a volume flux as a consequence of a concentration gradient across the membrane (osmotic flow). This osmotic action (osmotic pump) arises only from a pressure gradient which develops across the internal membrane as a consequence of pressure jumps at the membrane/solution interfaces. Therefore, the internal pressure, $\bar{P}$, is composed of the hydrostatic pressure and the pressure jumps, $\delta P'$ and $\delta P''$, at the two membrane/solution interfaces.

The boundary conditions at the solution/membrane interfaces require that the electrochemical potentials, η_i, of the co- and counterions at the interfaces and the chemical potential, μ_s, of the solute are equal. Therefore, the following boundary conditions have to be satisfied. The boundary conditions for the charged species are as follows (i = co or g):

$$\eta'_i = \mu'_{io} + RT \ln a'_i + z_i F \varphi'$$
$$= \bar{\eta}'_i = \bar{\mu}'_{io} + RT \ln \bar{a}'_i + z_i F \bar{\varphi}' \tag{4.20a}$$

$$\eta''_i = \mu''_{io} + RT \ln a''_i + z_i F \varphi''$$
$$= \bar{\eta}''_i = \bar{\mu}''_{io} + RT \ln \bar{a}''_i + z_i F \bar{\varphi}'' \tag{4.20b}$$

where

η_i = electrochemical potential of the co- or counterion (J/mol)
μ_i = chemical potential of the co- or counterion (J/mol)
μ_{io} = standard chemical potential of the co- or counterion (J/mol)
a_i = ion activity = $\gamma_i c_i$ (mol/l)

γ_i = ion activity coefficient
φ = electrical potential of the corresponding phase (mV)

Similarly, the boundary conditions for a solute are:

$$\mu'_s = \mu'_{s0} + RT \ln a'_s = \bar{\mu}'_s = \bar{\mu}'_{s_0} + RT \ln \bar{a}'_s \tag{4.21a}$$

$$\mu''_s = \mu''_{s_0} + RT \ln a''_s = \bar{\mu}''_s = \bar{\mu}''_{s0} + RT \ln \bar{a}''_s \tag{4.21b}$$

where ′ and ″ denote the two external phases and the bar designates the corresponding membrane phase at the membrane/solution interface. In this, the pressure-dependent part of the chemical potential of the ions or solute, $\bar{V}_i \Delta P$, was neglected; i = co, g, s. In addition, the boundary conditions of water can be written:

$$\mu'_w = \mu'_{w0} + RT \ln a'_w + \bar{V}'_w \Delta P' = \bar{\mu}'_w = \bar{\mu}'_{w0} + RT \ln \bar{a}'_w + \bar{V}'_{wm} \Delta \bar{P}' \tag{4.22a}$$

$$\mu''_w = \mu''_{w0} + RT \ln a''_w + \bar{V}''_w \Delta P'' = \bar{\mu}''_w = \bar{\mu}''_{w0} + RT \ln \bar{a}''_w + \bar{V}''_{wm} \Delta \bar{P}'' \tag{4.22b}$$

Assuming $\bar{\mu}_{w0} = \mu_{w0}$ as well as $\bar{V}_{wm} = \bar{V}_w$ and taking into account the relation $RT \ln a_w = -V_w \bar{\Pi}$ where $\bar{V}_{wm}$ is the partial molar volume of water within the membrane and Π denotes the osmotic pressure, the local equilibrium conditions for water, equations (4.22a) and (4.22b), yield:

$$\delta P' = \Delta \bar{P}' - \Delta P' = \bar{P}' - P' = \delta \Pi' = \bar{\Pi}' - \Pi' \tag{4.23a}$$

$$\delta P'' = \Delta \bar{P}'' - \Delta P'' = \bar{P}'' - P'' = \delta \Pi'' = \bar{\Pi}'' - \Pi'' \tag{4.23b}$$

where the pressure jumps at the solution/membrane interfaces, $\delta P'$ and $\delta P''$, correspond to a kind of swelling pressure. Replacing activities by the corresponding concentrations and using $x_w + \sum x_i = 1$ as well as the approximations $c_i \bar{V}_w \simeq x_i$ and $C_i \bar{V}_w \simeq \bar{x}_i$ (x_i = mol fraction), the following relationship for δP results:

$$\begin{aligned}\delta P &= (RT/\bar{V}_w) \ln(c_w/\bar{c}_w) = (RT/\bar{V}_w) \ln(x_w/\bar{x}_w) \\ &= (RT/\bar{V}_w) \ln\left[\left(1 - \sum_i x_i\right) \Big/ \left(1 - \sum_i \bar{x}_i\right)\right]\end{aligned} \tag{4.23c}$$

Using the Taylor expansion of $\ln(1-x) \approx -x$ and terminating after the first term, finally yields:

$$\delta P = \delta \Pi = RT\left[\sum_{i=1}^{n-1} (C_i - c_i)\right] \tag{4.23d}$$

where $c_i \approx x_i/\bar{V}_w$ and $C_i \approx \bar{x}_i/\bar{V}_w$ (i = co, g, s).

The local equilibrium conditions of ions lead to the corresponding partition coefficients, K_i (i = co, g) [17, 18]:

$$K_i = K_{i0}(\gamma_i/\bar{\gamma}_i) \exp(z_i F \Delta \varphi_{D0}/RT) \tag{4.24}$$

where

$K_i = C_i/c_i$ = individual ion partition coefficient

$K_{i0} = \exp(\Delta\mu_{i0}/RT)$ = individual standard ion partition coefficient

$\Delta\mu_{i0} = \mu_{i0} - \bar{\mu}_{i0}$ = standard chemical potential difference of ions

$\Delta\varphi_{D0} = \varphi - \bar{\varphi}$ = Donnan potential at the solution/membrane interface

γ_i = ion activity coefficient

For dilute solutions ($\bar{\gamma}_i, \gamma_i \approx 1$) and if, in addition $K_{i0} = 1$ is assumed, equation (4.24) simplifies and yields:

$$(C_i/c_i)^{1/z_i} = \exp(F\Delta\varphi_{D0}/RT) \tag{4.24a}$$

As is obvious from this equation, the term $(C_i/c_i)^{1/z_i}$ is independent of the ion species since the right hand side of equation (4.24a) does not contain ion specific terms.

For a neutral solute, equation (4.24) reduces to:

$$K_s = K_{s0}(\gamma_s/\bar{\gamma}_s) \tag{4.24b}$$

where the subscript s stands for solute. Therefore, $K_s \approx K_{s0}$ for dilute solutions where $\bar{\gamma}_s \approx 1$ and $\gamma_s \approx 1$. In this limiting case, the partition coefficient is thus determined only by the change in the standard chemical potential of the solute, $\Delta\mu_{s0}$. It should be noted that instead of $\bar{\gamma}_s \approx 1$ and $\gamma_s \approx 1$, $\bar{\gamma}_s/\gamma_s \approx 1$ is sufficient for $K_s \approx K_{s0}$ to hold. In this connection, it should also be mentioned that many authors, including Lewis & Randall [19, 20] assume that the standard chemical potentials in the two phases are identical thus including all existing differences in the activity coefficient: $\gamma_s^* = \gamma_s \exp(\Delta\mu_{s0}/RT)$.

The basic transport relationships of the solute fluxes, Φ_i, are nonlinear differential equations due to the term

$$\varepsilon D^{\mathrm{v}}_{\mathrm{im}}(z_i F C_i/RT)(\mathrm{d}\bar{\varphi}/\mathrm{d}x)$$

Therefore, no general complete integral of equations (4.18) exists. The flux equations can thus only be integrated for special cases. The most simple case is that with uncharged species as then the electrical term disappears. If only uncharged species are present in the solution/membrane system, equations (4.18a) and (4.18b) reduce to a system of linear differential equations. When dealing with the transport of a binary solution across a membrane, the corresponding differential equations for Φ_s and q are then easily integrated.

Considering a dialysis/osmosis system where a homogeneous membrane separates two solutions of equal solute or salt concentrations ($c_s' = c_s''$), it is easy to find the integral of the corresponding differential equations (4.18a) and (4.18b). Since the solute concentration is the same on both sides of the membrane, $C_i(x) = C_i$ = independent of x. The differential equations then reduce to algebraic linear equations. Since no term of equations (4.18a) and (4.18b) depends on x, $\mathrm{d}\bar{\varphi}/\mathrm{d}x$ is also a constant independent of x. Thus $\mathrm{d}\bar{\varphi}/\mathrm{d}x = -\Delta\bar{\varphi}/\delta$ where $\Delta\bar{\varphi} = \bar{\varphi}' - \bar{\varphi}''$. The entire potential difference between two reversible Ag/AgCl electrodes, for instance, is composed of the potential drops at the electrode/solution interfaces, the Donnan potentials at the membrane/solution interfaces, and the internal potential difference $\Delta\bar{\varphi}$. Therefore,

$$\Delta E = \delta E' + \Delta\varphi'_{DO} + \Delta\bar{\varphi} + \Delta\varphi''_{DO} + \delta E''$$

where $\delta E'$ and $\delta E''$ are the potential jumps at the solution/electrode interfaces. Because of the symmetry of the solutions, $\delta E' = -\delta E''$ and $\Delta\varphi'_{DO} = -\Delta\varphi''_{DO}$ and thus $\Delta E = \Delta\bar{\varphi}$ results under the specific conditions. With these simplifications, if the specific boundary conditions $\delta\Pi' = \delta\Pi''$ and thus $\Delta\bar{P} = \Delta P$ are considered in addition, equations (4.18a), (4.18b) and (4.19) yield:

$$q = (d_h/\delta)(\Delta P - \omega F C_X \Delta E) \tag{4.25}$$

$$\Phi_i = (FC_i/RT\delta)(z_i \varepsilon D^v_{im} - d_h RT\omega C_X)\Delta E + (d_h C_X/\delta)\Delta P \tag{4.26}$$

where $\delta = td$ is the path length of the pores (cmM); t is the tortuosity factor. The electrical current crossing the membrane is given by:

$$j = F(z_g \Phi_g + z_{co}\Phi_{co}) \tag{4.27}$$

Comparing these linear relationships with the phenomenological ones, equations (4.6a)–(4.6c), by putting $\Delta\Pi = 0$, yields the following relations between phenomenological transport coefficients and those of the model-dependent relationships:

$$L_E = (F^2/\delta)[(1/RT)(z_g^2 \varepsilon C_g \bar{D}^v_g + z_{co}^2 \varepsilon C_{co} \bar{D}^v_{co}) + C_X^2 d_h] \tag{4.28}$$

$$L_{EP} = L_{PE} = -(F/\delta)\omega C_X d_h \tag{4.29}$$

$$L_P = d_h/\delta \tag{4.30}$$

As can be seen from equation (4.26), an electrical potential difference, ΔE, acts on the pore solution like a pressure difference $\Delta P = FC_X \Delta E$, and might thus be termed 'electrical tension'. Equation (4.25) yields under the boundary condition $\Delta P = 0$:

$$q = -(d_h/\delta)\omega F C_X \Delta E = -\omega C_X F L_P \Delta E \tag{4.31}$$

the well-known Schmid relation which can be used to estimate the fixed-charge concentration of a membrane by measuring the volume flux, q, caused by a potential difference, ΔE, if L_P is known from independent measurements. Applying this relationship to estimate the fixed-charge concentration C_X, it has to be kept in mind that the underlying model is based on a continuum theory and thus assumes a movement of the homogeneously charged pore fluid with the bulk velocity expecting the ions to move with the same speed as the water. However, as reported by Spiegler [21], this condition is not met on a molecular level because of the specific interactions of the moving ions with each other and with the fixed charges causing the electrophoretic effect, for instance [22]. These interactions accelerate or retard the co- or counterions.

respectively. The effects of the electrostatic interactions can be taken into account by means of a so-called conductivity factor, λ, introduced into equation (4.31). This conductivity factor was shown to possess a value of about 0.5. [23]

4.2.4 *Viscous flow membrane models (capillary models)*

Instead of the solution/diffusion model, pore flow models can be used to characterize transport across synthetic membranes. The most common one is the capillary model. When Hagen–Poisseuille's law is applied to the permeation of solutions through capillaries with uniform diameters (average pore diameter), the hydrodynamic permeability of the corresponding porous membrane is correlated with the geometric parameters of the membrane (e.g. pore radius, porosity (= water content), pore length) [24]:

$$l_p = (\varepsilon/8\eta\delta)\bar{r}^2 \qquad (4.32)$$

where

$\bar{r}$ = mean pore radius (cmM)
$\delta = td$ = pore length (cmM)
t = tortuosity factor
η = viscosity of the solution (cP; g/cm s)
$\varepsilon = w'$ = porosity of the membrane (cm^3 pore vol./cm^3 wet membrane)

Furthermore, a relationship between the effective diffusion coefficient of a solute within the membrane, $\varepsilon_s\bar{D}_s$, and its value in free solution, D_s, can be obtained. Assuming the effect of the membrane matrix on the transport of solute to be fully obstructive (i.e. steric obstruction) the following relationship is obtained [25]:

$$\varepsilon_s\bar{D}_s/D_s = [(1-\alpha)/(1+\alpha)]^2 = w'^2/(2-w')^2 \qquad (4.33)$$

where α is the relative proportion of the volume of the membrane polymer. Extraction of the effective solute diffusion coefficient from the corresponding experimental data then makes it possible to decide whether the effect of the membrane matrix is only obstructive or not. If the experimentally determined ratio $\varepsilon_s\bar{D}_s/D_s$ is in agreement with equation (4.33), one may conclude that no specific interaction between solute and membrane matrix exists and transport of solute takes place by free diffusion within the water phase of the membrane. Moreover, a nearly free diffusion of the solute in the membrane makes the existence of free water in the membrane phase plausible. Thus, water may possibly also be transported by free diffusion. In the presence of a superimposed pressure difference (superimposed on the concentration difference), water will be transported by both diffusion and convection (e.g. Hagen–Poisseuille flow = viscous flow).

4.2.5 *Evaluation of typical transport parameters*

From the transport relationships, summarized in the preceding subsections, a set of membrane parameters is available for the characterization of water and solute transport across synthetic membranes in which the transport parameters depend on the physicochemical state of the membrane system (e.g. temperature, pressure, solute concentration, viscosity of the solution). The hydrodynamic permeability l_p, for instance, is a function of these physicochemical parameters:

$$l_p = l_p(T, P, c_s, \eta) \qquad (4.34)$$

Using a Cuprophane PM-150 membrane (manufactured by the Bemberg Company, a subsidiary of ENKA, The Netherlands) and saccharose solutions of different concentrations, the volume flux, q, was measured as a function of

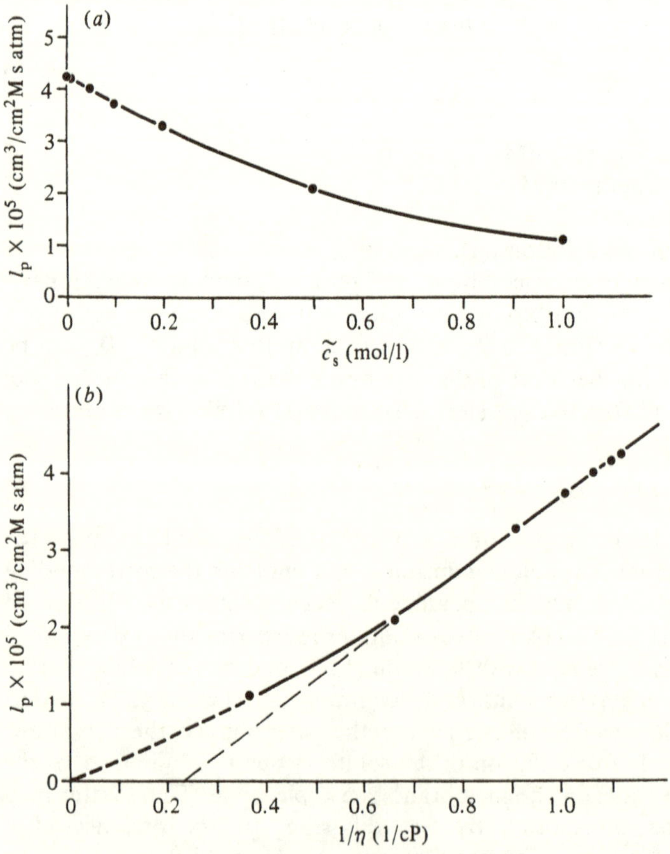

Fig. 4.3. Hydrodynamic permeability, l_p, as a function of (*a*) the mean external saccharose concentration $\tilde{c}_s$, and (*b*) the reciprocal viscosity of the external saccharose solutions $1/\eta$, using a Cuprophane PM-150 membrane at 25 °C. [2]

the saccharose concentration to demonstrate the effect of the viscosity η on the hydrodynamic permeability. Figures 4.3(*a*) and (*b*) show l_p as functions of the saccharose concentration and of $1/\eta$, respectively. As is obvious from figure 4.3(*b*), l_p depends linearly on $1/\eta$ up to saccharose concentrations of about 0.5 mol/l. To complete the picture, the reflection coefficient σ and the water content of the Cuprophane PM-150 membrane are also shown as functions of c_s in figures 4.4 and 4.5. As can be seen from these figures, the decay of the water content of the membrane with increasing saccharose concentration certainly contributes to the decrease of the hydrodynamic permeability since it is also correlated with the water content of the membrane.

In addition, using glucose, sodium chloride, and calcium chloride solutions, solute permeabilities $P_s = \varepsilon K_s \bar{D}_s$ of a Cuprophane PM-150 membrane were

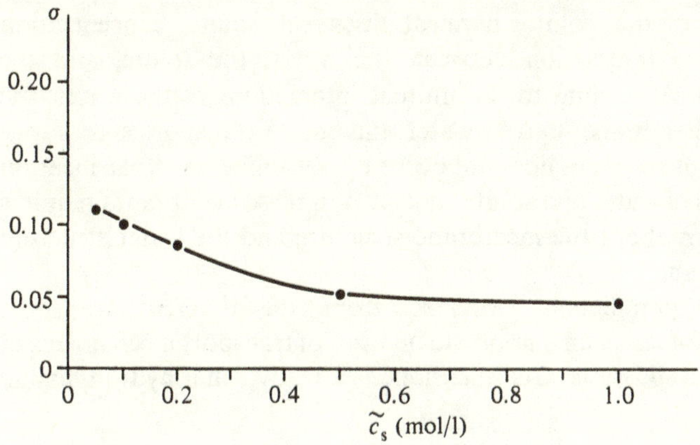

Fig. 4.4. Reflection coefficient, σ, of a Cuprophane PM-150 membrane for saccharose as a function of the mean saccharose concentration, $\tilde{c}_s$, at 25 °C.

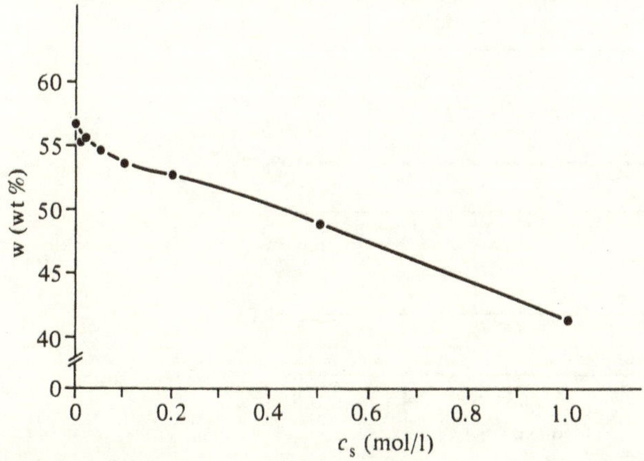

Fig. 4.5. Equilibrium water content, w, of a Cuprophane PM-150 membrane as a function of the external saccharose concentration, c_s, at 25 °C.

measured as functions of solute concentration. The corresponding experimental results are shown in figure 4.6. Further experimentally determined solute permeabilities of a Cuprophane PM-150 membrane, measured at 37 °C, are listed in table 4.1. As is obvious from figure 4.6, the solute permeabilities are constant, independent of the respective solute concentration. Estimation of the ratio $\varepsilon_s \bar{D}_s / D_s$ for two solutes (e.g. sodium chloride, vitamin B_{12}) gives a value of 0.15 which is in fairly good agreement with the value of 0.13 obtained from equation (4.33). This indicates that essentially steric hindrance by the membrane matrix is responsible for the reduction of the solute diffusion coefficients within the membrane.

For comparison, hydrodynamic and solute permeabilities of different ion-exchange membranes and a homogeneous cellulose acetate (CA) membrane are shown in figures 4.7–4.9, again as functions of solute concentration. As can be seen from these figures, there is a drastic dependence of both the hydrodynamic and the solute permeabilities on solute concentration indicating a strong interaction between the water, the solute, and the membrane matrix. According to the mutual interactions of the water, the solutes, and the membrane matrix, water and solute transport across ion-exchange and CA membranes does not occur by free diffusion. A clarification of the mechanisms of water and solute transport in these membranes requires further information about the membrane structure and the water structure within the membrane.

The transport parameters, obtained from the different transport relationships, do not make an *a priori* elucidation of transport mechanisms of water and solutes possible. As mentioned already, the hydrodynamic

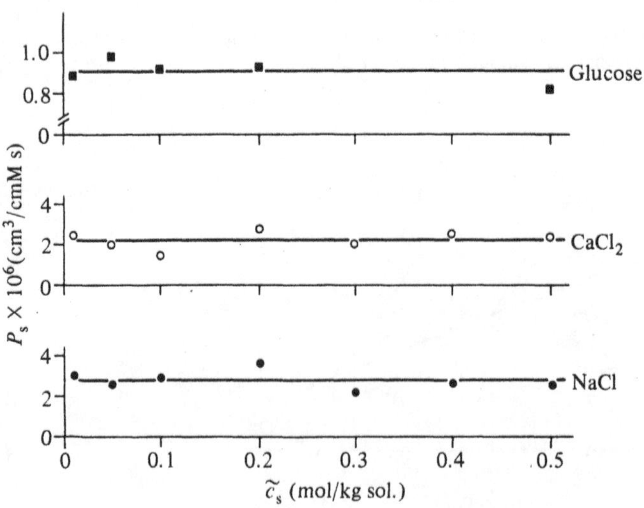

Fig. 4.6. Solute permeabilities, P_s, of a Cuprophane PM-150 membrane as functions of the external mean solute concentration, $\tilde{c}_s$, at 37 °C using different solutes.

Table 4.1 *Solute permeabilities P_s of Cuprophane PM-150 flat sheet membranes at $T = 37\,°C$*

Test substance	$(P_s/d) \times 10^4$ (cm³/cm²M s)	$P_s \times 10^6$ (cm³/cmM s)
Sodium chloride	11.2	2.60
Urea	10.4	2.34
Potassium chloride	9.65	2.24
Calcium chloride	9.28	2.13
Creatinin	6.12	1.43
Phenylalanin	3.23	0.75
Glucose	3.87	0.91
Saccharose	2.26	0.53
Raffinose	1.60	0.37
Vitamin B_{12}	0.72	0.18
Inulin	–[a]	–[a]

[a] After an experimental run of about three weeks, no variation of the inulin concentration in the external phases was measurable.

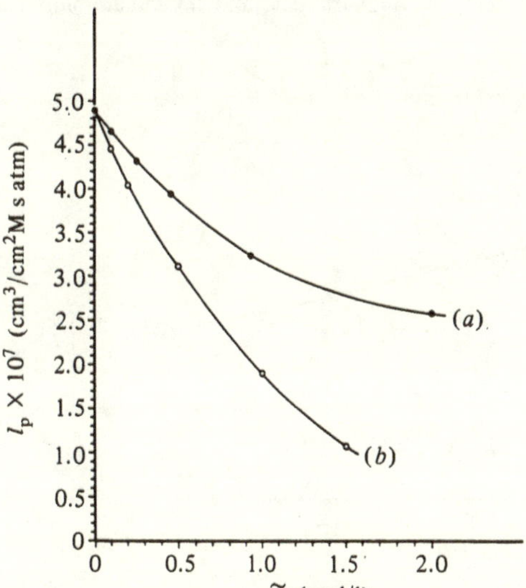

Fig. 4.7. Hydrodynamic permeability, l_p, of an anion–exchange membrane (manufactured by the Bayer Company, Germany) as a function of the external mean salt concentration, $\tilde{c}_s$, at 20 °C using (*a*) magnesium chloride and (*b*) aluminium chloride solutions.

permeability, l_p, obtained from the phenomenological relationships, characterizes both convective (viscous flow) and diffusive transport of water across the corresponding membrane. To some extent, the same holds true for the hydrodynamic permeability coefficient, d_h, obtained from the fine porous membrane model relationships, although the underlying assumption of the fine porous membrane model is that q characterizes convective flow through narrow pores of the membrane. However, to prove that q is mainly due to

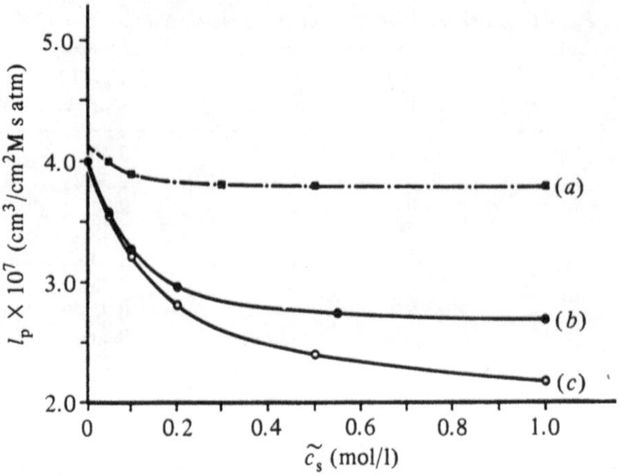

Fig. 4.8. Hydrodynamic permeabilities, l_p, of an anion–exchange membrane CA 1 (manufactured by the Asahi Chemical Company, Tokyo, Japan) as functions of the mean salt concentration, $\tilde{c}_s$, using (*a*) hydrochloric acid, (*b*) sulphuric acid and (*c*) sodium sulphate solutions.

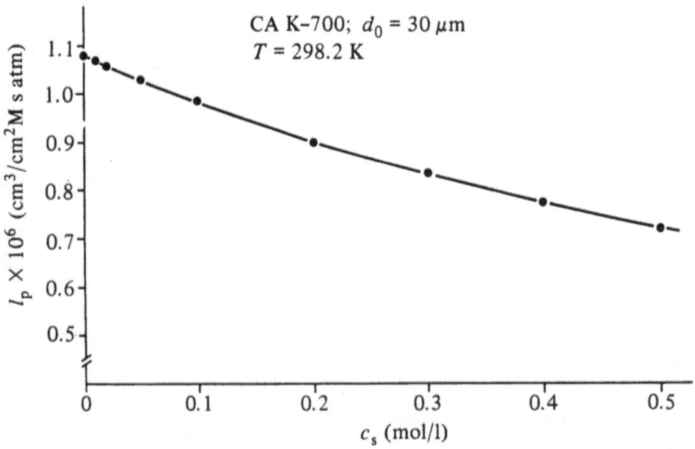

Fig. 4.9. Hydrodynamic permeability, l_p, of a homogeneous CA K-700 membrane as a function of the external sodium chloride concentration, c_s, at 25 °C.

convective pore flow, it would be necessary to make sure that the model assumptions match the properties of the real membrane. On the other hand, diffusion coefficients of water and solutes are obtained when the transport relationships of the solution/diffusion model are applied. Thus, the questions arise which kind of transport relationship should be applied for the characterization of a specific membrane and how could the basic transport mechanisms of water and solutes be elucidated. The answer to the first question requires a detailed knowledge of the membrane structure (e.g. coarse porous, fine porous, dense) and of the membrane organization (e.g.

homogeneous, asymmetric, composite) which can only be obtained from extensive electron microscopic investigations. On the basis of this information, appropriate relationships can then be selected for the characterization of water and solute transport across the membrane under consideration. In addition, a physicochemical interpretation of water and solute transport might be made possible by evaluating meaningful permeability coefficients from the measured transport parameters.

The answer to the second question requires, in addition to detailed information about the membrane structure and organization, knowledge of the water structure within the membrane and possibly also knowledge of the interactions between water, solutes, and the membrane matrix. To demonstrate the full procedure, the results of electron microscopic investigations of typical membranes will be summarized first. Subsequently, investigation and elucidation of the water structure within synthetic membranes will be discussed.

4.3 Electron micrographs of synthetic membranes

Exhaustive information on membrane morphology and organization can only be obtained by electron microscopic investigations of appropriate membrane samples. Scanning (SEM) and transmission (TEM) electron microscopic pictures of typical membranes are shown in figures 4.10–4.13. Figures 4.10(*a*)–4.10(*c*) show top and bottom surfaces of coarse porous membrane filters made from (*a*) cellulose, (*b*) cellulose triacetate (CTA), and (*c*) cellulose nitrate (CN). Further coarse porous membranes are shown in figures 4.11(*a*)–4.11(*d*). There are two kinds of coarse porous membranes: one group has a rather broad pore size distribution (figures 4.10(*a*)–4.10(*c*), 4.11(*a*) and 4.11(*b*), 'sponge structure') whereas the other group has pores of nearly uniform diameter (narrow pore size distribution; figures 4.11(*c*) and 4.11(*d*), 'sieve structure').

In contrast to coarse porous membranes, fine porous membranes do not exhibit pores under the electron microscope. Cross-sections of typical fine porous membranes are reproduced in figures 4.12(*a*)–4.12(*d*) where cross-sections of Cuprophane, cellophane, polycarbonate, and polyacrylonitrile membranes are shown. These membranes are mainly used for dialysis and hemodialysis (artificial kidney). The Cuprophane and cellophane membranes are not completely homogeneous but rather possess surface layers clearly differing in structure from the internal homogeneous membrane matrix. Cellulose membranes apparently consist of a fine porous matrix located between two dense (nonporous) surfaces of *ca.* 100–300 nm thickness.

Figures 4.13(*a*)–4.13(*d*) show SEMs of cross-sections of typical asymmetric and composite membranes. All these membranes consist of an extremely thin film (active layer) on top of a more or less porous supporting framework (matrix). In asymmetric membranes, the porosity of the supporting matrix

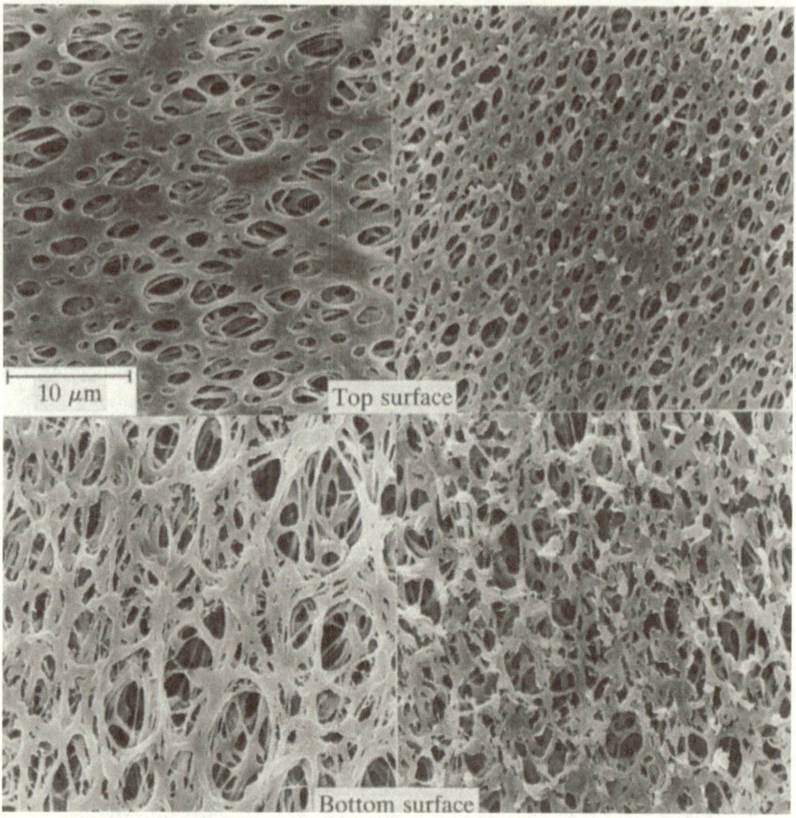

(a) (b)

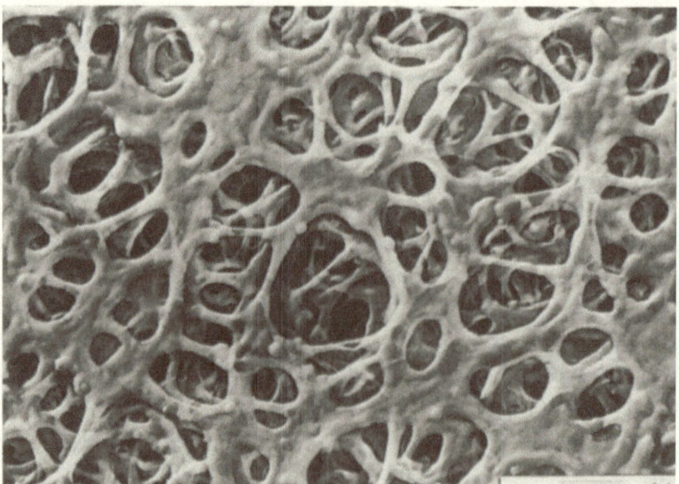

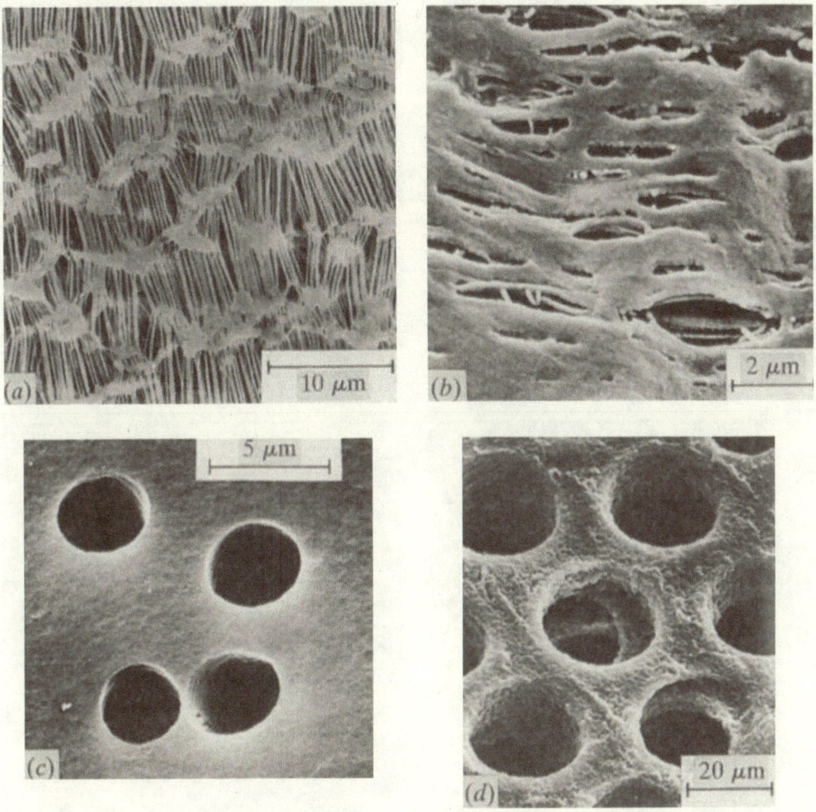

Fig. 4.11. SEMs of the top surface of (*a*) PTFE (Sartorius), (*b*) polypropylene (Celgard 2400), (*c*) polycarbonate (Nuclepore), and (*d*) an oriented gel membrane (Thiele membrane). [79]

Fig. 4.10. SEMs of top and bottom surfaces of membrane filters made from (*a*) cellulose (Gelman cellulose-α-450), (*b*) cellulose triacetate (Gelman), and (*c*) cellulose nitrate (Sartorius GmbH). [79, 82]

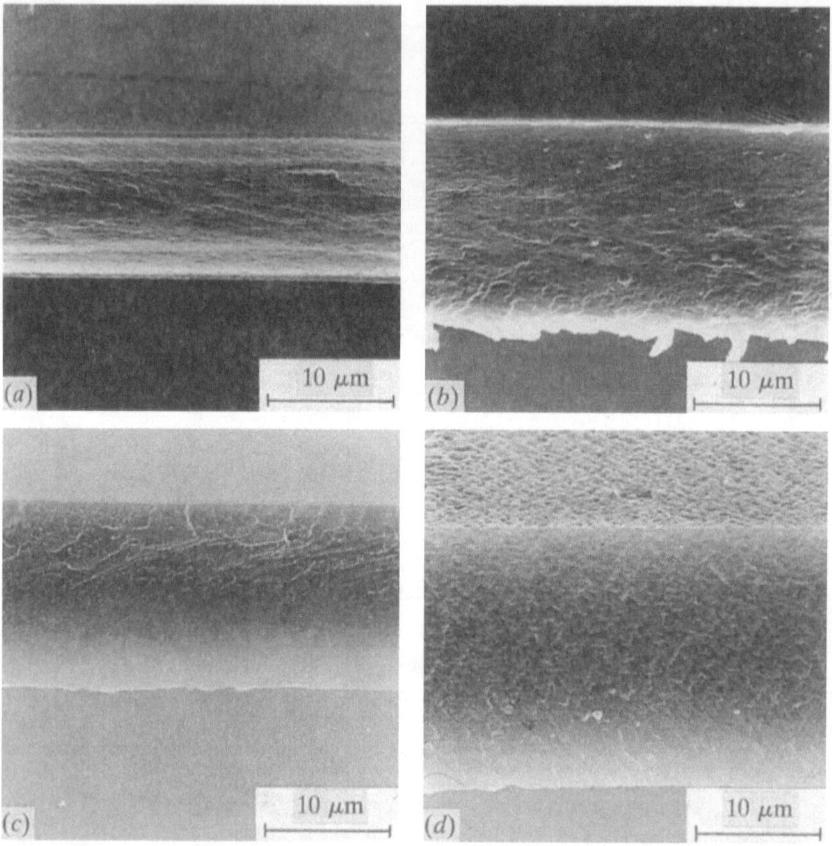

Fig. 4.12. SEMs of a cross-sectional view (with a partial top view of one surface) of (*a*) a Cuprophane, (*b*) a Nadir, (*c*) a polyacrylonitrile, and (*d*) a polycarbonate membrane. [79]

Fig. 4.13. SEMs of cross-sections of asymmetric (*a*) CA, (*b*) PA, (*c*) PS, and (*d*) and (*e*) composite membranes where the supporting layer of the composite membranes is made from PS. The composite membranes are manufactured by UOP ((*d*) RC-100), San Diego, California and Desalination Systems ((*e*) Desal-LP; LP = low pressure), Los Angeles, California. [79]

(a)

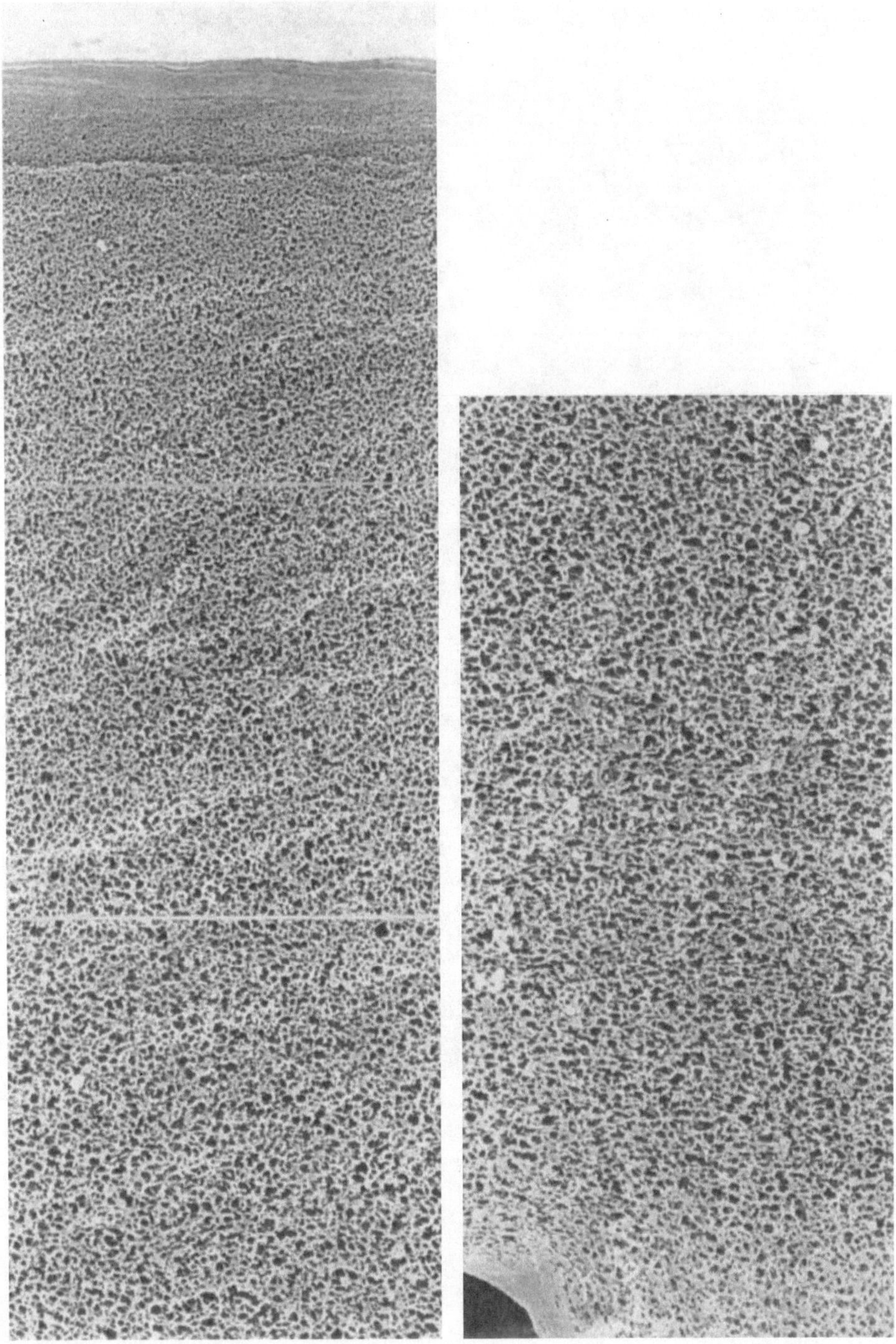

(b) (c)

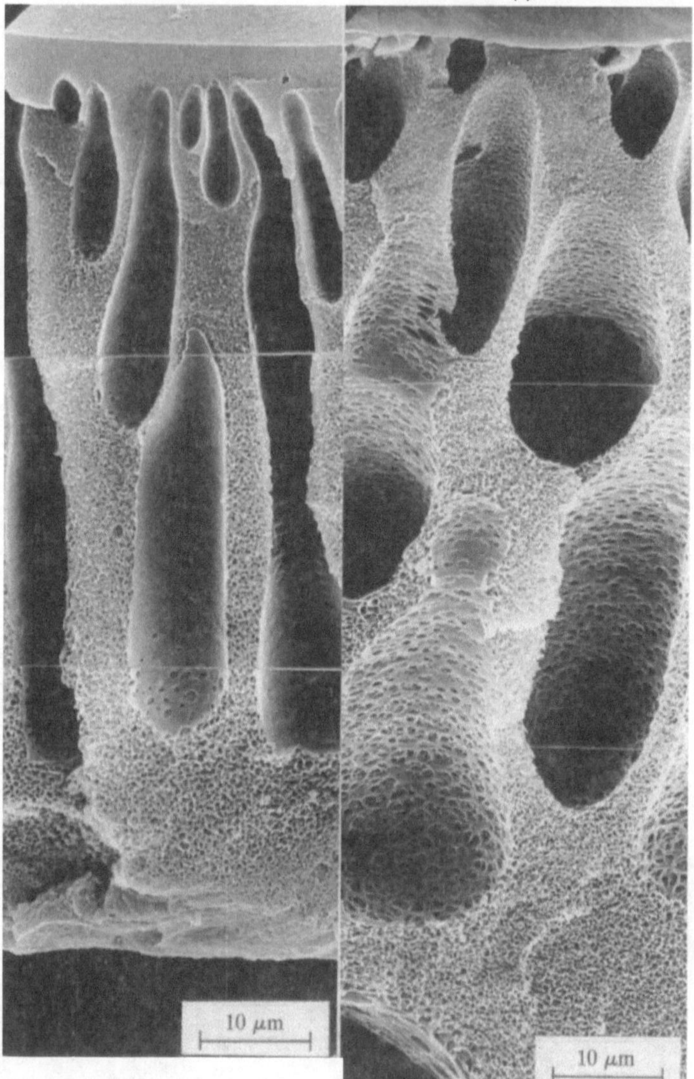

(*d*) (*e*)

increases across the membrane from top to bottom. The pore diameters, the shape of the pores, and the porosity of the supporting framework can be varied within certain limits by adjusting the composition of the casting solution and conditions of membrane fabrication.

The transport properties of all these membranes can be fully characterized by use of the phenomenological relationships (equations (4.4)–(4.10)). [26, 27, 28, 29] However, no information on the transport mechanism of water and solutes can be obtained from these transport coefficients without using, in addition, membrane model-dependent transport relationships and applying the information obtained from electron microscopic investigations. When electron microscopic investigation has proved the membrane to be a coarse porous one, pore–flow model relationships are appropriate for the evaluation of corresponding transport parameters (e.g. pore radii). In this case, transport of matter across the membrane will take place by diffusion if a concentration gradient is the only driving force. In the presence of an additional pressure gradient, convective transport (viscous flow) will predominate. On the other hand, if no pores are revealed under the electron microscope, one might apply both the solution/diffusion and the fine porous membrane model relationships. Using the measured transport coefficients, one can then estimate the diffusion coefficients of water and solutes within the membrane. These coefficients might yield at least some indication of the possible kinds of transport mechanisms. To quote an example, typical hydrodynamic permeabilities, l_p, of a variety of homogeneous membranes are summarized in table 2 together with the membrane thicknesses, d, their water contents, w′, and the water diffusion coefficients obtained from equation (4.14) using $\varepsilon = w'$.

Inspection of table 4.2 demonstrates that the diffusion coefficients, estimated from equation (4.14), decrease with decreasing water content of the membrane. Membranes possessing comparatively high water contents (e.g. cellulose and carboxy-methyl cellulose membranes) exhibit water diffusion coefficients which are much larger than the diffusion coefficient of bulk water ($D_w \approx 2.5 \times 10^{-5}\,cm^2/s$) suggesting that water transport across such membranes might be characterized by pore flow models rather than by solution/diffusion models. This would also be consistent with the values of the mean pore diameters estimated for these membranes by means of the capillary model and shown in the last column of table 4.2.

On the other hand, the water diffusion coefficients within the ion-exchange membranes used are of the order of the water diffusion coefficient of bulk water. Thus, one would be inclined to assume that water transport across these membranes might be purely diffusive in nature. As is obvious from the last column of table 4.2, however, the mean pore radii range from 4 to 6 Å. Taking into account an additional tortuosity factor of the order of 2–4, the estimated pore diameters would range from 12 to 25 Å which seems quite realistic and suggests a substantial pore-flow contribution to water and solute transport.

Table 4.2 *Measured hydrodynamic permeabilities, l_p, membrane thicknesses, d, and water contents, w′, of various synthetic membranes as well as diffusion coefficients within the membranes, $\bar{D}_w$, and average pore radii, $\bar{r}$, estimated by use of transport parameters of appropriate transport relationships.*

Membrane	$l_p \times 10^7$ (cm³/cm²M s atm)	d (μm)	w′ (vol %)	$\bar{D}_w \times 10^5$ (cm²M/s)	$\varepsilon_s\bar{D}_w \times 10^5$ (cm²/s)	$\bar{r}$ (Å)
CMC	1000	30	62.9	65	41	19
PM-150	421	20	66.2	17	11	10
PM-325	253	47	65.3	25	16	12
PM-600	182	75	66.7	28	19	13
BAYER	4.8	350	42.0	5.4	2.3	6
CA-1	4.1	225	28.0	4.5	1.3	5
CK-1	2.5	225	36.0	2.1	0.77	4
A-101	8.5_1	173	24.6	8.1	2.0	7
Nafion-290	2.5	334	26.8	4.2	1.1	5
CA E-320	11.3	5.6	26.1	0.33	0.08_6	1.4
CA L-700	4.84	6.0	19.2	0.21	0.04_0	1.1
CA K-700	0.77	25.0	17.4	0.15	0.02_6	0.9
CA K-700	3.96	4.5	17.4	0.14	0.02_4	0.9
CA K-700	1.06	30.0	17.4	0.25	0.04_3	1.2
CA F-700	2.74	6.0	15.6	0.14	0.02_2	0.9

CMC = carboxy-methyl cellulose; PM-150–PM-600 = Cuprophane membranes of different thicknesses; Bayer = anion exchange membrane; CA-1, A-101, and CK-1 = anion (A) and cation (K) exchange membranes, respectively, manufactured by Asahi Chemical; Nafion membrane made from sulphonated polytetrafluoroethylene and manufactured by Du Pont De Nemours, Wilmington (Delaware); CA = homogeneous cellulose acetate membranes prepared from CA materials possessing different acetyl contents (E-320 = 32.0; L-700 = 33.2; K-700 = 39.1; F-700 = 40.0 wt %); ε_s = surface porosity (cm² pore area/cm² wet membrane); $\varepsilon_s\bar{D}_w$ is the diffusion coefficient of water related to the pore volume in the membrane.

Finally, the estimated water diffusion coefficients within the homogeneous CA membranes are much smaller than the diffusion coefficient of bulk water suggesting that water transport across these membranes might occur solely by hindered diffusion. But a consideration of the pore diameters, estimated to be of the order of 3–6 Å when a tortuosity factor of 2–4 is again taken into account, shows that a substantial pore-flow contribution to water transport cannot be completely ruled out.

Summarizing, it might be concluded that it is nearly impossible to decide the nature of the transport mechanisms of water (and solute) in fine porous membranes solely on the basis of hydrodynamic permeabilities when the membranes possess hydrodynamic permeabilities which are of the same order of magnitude as diffusive permeabilities. This conclusion was reached by Kuhn [30] who showed that pore-flow and solution/diffusion models merge at pore diameters comparable to the dimensions of the permeating species and that the distinction between pore-flow and diffusive flow becomes to some extent a question of semantics. Further information is thus necessary to allow a more certain clarification of transport mechanisms. This information can be obtained from an investigation of the water structure within membranes.

4.4 Water structure in synthetic membranes

The structure of water in synthetic membranes might differ substantially from that of bulk water as a consequence of interactions of the water with the membrane matrix and a lack of space within the polymer network where the distances between neighbouring polymer chains or segments are of the order of molecular dimensions. Information on the structure of water in membranes can be obtained from IR-spectroscopic [31, 32], NMR-spectroscopic [33], calorimetric [34, 35], and water sorption measurements [36, 37]. In general, IR- and NMR-spectroscopic investigations yield information about the interaction of the water molecules with each other and with the polymer matrix and on the movement of the molecules (NMR-spectroscopic relaxation times) whereas calorimetric and sorption measurements yield information about the thermodynamic state of the water/membrane system. Information about the structure of water can then be obtained only by additional assumptions based on model considerations.

The polymer phase of the membrane shares a comparatively large interface with the microphase of the 'membrane water'. If the phase boundary can be considered as a structural boundary comparable to a pore wall or an external surface, adsorption of water onto these surfaces may exist. However, in many cases the water is finely dispersed in the membrane phase so that it is absorbed (occluded) rather than adsorbed. Any intermediate stage between adsorption and absorption may exist, for instance, when the pore diameters or interstices are of molecular dimensions. For this reason, McBain [38] suggested the use of the term 'sorption' when a clear distinction between ad- and absorption is not possible. This terminology will be used in the following.

For sorption of gases and vapours by polymer systems, which are subject to swelling during the sorption process, possible changes of the substrate (polymer) must be taken into account (e.g. reorientation of polymer chains or segments). Thus, sorption can be treated in a similar way to the mixing of two components, $n_1\mathrm{S} + n_2\mathrm{A} \leftrightarrow (\mathrm{AS})$ where S is the substrate, A the sorbed component, and n_1 and n_2 are the mol numbers of components 1 and 2, respectively. This is the formulation used when sorption is treated by the so-called mixing models. [39, 40] The thermodynamic treatment thus includes both the sorbed component and the substrate. The system resulting from the sorption process is treated as one phase.

4.4.1 *Thermodynamic functions of water sorption*

The thermodynamic functions of water sorption (e.g. $\Delta\bar{\mu}_w$, $\Delta\bar{H}_w$, $\Delta\bar{S}_w$) are obtained from water sorption isotherms of membranes, measured at different temperatures. Since sorbed water is at thermodynamic equilibrium, the equilibrium conditions can be applied to the water vapour/synthetic membrane system. At first, the chemical potential of water in the membrane must be equal to the chemical potential of water (vapour) in the external phase

(vapour phase)

$$\bar{\mu}_w(T, w) = \mu_v(T, p_v) \tag{4.35}$$

where

p_v = water vapour pressure (atm)
$\mu_v(T, p_v)$ = chemical potential of water vapour (J/mol)
$\bar{\mu}_w(T, w)$ = chemical potential of water within the membrane (J/mol)
$w = w(T, p_v)$ = water content of membrane (g H_2O/g wet membrane)

Dividing the equilibrium condition by T and forming the total differential at constant water content w, yields in the usual way

$$d[\bar{\mu}_w(T, w/T]_w = d[\mu_v(T, p_v)/T] \tag{4.36a}$$

$$-[(\bar{H}_w - \bar{H}_v)/RT^2]dT = (\bar{V}_v - \bar{V}_w)dp_v \tag{4.36b}$$

where

$\bar{H}_v$ = partial molar enthalpy of water vapour at T and p_v (J/mol)
$\bar{H}_w$ = partial molar enthalpy of sorbed water in the membrane (J/mol)
$\bar{V}_v$ = molar volume of water vapour at T and p_v (cm^3/mol)
$\bar{V}_w$ = molar volume of sorbed water (cm^3/mol)

Neglecting $\bar{V}_w$ in comparison to $\bar{V}_v$ and assuming ideal gas behaviour of water vapour ($p_v\bar{V}_v = RT$), equation (4.36b) yields a relationship equivalent to the Clausius–Clapeyron equation:

$$(d \ln p_v/dT)_w = -\Delta\bar{H}_v/RT^2 \tag{4.36c}$$

where the index w at the differential indicates that the derivation has to be performed at constant water content w. The partial molar heat of sorption, $\Delta\bar{H}_v = \bar{H}_w(T, w) - \bar{H}_v(T, p_v)$, equals the enthalpy change occurring when 1 mol of water vapour at T and p_v is sorbed by an infinite amount of wet membrane of water content w, which is constant during the sorption process.

In general, one is essentially interested in the differences of the thermodynamic functions as compared with liquid water since these differences supply information on the state of sorbed water as opposed to bulk water. In order to obtain the corresponding differences of the thermodynamic functions of water sorption, it is common to use water at the same temperature T and at saturation vapour pressure p_0 as the reference state. Subtracting the chemical potential of water at saturation, $\mu_w(T, p_0) = \mu_v(T, p_0)$, from both sides of equation (4.35) yields:

$$\bar{\mu}_w(T, w) - \mu_w(T, p_0) = \Delta\bar{\mu}_w^* = \mu_v(T, p_v) - \mu_v(T, p_0) = RT \ln(p_v/p_0) \tag{4.37}$$

Since the right hand side of this equation is known, one obtains the chemical potential difference between sorbed water and water at saturation vapour pressure by measuring the amount of sorbed water as a function of water vapour pressure (the sorption isotherm). The temperature dependence of the chemical potential difference then yields the partial molar heat of sorption, $\Delta\bar{H}_w^*(T, w)$, referred to water at p_0:

$$[\partial(\Delta\bar{\mu}_w^*/T)/\partial(1/T)]_w = R\partial[\ln(p_v/p_0)/\partial(1/T)]_w = \Delta\bar{H}_w^* \tag{4.38a}$$

where the index w indicates again that the derivation is performed at constant water content w. The partial molar heat of sorption, $\Delta\bar{H}_w^* = \bar{H}_w(T, w) - \bar{H}_w(T, p_0)$, is thus obtained from the slope of $\ln(p_v/p_0)$ vs. $1/T$ and the corresponding partial molar entropy difference of water sorption, $\Delta\bar{S}_w^*$, again referred to water at saturation vapour pressure, is obtained by use of the following thermodynamic relationship:

$$\Delta\bar{S}_w^* = \bar{S}_w(T, w) - \bar{S}_w(T, p_0) = (\Delta\bar{H}_w^* - \Delta\bar{\mu}_w^*)/\mathrm{T} \tag{4.38b}$$

In contrast to $\Delta\bar{H}_v$, the partial molar heat of sorption, $\Delta\bar{H}_w^*(T, w)$, is equal to the enthalpy change occurring when 1 mole of water at p_0 instead of water vapour at p_v, is sorbed by an infinite amount of wet membrane with the water content w, so that its water content remains constant during the sorption process. Thus, the following equation holds:

$$\Delta\bar{H}_v[T, w(p_v)] = \Delta\bar{H}_w^*[T, w(p_v)] - L_p(T, p_0) - RT\ln(p_v/p_0) \tag{4.39}$$

where $L_p(T, p_0)$ is the positive molar heat of evaporation of water at p_0.

In addition to the partial molar heat of sorption, the integral heat of sorption, $\Delta H(T, w)$, is also of interest. $\Delta H(T, w)$ corresponds to the enthalpy change taking place when an amount of water vapour at T and p_v is sorbed by an amount of dry membrane. In general, $\Delta H(T, w)$ will depend not only on the final water content but also on the initial water content and should thus be denoted as $\Delta H(T, w', w'')$. Thus $\Delta H(T, w)$ denotes the integral heat of sorption obtained when dry polymer sorbs water vapour up to the water content w. First, the enthalpy of a binary system is defined:

$$H(T, w) = n_w\bar{H}_w(T, w) + n_p\bar{H}_p(T, w) \tag{4.40}$$

where n_w and n_p are the mol numbers of the water vapour and the membrane polymer. Instead of the mol number, thermodynamic functions of the polymer are usually expressed in terms of the polymer weight since the molecular weight of the polymer is unknown although an average polymer weight might be available. Adding n_w mol of water vapour to n_p mol of dry polymer, the following enthalpy change occurs:

$$\Delta H(T, w) = n_w\bar{H}_w(T, w) + n_p\bar{H}_p(T, w) - n_w\bar{H}_v(T, p_v) - n_p\bar{H}_{p0}(T) \tag{4.41a}$$

where $\bar{H}_p(T, w)$ and $\bar{H}_{p0}(T)$ are the enthalpies of the wet and dry polymer referred to a mol of polymer at a given temperature and pressure. Referring the integral heat of sorption, $\Delta H(T, w)$, to the amount of water in the membrane, the following integral molar heat of sorption is obtained:

$$\Delta\bar{H}(T, w) = \Delta H/n_w = \bar{H}_w - \bar{H}_v + (n_p/n_w)(\bar{H}_p - \bar{H}_{p0}) \tag{4.41b}$$

Table 4.3. *Several physicochemical properties of the employed synthetic membranes*

Membrane	Acetyl content (wt %)	Degree of substitution α	Water content v* at saturation (T = 298 K) (wt %)	Fixed charge concentration (eq/g wet membrane)
Nafion-117				9.10×10^{-4}
Nafion-290			18.3	3.63×10^{-3}
Nafion 901			7.4	6.75×10^{-4}
CMC			122.7	2.76×10^{-5}
Cuprophane (PM-150)	0	0	125.7	1.11×10^{-5}
CA E-320	32	1.8	27.0	2.12×10^{-6}
CA L-700	38	2.3	16.7	1.43×10^{-6}
CA K-700	39	2.4	15.4	1.31×10^{-6}
CA F-700	40	2.5	13.2	1.14×10^{-6}
CA T-900	44	2.9	9.8	0.89×10^{-6}

Furthermore, the following relations hold with $\Delta\bar{H}_v = \bar{H}_w - \bar{H}_v$ [40]:

$$(\partial \Delta H/\partial n_w)_{n_p} = \Delta\bar{H}_v - n_w(\partial\bar{H}_v/\partial p_v)_T(\partial p_v/\partial n_w)_{n_p} \tag{4.41c}$$

$$n_w(\partial \Delta\bar{H}/\partial n_w)_{n_p} = \Delta\bar{H}_v - \Delta\bar{H} - n_w(\partial\bar{H}_v/\partial p_v)_T(\partial p_v/\partial n_w)_{n_p} \tag{4.41d}$$

where the index n_p means derivation at constant polymer weight.

It should be pointed out that only the integral molar heat of sorption, $\Delta\bar{H}(T, w)$, is directly measurable by means of calorimetry. The partial molar heats of sorption, $\Delta\bar{H}_v$ and $\Delta\bar{H}^*_w$, cannot be determined directly by calorimetric measurement but can only be obtained from the temperature dependence of quantities such as the vapour pressure which characterize equilibrium states.

4.4.2 *Determination of membrane water content*

The water content of a membrane depends on the water vapour pressure (water activity) of the surroundings. Measuring the water content of a membrane as a function of water vapour pressure at a constant temperature yields the water sorption isotherm at this temperature. One asymmetric CA membrane, prepared from CA K-700, and several homogeneous CA membranes, prepared from Bayer Cellit T-900, F-700, K-700, L-700, and from Eastman Kodak CA E-320, as well as commercially available cellulose membranes and Nafion membranes (117, 290, 901), were used to determine the corresponding sorption isotherms at different temperatures and to extract the thermodynamic functions of water sorption. Some physicochemical properties of the membranes used are summarized in table 4.3.

Casting solutions of the CA membranes were prepared by dissolving the respective CA material in acetone (T-900), in a mixture of dioxan with methylacetamide (9:1 by volume; T-900), in pure methylglycolacetate (F-700, K-700, L-700), or in a mixture of methylglycol with methylglycolacetate (1:1

by volume; E-320). The filtered and degassed casting solutions were then poured onto a glass plate which was located in an air-conditioning system to control the temperature (50 °C) and to maintain a very low relative humidity ($\leqslant$10% rh).

To perform the experimental determination of the water content of synthetic membranes as a function of water vapor pressure, appropriate membrane samples of about 100 mg each are equilibrated with water vapour of the corresponding vapour pressure (humidity) by placing the membrane sample in an air-conditioning system and controlling the humidity ($\pm$0.5%) and the temperature ($\pm$0.2 °C). After 2–5 days, equilibrium is considered to have been reached. The membrane samples are then processed in the following way. Each membrane sample is put into a weighing bottle of known weight and the closed bottle, containing the wet membrane sample, is weighed. Depending on the temperature resistance of the membrane material, the bottle is then reopened and placed in either an appropriate oven or a drying pistol containing phosphorous pentoxide, for instance. Using an oven, the membrane sample is dried at atmospheric pressure or under vacuum at temperatures ranging from 110 °C to 130 °C, or using a drying pistol, the membrane sample is dried under vacuum, at temperatures ranging from 50 °C to about 70 °C. The bottles are repeatedly weighed until a constant weight has been reached. Then the weighing bottles are taken out of the evaporation

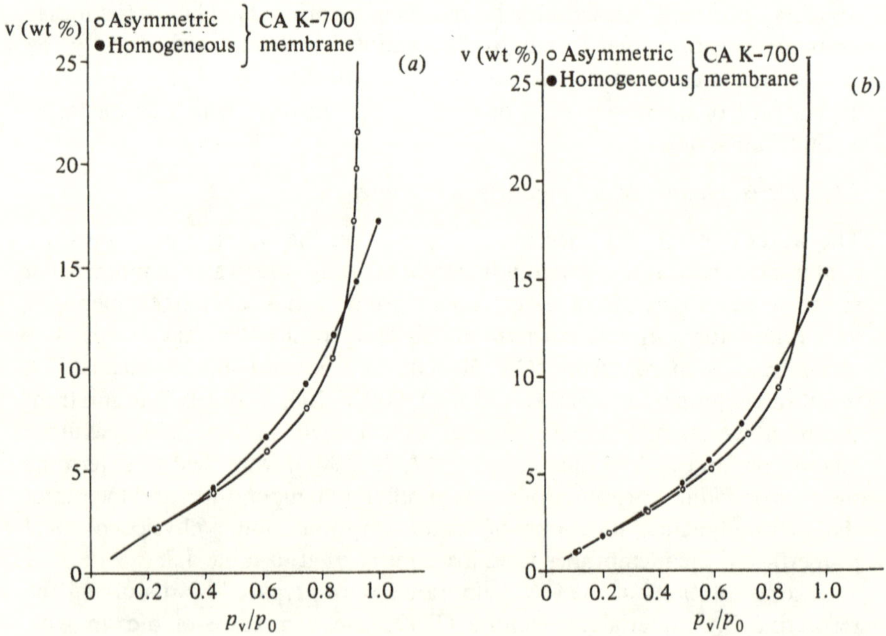

Fig. 4.14. Desorption isotherms of a homogeneous and an unannealed asymmetric CA K-700 membrane at (*a*) 35.1 °C and (*b*) 45.1 °C. [36]

(heating) system, closed again, and cooled in a desiccator. The cooled bottles are then weighed again. The difference in weight between the wet membrane sample and the dry one is taken as the water content of the membrane. This can be referred to either the wet membrane weight (w) or the dry membrane weight (v). Typical sorption isotherms are shown in figures 4.14–4.17. Since the amount of sorbed water depends on the previous history of the membrane, mainly desorption isotherms were measured. However, sometimes both absorption and desorption isotherms were measured.

In general, the amount of water, sorbed by homogeneous CA membranes, increases with decreasing acetyl content of the membrane (increasing number of OH groups per glucose unit). The desorption isotherm of the CA E-320 membrane differs from the corresponding isotherm of Cuprophane PM-250 only at relative water vapour pressures beyond $p_v/p_0 > 0.85$. Cuprophane sorbs substantially more water at saturation than the CA membrane. The water uptakes of the L-700, K-700, and F-700 membranes are quite similar whereas the water uptake of the T-900 membrane is obviously less.

As can be seen from figures 4.14(*a*) and 4.14(*b*), at higher water activities the asymmetric CA K-700 membrane sorbs more water than the homogeneous one. Since capillary structures have been suggested by electron microscopy for asymmetric membranes [41, 42, 43], one might reasonably assume that this 'excess' water in the asymmetric membrane is confined to well-defined capillaries. The water sorption isotherm for the asymmetric membrane is nearly identical to the isotherm which characterizes equilibrium sorption in a homogeneous membrane at very low relative humidities; however, 'excess' water is dissolved in the homogeneous membrane at intermediate humidities.

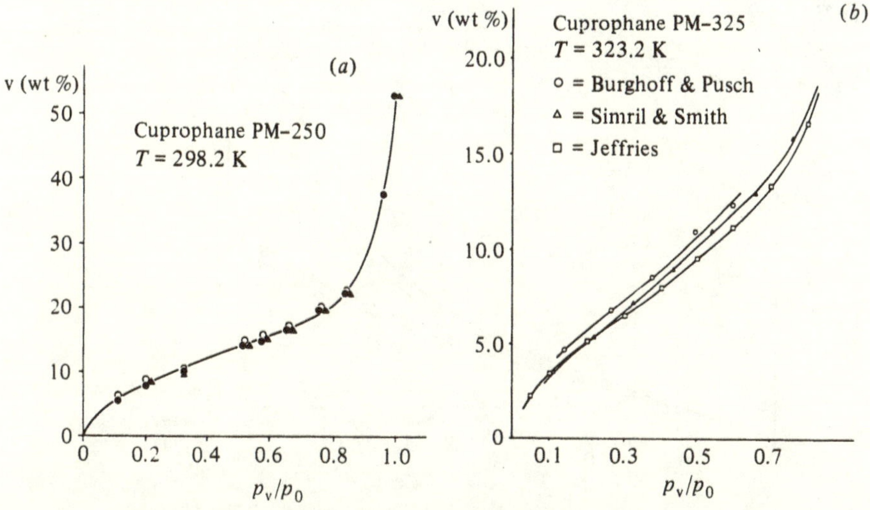

Fig. 4.15. Desorption isotherm of (*a*) Cuprophane PM-250 membrane at 298.2 K [81] measured with three different sample membranes and (*b*) Cuprophane PM-325 at 323.2 K.

This intriguing observation might be considered to be a consequence of condensation at intermediate humidities in the homogeneous membranes in capillaries bound by quite specifically defined pore radii. Although one might reasonably expect some significant asymmetric distribution of pores in a 'homogeneous' membrane (e.g., smaller pores in the air-dried surface), these equilibrium experiments do not discern between symmetric and asymmetric structures. Assuming cylindrical pores, the specific pore size corresponding to the relative humidity values in question is given by:

$$RT\ln(p_v/p_0) = -(2\bar{V}_w\gamma/r)\cos\theta \tag{4.42}$$

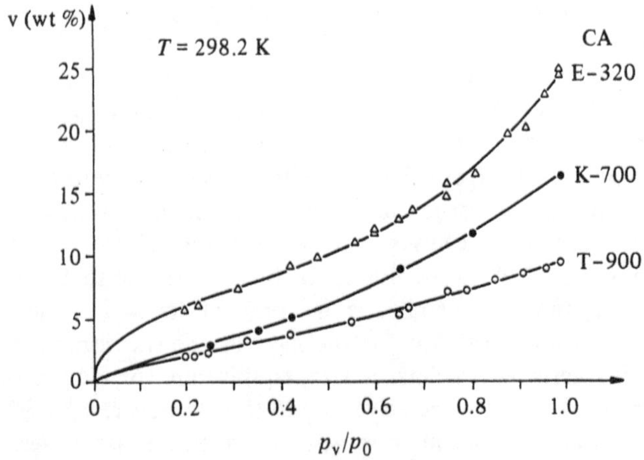

Fig. 4.16. Desorption isotherms of homogeneous CA membranes with different acetyl contents at 25 °C. [83]

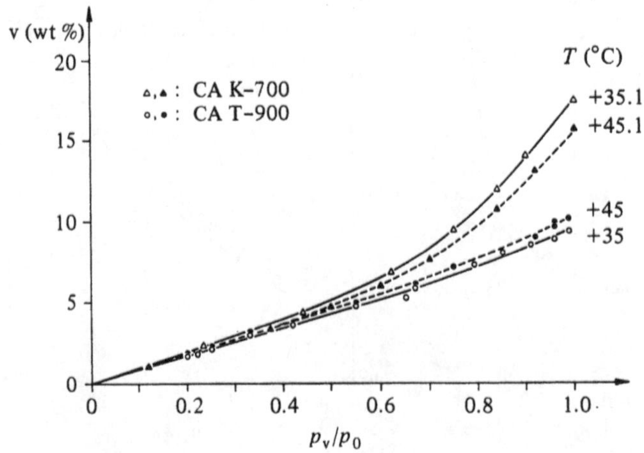

Fig. 4.17. Desorption isotherms of a homogeneous CA K-700 and CA T-900 membrane at two different temperatures. [82]

where

γ = surface tension between water and air (erg/cm^2)
θ = contact angle between capillary wall, liquid, and vapour phase
r = pore radius (cm)

The pore size distribution may, therefore, be calculated by combining the results of equation (4.42) with the geometric constraints and the measured excess water sorbed at the corresponding humidity. With $\bar{V}_w = 18\ \text{cm}^3/\text{mol}$ and $\cos\theta = 1$ ($\theta = 0°$) one, therefore, obtains V_r, the excess water corresponding to the capillary radius which would satisfy equation (4.42) at various humidities as a function of the corresponding pore radius. These results are shown in figure 4.18. These estimations suggest that the so-called 'homogeneous' CA membrane may, in fact, contain capillary pores which contribute significantly to the observed sorption. For relative humidities which correspond to pore radii up to 0.004 μm, the homogeneous membrane sorbs more water than the asymmetric one. Assuming a strict capillary condensation model, these data would suggest that the homogeneous membrane may possess pores up to 0.004 μm (= 40 Å), whereas the asymmetric membrane may possess pores with radii mainly in the range 0.01–0.3 μm. However, pore diameters, deduced from water sorption isotherms, are larger by an order of magnitude than the average pore diameter obtained from hydrodynamic permeabilities (table 4.2). One might conclude,

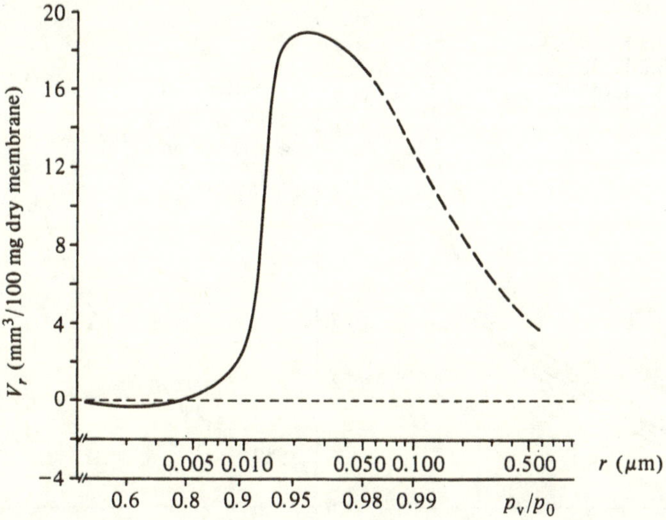

Fig. 4.18. Excess water sorbed by an unannealed asymmetric CA K-700 membrane as a function of water vapour activity p_v/p_0. The corresponding calculated pore radius r is given as a second abscissa. Negative V_r corresponds to excess water in the homogeneous CA K-700 membrane. [36] The dashed part of the line is an extrapolation and to some extent uncertain.

therefore, that mechanisms other than capillary condensation are responsible for the different water uptake of homogeneous and asymmetric CA membranes at intermediate humidities. [44]

Sorption isotherms of a Nafion-290 and a Nafion-901 are shown in figures 4.19(*a*) and 4.19(*b*). The isotherms of the Nafion-290 membrane resemble to some degree those of asymmetric CA membranes. All the isotherms coincide at very low water activities ($p_v/p_0 < 0.05$) and at very high water activities ($p_v/p_0 > 0.8$). As the Nafion-901 membrane sorbs much less water than the Nafion-290 one, its sorption isotherms do not exhibit the branch at high humidities. The water sorbed by the Nafion-290 membrane at high humidities might be due to capillary condensation. Since this membrane contains a polymer network for reinforcement where the contact between the membrane polymer and the network is poor, there is a gap around the fibres of the network which acts like a capillary. At high enough humidities, water might thus condense in these capillary-like interspaces, and this capillary condensation is nearly independent of the temperature. At intermediate water activities the amount of sorbed water depends on the temperature; decreasing with increasing temperature which indicates a negative partial heat of sorption.

Plotting the logarithm of the normalized water activity, $\ln(p_v/p_0)$, at different but constant water contents, w, as a function of $1/T$, straight lines are

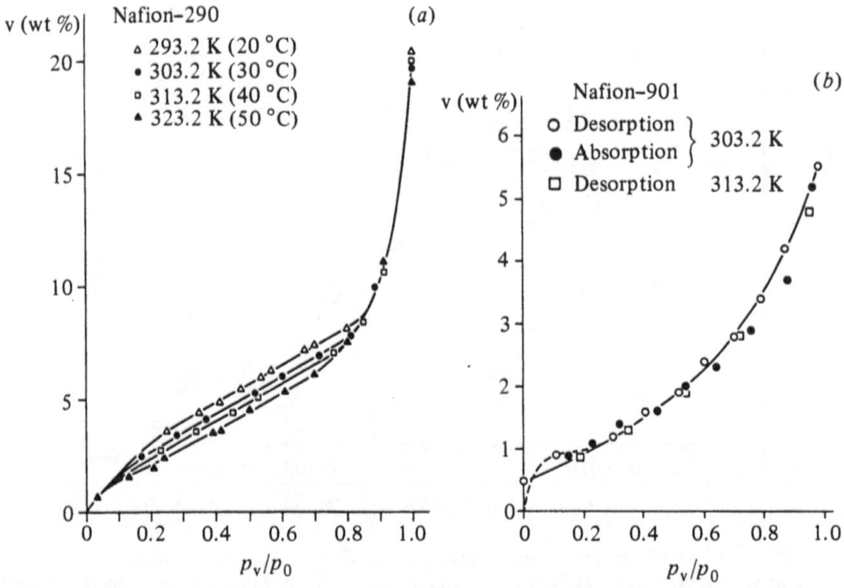

Fig. 4.19. Desorption and absorption isotherms of (*a*) Nafion-290 and (*b*) Nafion-901 membranes at different temperatures. [84] The dashed line of (*b*) is an extrapolation.

obtained. Typical $\ln(p_v/p_0)$ versus $1/T$ straight lines are plotted in figures 4.20(*a*) and 4.20(*b*) for the CA K-700 and the Nafion-290 membranes and demonstrate a linear relationship. Partial molar enthalpies of sorption calculated from the slope of the corresponding straight line are included in figure 4.20(*b*).

The calculated values of the thermodynamic functions of water sorption for the different membranes used are reproduced in figures 4.21(*a*)–4.21(*c*) and 4.22(*a*) and 4.22(*b*). Where necessary, these overall values of the thermodynamic functions may be considered as linear combinations of the partial molar quantities which describe dissolution and capillary

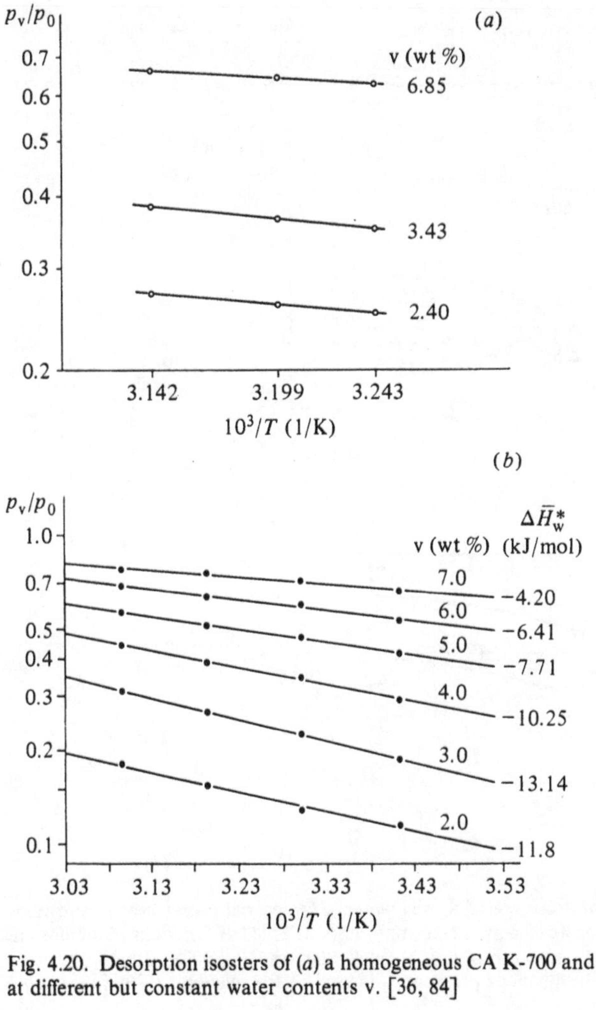

Fig. 4.20. Desorption isosters of (*a*) a homogeneous CA K-700 and (*b*) a Nafion-290 membrane at different but constant water contents v. [36, 84]

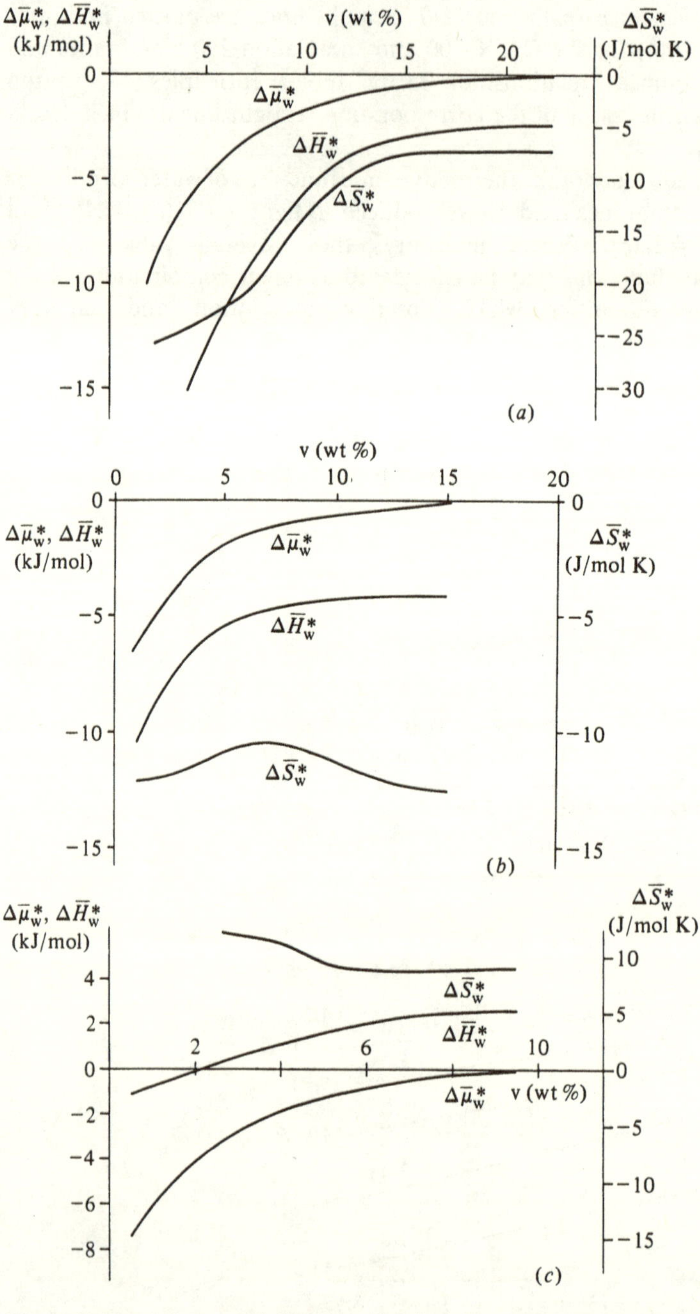

Fig. 4.21. Chemical potential difference of sorbed water, $\Delta\bar{\mu}_w^*$, partial molar heat of sorption, $\Delta\bar{H}_w^*$, and partial molar entropy difference of sorption, $\Delta\bar{S}_w^*$, at 25 °C as functions of membrane water content, v, using (*a*) homogeneous E-320, (*b*) K-700, and (*c*) T-900 cellulose acetate membranes. Water at saturation vapour pressure p_0 is used as the reference state. [44, 83]

condensation, respectively. For $\Delta\bar{G}_w = \Delta\bar{\mu}_w$, for instance, the following linear combination holds:

$$\Delta\bar{G}_w = x_d\Delta\bar{G}_d + x_c\Delta\bar{G}_c = x_d\Delta\bar{\mu}_d + x_c\Delta\bar{\mu}_c \tag{4.43}$$

where x_d is the mol fraction of dissolved water, x_c is the mol fraction of water sorbed by capillary condensation, $\Delta\bar{G}_d$ is the partial molar Gibbs free energy difference of dissolution, $\Delta\bar{G}_c$ is the molar Gibbs free energy difference of capillary condensation, $\Delta\bar{\mu}_d$ is the chemical potential difference of dissolution, and $\Delta\bar{\mu}_c$ is the chemical potential difference of capillary condensation.

The contribution of capillary condensation to the enthalpy and free energy changes associated with sorption only becomes significant, however, at

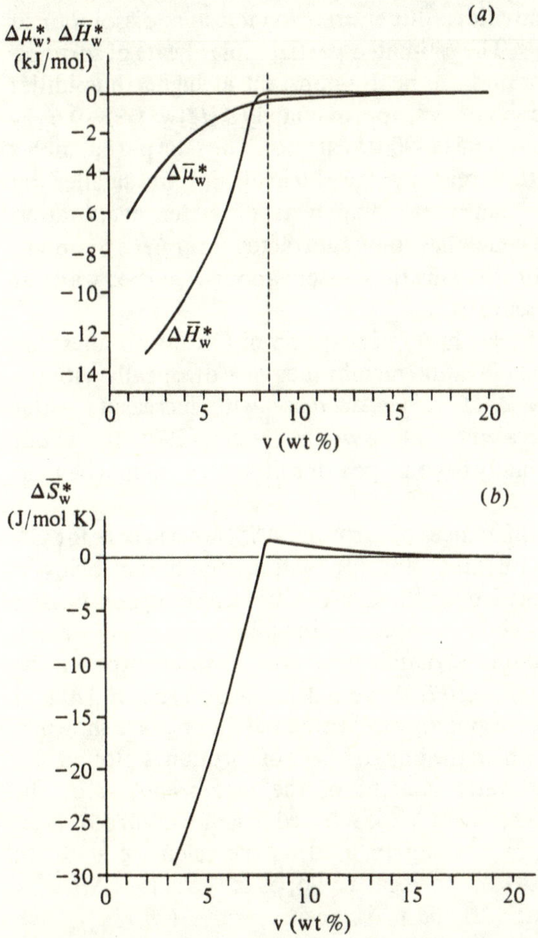

Fig. 4.22. (*a*) Chemical potential difference of sorbed water, $\Delta\bar{\mu}_w^*$, partial molar heat of sorption, $\Delta\bar{H}_w^*$, and (*b*) partial molar entropy difference of sorption, $\Delta\bar{S}_w^*$, at 25 °C using a Nafion-290 membrane, H^+-form at 40 °C. [84]

extremely small pore radii since the values of these thermodynamic functions are inversely related to the pore radii through equations (4.36c), (4.37), (4.38a) and (4.38b). At high relative humidities (large equivalent pore radii), the contribution of capillary condensation to the numerical values of these functions (as defined) approaches zero. At extremely low water activities, however, the fraction of water taken up by capillary condensation is very small compared to the dissolved water and can, therefore, be neglected to a first approximation.

Thus, in the limiting cases, the calculated values of the thermodynamic functions are described entirely by dissolution. Since only pores of $r < 0.01\ \mu m$ are presumed to exist in the homogeneous CA membranes, no significant capillary condensation would occur at high water activities ($p_v/p_0 > 0.9$). Increasing regain at higher activities is due entirely to molecular dissolution in homogeneous CA membranes. The estimated partial molar heats of sorption for homogeneous CA membranes are nearly constant at higher humidities and are, at saturation water content, w*, approximately $\Delta\bar{H}_w^*(w^*) \approx -0.6$ (E-320), ≈ -1.0 (K-700), and $\approx +0.65$ (T-900) kcal/mol. Thus the partial molar heats of water sorption are larger by 0.6–1.0 kcal/mol or smaller by 0.65 kcal/mol, respectively, than the molar heat of water evaporation ($L_p = 10.5$ kcal/mol at 25 °C) which has to be subtracted from $\Delta\bar{H}_w^*$ to obtain the partial enthalpy change for the transition water vapour → sorbed water at saturation water vapour pressure p_0.

In contrast to the partial molar heats of sorption of CA membranes, the partial molar heats of sorption of Nafion membranes vary drastically with the water content. The negative $\Delta\bar{H}_w^*$ values decrease with increasing water content from about -3 kcal/mol at 2 wt % (Nafion-290) to about -1 kcal/mol at 7 wt % and finally become positive at water contents beyond about 8 wt %.

The partial molar entropy differences of sorption, ΔS_w^*, are negative for CA membranes with lower acetyl content and for Nafion membranes below a specific water content and positive for the CA T-900 membrane and Nafion membranes at water contents above a specific value (about 8 wt % of water).

In addition to the partial thermodynamic functions of water sorption, the integral molar heats of sorption, $\Delta\bar{H}(T, v)$, were determined using E-320 and T-900 CA membranes. The corresponding experimental findings are shown in figures 4.23(*a*) and 4.23(*b*) in which the integral heat of sorption is plotted as a function of either the final water content of the membrane, v, or the normalized water content, v/v*, where $\Delta\bar{H}$ is referred to a mol of sorbed water and water vapour at pressure p_v is again used as the reference state. In accordance with equation (4.41d), the negative slopes of the curves $\Delta\bar{H}$ vs. v and $\Delta\bar{H}$ vs. v/v* mean that $(\partial\bar{H}_v/\partial p_v)_T$ is positive since $(\partial p_v/\partial n_w)_{n_p} > 0$, $\Delta\bar{H}_v < 0$, $\Delta\bar{H} < 0$ and $|\Delta\bar{H}_v| < |\Delta\bar{H}|$.

The calculated values of the thermodynamic functions, $\Delta\bar{\mu}_w^*$, $\Delta\bar{H}_w^*$, and $\Delta\bar{S}_w^*$, presented in figures 4.21(*a*)–4.21(*c*), 4.22(*a*) and 4.22(*b*) and listed to some extent in table 4.4, are quite consistent with the assumption that the water in the membranes used interacts with the membrane matrix (polymer). The negative partial enthalpy changes of the water sorbed by Nafion membranes at lower water contents might be considered as the energy of hydration of the fixed charge groups (SO_3H groups) which would also be consistent with the

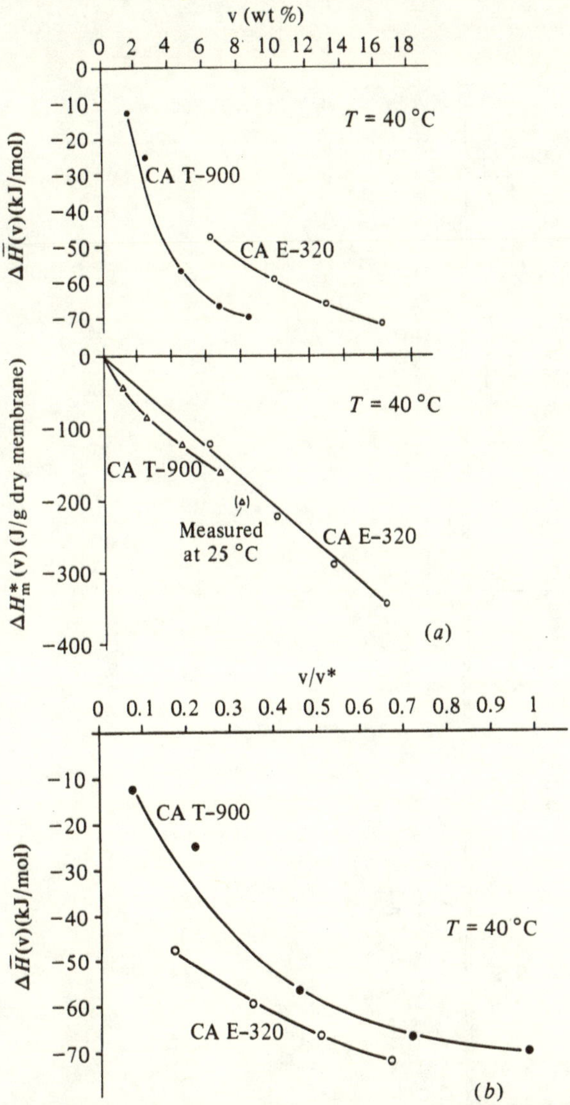

Fig. 4.23. Integral molar heats of sorption, $\Delta\bar{H}(v)$, for homogeneous E-320 and T-900 CA membranes as a function of (*a*) the water content, v, and (*b*) the normalized water content, v/v*. [86] Different membrane samples were used to obtain the data shown in (*a*) and (*b*), respectively. It is common experience that the saturation water content v* might vary with the membrane sample.

Table 4.4 *Water contents* $v^*(T)$, *integral heats of sorption* $\Delta\bar{H}^*(T, v^*)$, *partial enthalpy differences* $\Delta\bar{H}_w^*(T, v^*)$ *and* $\Delta\bar{H}_p(T, v^*)$ *of CA membranes for water at* 30 *and* 40°C *at saturation water content* v^*

Membrane	T (°C)	v^* (g H_2O/g CA)	$\Delta\bar{H}^*(T, v^*)$ (J/g CA)	$\Delta\bar{H}^*(T, v^*)$ (kJ/mol H_2O)	$\Delta\bar{H}_w^*(T, v^*)$ (kJ/mol H_2O)	$\Delta\bar{H}_p(T, v^*)$ (J/g CA)
E-320	30	0.24	−573	−43.0	−2.51	−31.9
	40	0.20	−417	−37.5	−2.51	−20.2
L-700	30	0.18	−361	−36.1		
	40	0.15	−287	−34.4		
F-700	30	0.16	−284	−31.9		
	40	0.16	−267	−36.9		
T-900	30	0.10	−150	−27.1	+2.72	−33.5
	40	0.11	−146	−23.9	+2.72	−35.1

$v = m_w/m_p$; $w = m_w/(m_w + m_p)$; $v = w/(1-w)$
$m_w = g\,H_2O$; $m_p = g$ dry membrane (polymer)
$\Delta\bar{H}^*(T, v^*) = \Delta\bar{H}(T, v^*) + L_p(T, p_0)$; $L_p > 0$

negative partial molar entropy changes of sorption (comparable to the negative partial molar entropy differences of hydration of ions) indicating a higher degree of order of the corresponding water (or less statistical degrees of freedom).

The 'sorption' sites for water in CA membranes are most likely the OH groups of the glucose rings, possibly the oxygen atom of the glycosidic bond (acetic groups), the C=O groups of the acetyl residues, and last, but not least, the fixed charges (COOH groups) which are present as the end groups of the cellulose chains. There are also van der Waals interactions with the polymer chains.

4.4.3 *Calorimetric measurements with CA membranes*

Calorimetry has been proved to be an additional sensitive technique for investigating the state of sorbed water. Taniguchi & Horigome [45] studied melting endotherms of water in CA membranes by differential scanning calorimetry. They detected no phase transition at all in homogeneous membranes. With asymmetric membranes, phase transitions appeared only at water contents higher than 13 wt % within a temperature range of −15 to 0 °C. Yasuda *et al.* [46] specifically measured the heat of fusion in water–glycerol monomethacrylate systems and found no phase transition at water contents up to 0.36 g water/g dry polymer. It was thus assumed that up to this water content the water is present as so-called bound water only, whereas any water sorbed in excess of 0.36 g/g dry polymer is supposed to be free water, which freezes with the normal heat of crystallisation. Other authors [47, 48, 49] have also reported that no phase transition was observed when the specific heat of water in membranes [47] or biological systems [48, 49] were measured. Plooster & Gittlin [50] reported results similar to those of Taniguchi & Horigome for the state of water adsorbed on silica surfaces. They estimated the heat of melting of ice which melts at temperatures below 0 °C and found values which are smaller than the normal heat of fusion of ice at 0 °C. Burghoff & Pusch [51] and Lukas & Pusch [52] measured the specific heat of CA membranes with different water contents as a function of temperature over the temperature range −40 to +30 °C. They did not detect a phase transition in any of the homogeneous CA membranes used.

Measured heat capacities c_m of homogeneous and asymmetric CA membranes at various water contents as functions of temperature are presented in figures 4.24(*a*)–4.24(*c*). The water content v of a membrane is given in wt % referred to the dry membrane material. With homogeneous CA membranes of different acetyl contents the heat capacity increases monotonically with temperature, with no detectable peaks. For a given water content of less than 15 wt %, no difference was observed between the heat capacity of a homogeneous and an asymmetric membrane (K-700 membranes). On the other hand, at a water content somewhat above 15 wt % in an asymmetric K-700 membrane, a broad maximum in the heat capacity–temperature curves appeared within the temperature range of −30 to −1 °C, peaking at −9 °C. At higher water contents the peak temperature increased continuously with increasing water content but never approached 0 °C.

To determine the heat of phase transition in the temperature range of diffused melting below 0 °C, the heat capacity of a sample within the transition region has to be known. Clearly, this heat capacity cannot be determined directly by measurement because of the superimposed melting energy of ice. However, one might obtain a calculated heat capacity of a sample by knowledge of the partial quantities of heat capacity and masses of both water and membrane material within the sample. Experimental results were used to derive the partial heats by the following analysis.

The partial heat capacities of both water, $\bar{c}_w$, and the CA membrane material, $\bar{c}_p$, can be calculated from the measured heat capacity, c_m, of a water-containing membrane sample by using the following relationship:

$$c_m = (m_w\bar{c}_w + m_p\bar{c}_p)/(m_w + m_p) \quad (4.44a)$$

where m_w and m_p are the weights of water and membrane, respectively. Dividing numerator and denominator by m_p, the relative water content $v' = m_w/m_p$ is obtained. Thus, equation (4.44a) might be rewritten to yield:

$$(1 + v')c_m = \bar{c}_p + v'\bar{c}_w \quad (4.44b)$$

The slope of a tangent at any point of the curve $(1 + v')c_m$ vs. v' corresponds to $\bar{c}_w$ and its intercept with the ordinate to $\bar{c}_p$. Such plots of the experimental values at different temperatures are presented in figures 4.25(*a*)–4.25(*d*) for homogeneous CA membranes with different acetyl contents. In figure 4.26 the

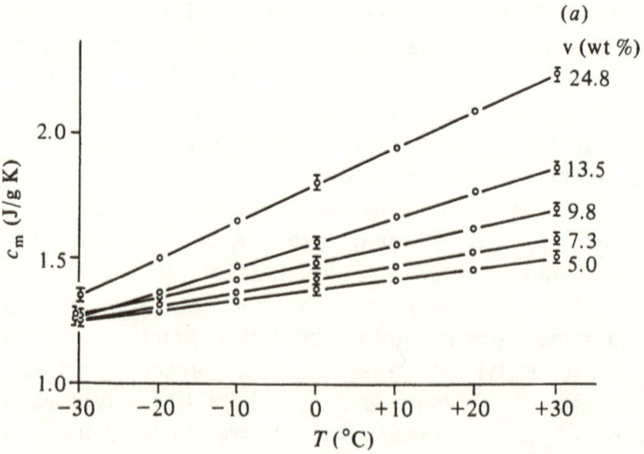

Fig. 4.24. Heat capacities, c_m, of (*a*) homogeneous E-320, (*b*) homogeneous and unannealed asymmetric K-700, and (*c*) homogeneous T-900 CA membranes as functions of temperature, *T*, at various water contents, v. [44, 83]

same plot is reproduced for an asymmetric CA K-700 membrane at water contents up to saturation (beyond about 16 wt %). The extracted partial specific heats are listed in table 4.5.

While the heat capacity data obtained with homogeneous CA membranes yield straight lines up to the saturation water content, two straight lines are obtained when plotting the heat capacity data of an asymmetric CA K-700 membrane as a function of its water content. The heat capacities of this

(*b*)

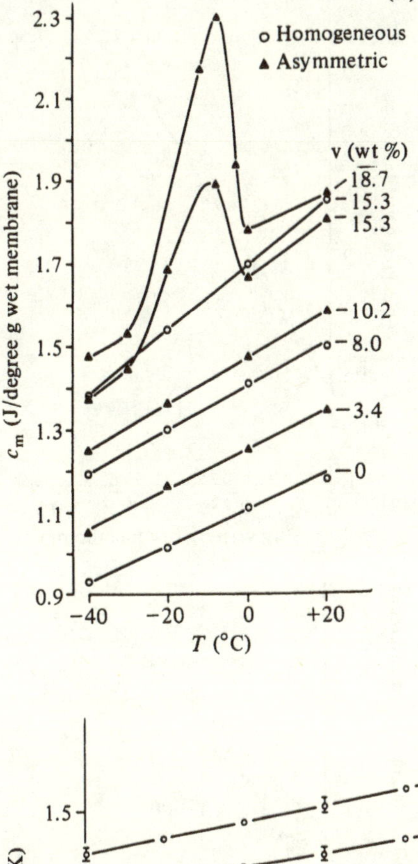

(*c*)

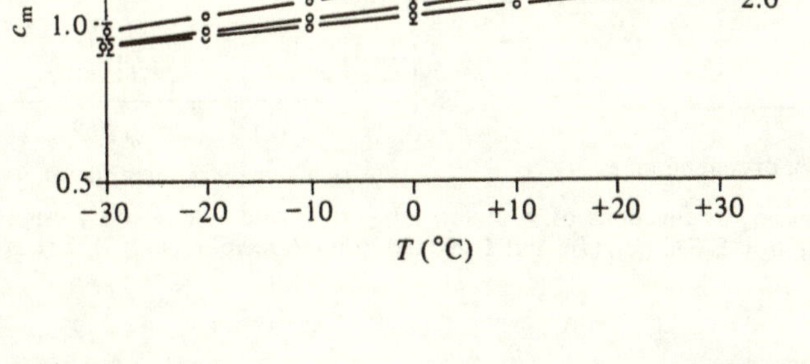

membrane, measured at water contents up to the saturation water content of the corresponding homogeneous membrane, yield a straight line with the same slope as that for the homogeneous membrane. On the other hand, another straight line, with a smaller slope, is obtained at water contents above the saturation content of the homogeneous membrane, as can be seen from figure 4.26. This indicates an abrupt change in the partial heat capacity of water in the asymmetric CA K-700 membrane at a water content of about 16 wt %. The smaller slope of this second straight line yields a partial heat capacity of the extra water in asymmetric membranes, essentially identical

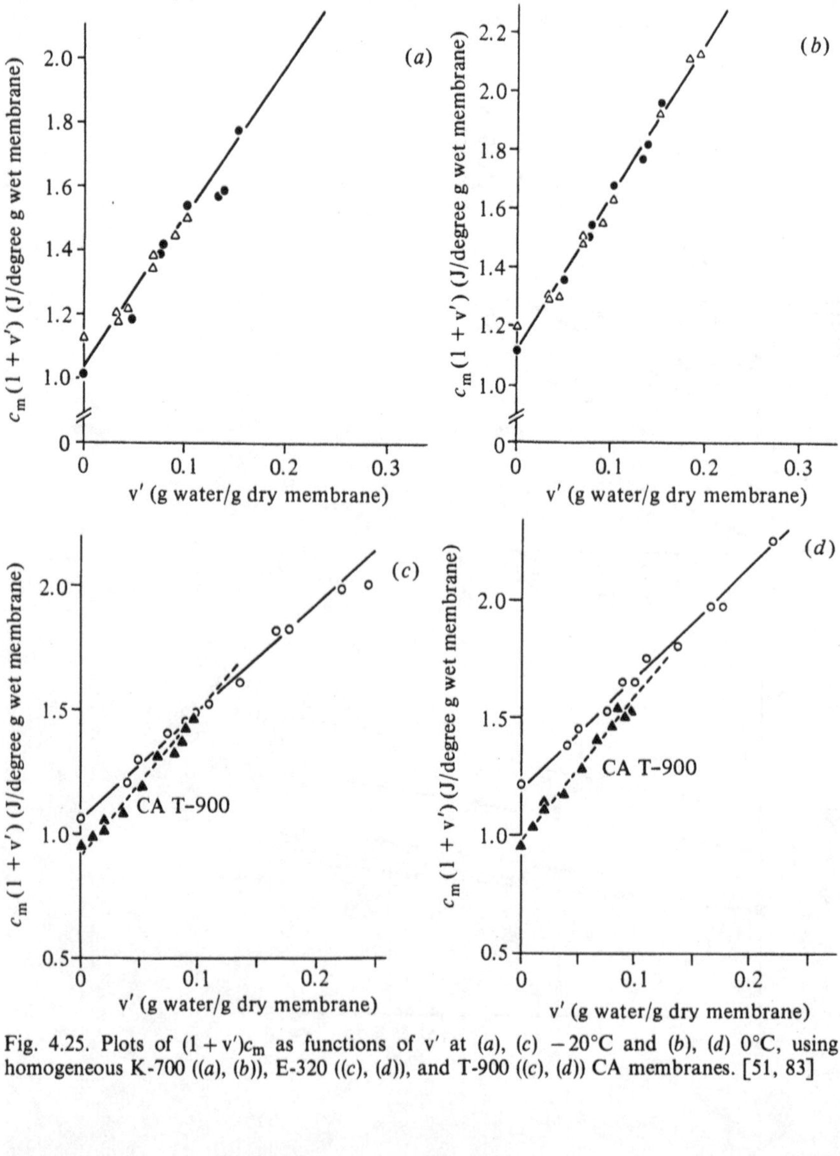

Fig. 4.25. Plots of $(1 + v')c_m$ as functions of v' at (a), (c) −20°C and (b), (d) 0°C, using homogeneous K-700 ((a), (b)), E-320 ((c), (d)), and T-900 ((c), (d)) CA membranes. [51, 83]

Table 4.5. *Partial specific heats of water, $\bar{c}_w$ in different CA membranes and of the corresponding membrane material, $\bar{c}_p$ at −20.0 and +20 °C*

Membrane	*T* (°C)	c_p (J/g degree)	$\bar{c}_w$ (J/g degree)
E-320	−20	1.06 ± 0.02	4.28 ± 0.41
	0	1.19 ± 0.02	4.99 ± 0.32
	20	1.32 ± 0.02	5.07 ± 0.27
K-700	−20	1.03 ± 0.02	4.70 ± 0.24
	0	1.11 ± 0.02	5.25 ± 0.26
	20	1.19 ± 0.02	5.80 ± 0.29
T-900	−20	0.91 ± 0.03	5.60 ± 0.60
	0	0.97 ± 0.01	6.00 ± 0.49
	20	1.02 ± 0.01	6.35 ± 0.34

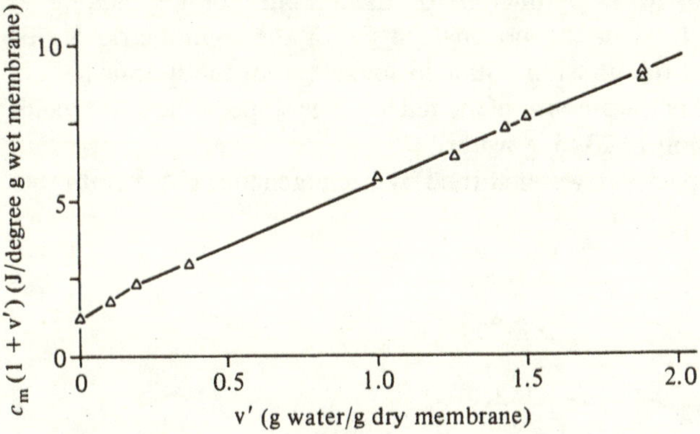

Fig. 4.26. Plot of $(1+v')c_m$ as a function of v' at 20°C and at water contents up to saturation using an unannealed asymmetric CA K-700 membrane. [51]

with either the heat capacity of ice or bulk water depending on the temperature. On the other hand, it is obvious that $\bar{c}_w$ and $\bar{c}_p$ are both identical for the two types of CA K-700 membranes below 16 wt % water content. In addition, it might be concluded that the partial heat capacities, $\bar{c}_w$ and $\bar{c}_p$, are independent of the water content for all CA membranes up to the saturation water content of the respective homogeneous CA membrane.

No distinction can also be made between homogeneous and asymmetric CA membranes with respect to the temperature dependence of the partial heat capacities of water and membrane materials at water contents less than the water content at saturation of the respective homogeneous membrane. As can be seen from figure 4.27, where the heat capacities of ice, bulk water, and saturated vapour are drawn in for comparison, the partial heat capacity of dissolved water is always larger than those of ice or liquid water at the same temperature and it increases with increasing acetyl content of the CA membrane. Furthermore, the partial heat capacities of the membrane materials are directly proportional to the temperature. Haly & Snaith [53] have also proved a similar relationship for wool. As shown by Tarasov & Yunitskii [54], this linear relationship results from a theoretical treatment using the one-dimensional Debye term for the vibrations of the polymer chains.

The overall heat capacity of a 1 g sample of the asymmetric CA K-700 membrane might be considered as the linear combination of the partial quantities describing the membrane material and the two states of dissolved water. The heat of fusion at temperatures below 0 °C is given by the difference between the total heat added to the system within the temperature range of −30 to 0 °C and the energy necessary to heat up the system within the temperature range of −30 to 0 °C calculated by means of the overall heat capacity of the sample. The results are shown in figure 4.28. The heat of melting, ΔH_s, is plotted as a function of the amount of ice melting at temperatures below 0 °C in the porous matrix of the asymmetric K-700 membrane. The linear relationship obtained suggests that the specific heat of fusion is independent of the amount of melted ice. The slope of the curve yields a specific heat of fusion of 231 J/g water.

The larger heat capacity of water sorbed by homogeneous CA membranes

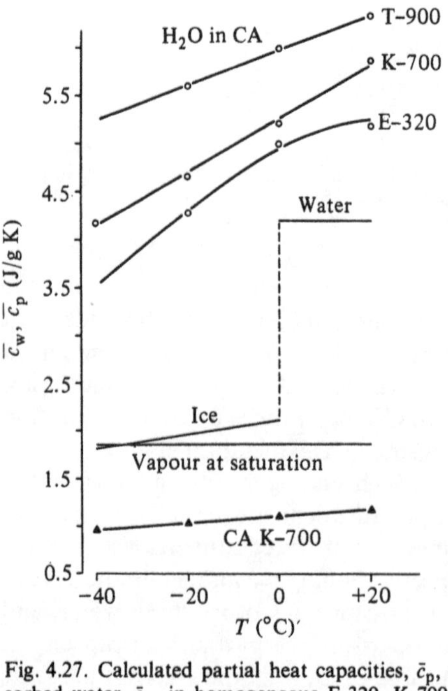

Fig. 4.27. Calculated partial heat capacities, $\bar{c}_p$, of a cellulose acetate K-700 membrane and sorbed water, $\bar{c}_w$, in homogeneous E-320, K-700, and T-900 CA membranes as functions of temperature. In addition, heat capacities of ice, bulk water, and saturated water vapour are presented for comparison. [51, 82]

and by asymmetric ones at low water contents may be explained by introducing a simple model. The heats of sorption connected to the sorption of water by CA membranes indicate, in agreement with results obtained from IR-sorption measurements [31, 32], a reduced hydrogen bond interaction between sorbed water molecules while at the same time there is a stronger interaction between the sorbed water and the polymer, for example the formation of hydrogen bonds. In general, the sorption energy results in an excess heat capacity over the heat capacity of bulk water. This model together with the experimental findings leads to the conclusion that all the water sorbed by homogeneous CA membranes is completely sorbed as 'bound' water. Since the heat capacity c_p of CA membranes is independent of the water content, the membrane structure should be essentially unchanged by sorption of water molecules, unlike the solvation process which would take place in homogeneous and asymmetric CA membranes at larger water contents in the presence of good organic solutes such as phenol. The fact that the smaller partial heat capacity of water observed with asymmetric CA membranes at larger water contents equals the heat capacity of bulk water indicates that there is almost no interaction between the additional water and the polymer backbone. Furthermore, the decrease in the partial heat capacity of water coincides with the onset of the rapid increase in water content in these asymmetric membranes as a function of regain. Since it is well known that this rapid increase in water content is due to capillary condensation of water in the porous substructure of the asymmetric membranes, this heat capacity observation supports the idea of well-defined pores existing in the matrix of asymmetric CA membranes. [36, 55] The additional water sorbed by capillary condensation can thus be assumed to be a second state of water within asymmetric membranes.

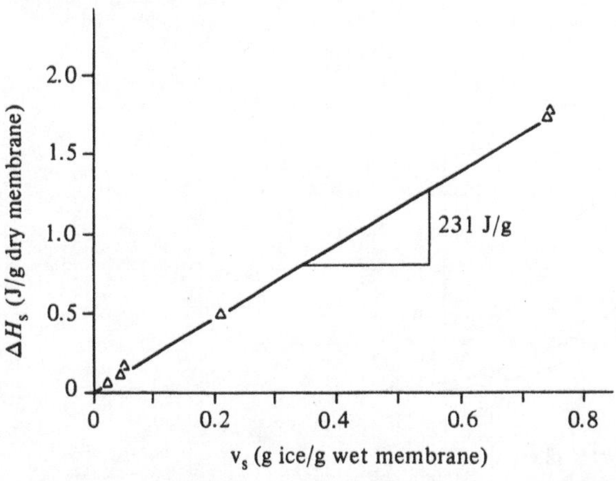

Fig. 4.28. Heat of fusion, ΔH_s, of ice melting in asymmetric CA K-700 membranes at temperatures below 0 °C as a function of the amount of melting ice, v_s. [51]

The correlation of the decrease in heat capacity to capillary condensation is quite consistent with the phase transition phenomena occurring concurrently at temperatures below 0 °C since the thermodynamic treatment of capillary phenomena predicts melting point depression and a reduced heat of fusion for water condensed in small pores, whereas the heat capacity of pore water is quite similar to that of bulk water [56–8]. It may therefore be suggested that the broad maxima in the heat capacity–temperature curves of asymmetric membranes are due to first-order transitions of water in non-uniform-sized pores.

In summary, the calorimetric data can be interpreted by assuming two different states of water in CA membranes. Only 'bound' water without any phase transition is observed in homogeneous CA membranes. The highly ordered state of water might be due to water–polymer molecule interactions. The caloric characteristics of water in homogeneous and asymmetric CA membranes are identical at low water contents ($w \leqslant w^*$). A reduced heat capacity of water in asymmetric CA membranes at larger water contents ($w > w^*$) coincides with the onset of melting at temperatures below 0 °C. Taking into account the theoretical results of the thermodynamic treatment of capillary phenomena, these experimental findings identify this water of lower heat capacity to be capillary water in the porous matrix of the asymmetric CA membranes.

Further support of the assumption of strong water membrane interactions was obtained from sorption measurements with aqueous phenol solutions. Measuring phenol and water uptake of homogeneous and asymmetric CA membranes as a function of phenol concentration at 25 and 45 °C, Burghoff *et al.* [6, 59] were able to correlate the water content v to the phenol content *u* of a homogeneous CA K-700 membrane. The corresponding experimental

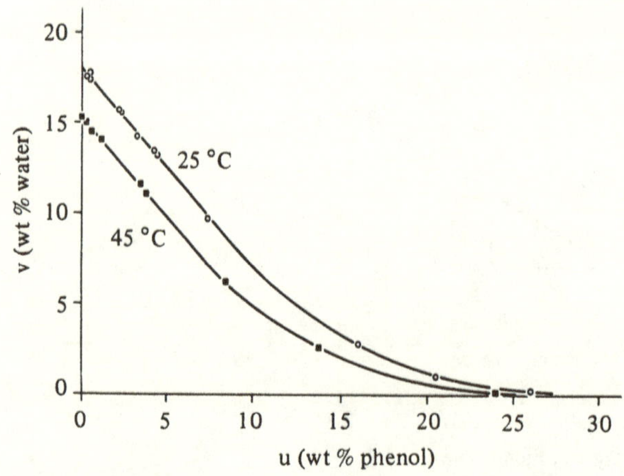

Fig. 4.29. Water content v as a function of the phenol content u of a homogeneous CA K-700 membrane. [85]

findings are shown in figure 4.29. Clearly, the water content is inversely related to the phenol content at phenol contents below about 10 wt % at 25 °C and below about 8 wt % at 45 °C. Both these phenol contents correspond to an external phenol concentration of $c_s = 3 \times 10^{-2}$ mol/l phenol. At higher external phenol concentrations, the water content of the membrane becomes much lower. It should be mentioned in this connection that CA membranes dissolve in aqueous phenol solutions beyond about 0.1 mol/l. Using the slopes of the linear portion of the v versus u curves (figure 4.29) and the molecular weights of water, $M_w = 18.01$ g/mol, and phenol, $M_{ph} = 94.11$ g/mol, the number of water molecules replaced by one phenol molecule can be calculated to be about five at both temperatures. Thus, one phenol molecule is assumed to occupy the volume of five water molecules, in agreement with the relative partial molar volumes of these substances. The attachment of a phenol molecule at a sorption site shields the attractive forces of that site, rendering attachment of a second layer more difficult because of steric hindrance.

4.5 Characteristics of water in microvoids or interstices

Water, which exists in small aggregates of several hundred or thousand molecules, exhibits specific properties since a substantial number of the water molecules exists near interfaces (surfaces). Intensive research in the field of the physical chemistry of water is devoted to the phase transition behaviour of thin water films on surfaces, of microemulsions, and of water in quartz capillaries. [60, 61] The phenomenon of supercooled water, i.e. water cooled down to about −40 °C without a transition to a solid phase, is explained in terms of nucleation kinetics. The absence of nucleation centres in small clusters of water prevents small water aggregates from heterogeneous nucleation at 0 °C. Also homogeneous nuclei formation cannot occur in aggregates containing a limited number of molecules because of the weaker cooperative effects in these aggregates. This may thus create a metastable state which persists even when the temperature is decreased further until a distinct limiting temperature is reached at which the small water aggregate phase abruptly turns to ice.

The precondition for the occurrence of a metastable state of supercooled water is the existence of small units of water in the vicinity of surfaces. The investigation of the physicochemical properties of water, present in such small systems, revealed the following characteristics [62, 63]:

No phase transition is observed at temperatures ranging from −40 to +20 °C.

Due to its metastable state, supercooled water exhibits an exceptionally large heat capacity which might be as large as 6.5 J/g K at −40 °C.

Ions are not dissolved by supercooled water.

The observed effects are also characteristic of the properties of the water of hydration in synthetic membranes. The similarity between water in synthetic membranes and water in small aggregates leads to the supposition that water in fine porous and dense synthetic membranes might exist as small units in the vicinity of interfaces just like supercooled water in corresponding systems. A distinct cluster morphology has already been proved for water occluded in Nafion membranes, for instance. [64] For this reason, Nafion membranes may be used for comparison with other membranes such as the CA membranes employed for calorimetric investigations.

The results of calorimetric investigations of water sorption by synthetic membranes suggest a mechanism of water sorption with a comparatively weak interaction between the polymer material and the initially sorbed water molecules. Subsequently, water uptake becomes strongly exothermic due to the mutual stabilization of the hydrogen bonds in the growing water clusters and due to stronger average interaction between water molecules in the clusters than in corresponding clusters in bulk water. Certainly, the close contact between sorbed water and polymer material, exhibited by the values of the partial molar thermodynamic functions $\Delta\bar{H}_w^*$, $\Delta\bar{S}_w^*$, and $\Delta\bar{\mu}_w^*$, plays an important part in determining the energy balance of the ab- and desorption process. Further discussion of the characteristics of sorbed water in synthetic membranes should thus be based on a model of sorption in intermicellar spaces. With this model, it is assumed that water, sorbed by certain polymer materials, accumulates in the interstices between the chains of the polymer. [65] Since the size of the voids or microvoids is of molecular dimensions, the properties of the occluded water and its interaction with the polymer matrix are governed by molecular mechanisms.

Submicroscopic interstices with diameters of the order of 2–5 nm (20–50 Å) exist, for instance, between the polymer bundles of raw cellulose. The presence of such cavities, called 'microvoids', which may be a consequence of the stiff side-groups of the polymer chains causing reduced crystallinity, was also proved for regenerated cellulose and cellulose derivatives. Gas permeation measurements with homogeneous CA membranes, for instance, actually suggested transport of gases through micropores. [66] A further indication of the existence of microvoids is the hygroscopic nature of specific cellulose derivatives which are completely substituted with hydrophobic groups. [67] In such systems, water might be sorbed by a kind of capillary action of the microvoids.

4.5.1 *Water in small systems*

From the ideas outlined above, it is assumed that water, sorbed by fine porous and dense synthetic membranes, might exist in a dispersed form as small aggregates. The water forms physical clusters due to its occlusion in the limited spaces formed by the chain segments of the membrane polymer, and due to the mutual interactions of the molecules. In the following, their probable size and thus the number of molecules per cluster will be estimated. The average number of water molecules per glucose unit for CA membranes at saturation water content, for instance, is 1.2–2.0. Since the distribution of the water within the membrane material will be strongly heterogeneous, water-containing domains, water-filled polymer chain interstices and microvoids will exist as well as the more or less crystalline regions which are inaccessible to water. Depending on the degree of crystallinity of the CA material, a cluster size of 1–4 water molecules per glucose ring is estimated if a homogeneous distribution of water over the membrane phase is assumed to exist. On the other hand, a maximum of 33 water molecules fit into a 1 nm (10 Å) cube. Although the volume of these water molecules might still range from 9.7 cm^3/mol, at close packing, to 22.5 cm^3/mol for the enveloping sphere at free rotation of the molecules [68], it does not change the order of magnitude of the size of the water clusters expected to consist of between ten and several hundred molecules. It is nearly impossible to treat such small systems, where the number of molecules is very small compared to Avogadro's number ($\approx 6 \times 10^{23}$ per mol), by classical thermodynamics. Stillinger [65] has already shown this and has developed formulations of the thermodynamics of sorption applicable to small systems. A more extended presentation of the thermodynamics of small systems can be found in the book by Hill. [69] The important implications of this theory are summarized below.

The Gibbs free energy, G, of small systems contains, in addition to the terms which are characteristic for macroscopic systems and which are proportional to the number of molecules, N, corrections for the surface energy of the system and further contributions to the energy of the system. For instance, the following formulation is given:

$$G = Ng(T, p) + N^{2/3}a(T, p) + b(T)\ln N + c(T, p) \tag{4.45a}$$

where $g(T, p)$, $a(T, p)$ and $c(T, p)$ are functions of temperature and pressure; $g = G/N$ is the specific Gibbs free energy of the equivalent macroscopic system, $N \to \infty$; $a(T, p)$ is the specific Gibbs free surface energy; $b(T)\ln N$ and $c(T, p)$ are the excess Gibbs free energy. It is further assumed that $\bar{G} = (G/N) \neq \mu = (\partial G/\partial N)_{T,p}$ and that the molar Gibbs free energy depends on the particle number, $(\partial \bar{G}/\partial N)_{T,p} = (\mu - G)/N$. The macroscopic behaviour is reached when N approaches infinity ($N \to \infty$).

In order to treat the special case of small water clusters in synthetic membranes by the thermodynamics outlined above, it is convenient to consider the large number of individual systems in the polymer matrix as an 'ensemble' in the sense of statistical mechanics. According to the fundamentals of statistical thermodynamics, the time averages of the thermodynamic quantities of the individual systems are then equal to the average of the ensemble where the small particle numbers of the individual systems lead to large fluctuations of the corresponding quantities with time. Simplifying, the water clusters are considered as M equivalent, distinguishable, independent, open systems which are contained in a volume V (volume of the microvoids) at constant T and p and which are in equilibrium with the external water phase which serves as a reservoir for water molecules. The chemical potential of the external water phase is designated as μ. The variation of the total entropy of the ensembles is then given by:

$$TdS_t = dU_t + p dV_t - \mu dN_t + \partial S_t/\partial M|_{U_t, V_t, N_t} dM \tag{4.45b}$$

where U_t is the total internal energy of the ensembles. The last term is of great importance. When the number of clusters increases at constant total particle number, constant total volume, and constant total internal energy of the ensembles, the total entropy will first increase since the N_t molecules can be distributed over a larger number of systems. Concurrently, however, the mean volume of an aggregate (cluster) must decrease and thus relatively more molecules will exist close to interfaces in contact with the polymer material. The entropy variations correlated with the generation of additional interfaces contribute to the variation of the overall entropy of the system.

Treating the properties of water sorbed by synthetic membranes according to the theory outlined above, it is possible to demonstrate that the absence of a sharp liquid–solid transition is characteristic for this sorbed water. Moreover, the thermodynamic preconditions for a liquid–solid phase transition are nonexistent for very small aggregates (e.g. clusters of a few water molecules). With regard to the observed larger partial molar specific heats of water in synthetic membranes, it should be noted that the heat capacity $C_p = (\partial H/\partial T)_{N,p}$ reflects the temperature dependence of the enthalpy of the system. With liquids it is commonly assumed that the different molecular degrees of freedom take up the energy supplied by the surroundings and thus determine the heat capacity. In addition, there exists a heat capacity of configuration in water since hydrogen bonds will form or break with decreasing or increasing temperature, respectively. A decrease in the average energy of hydrogen bonding (12 kJ/mol) by about 0.2% during a temperature increase of 1 °C causes an increase in the heat capacity of water of the order of 1.5 J/g K which corresponds to the increase in the heat capacity of water observed in CA membranes. Small clusters of about ten water molecules may take up the additional energy for rotation and/or vibration of the full aggregate thus still contributing to an increase of the specific heat of the system.

Further support for the existence of small water aggregates (clusters) in synthetic membranes might be obtained by means of a more detailed analysis of the water sorption isotherms. As is well known from the literature [70], sorption isotherms can be classified according to their shape as Langmuir-, Freundlich-, BET-type isotherms. The most general description of sorption isotherms is possible by means of the BET model. In many cases these models successfully describe the sorption process of specific systems. However, all these models fail in characterizing the sorption of water in polymers and biopolymers, especially at relative humidities near saturation. The reasons for this failure are specific to each model. As all of the experimentally determined isotherms of water–polymer systems are typically S-shaped, one would expect a multilayer model to give a best fit of the experimental data. Nevertheless, agreement between experimental data and the BET model, for instance, is obtained only at relative humidities below 40%. This might be due, for example, to the assumption of the BET model that sorption of gases and vapours at surfaces occurs by the formation of multilayers of gas molecules, in which no limitation on the number of layers is taken into account. However, bearing in mind the space restrictions present in polymeric membranes, one must apply sorption models which limit the number of layers of the sorbed gas (vapour) molecules.

4.5.2(*a*) *Monolayer model of sorption*

In water–polymer and many other similar systems, the density of the adsorbate is much higher than the density of the gas (vapour). Thus the amount of gas n in the gas phase (volume V_g) is negligible compared to the amount of adsorbate n_s, and the Gibbs surface excess of the gas is thus equal to the experimentally determined amount of adsorbate. Using the chemical potential of a gas at a temperature T below its critical temperature, equation (4.37), and substituting this into the following form of the Gibbs sorption isotherm for a binary system [71]:

$$-(\mathrm{d}\gamma)_T = \Gamma(\mathrm{d}\mu_v)_T \tag{4.46a}$$

where γ is the surface tension and $\Gamma = n_s - n_i$ is the Gibbs surface excess denoting the difference of the number of moles of component i between the surface (n_s) and the gas phase (n_i) per unit area, the following relationship results neglecting n_i in comparison to n_s:

$$-(\mathrm{d}\gamma)_T = n_s RT \mathrm{d}\ln(p_v/p_0)_T \tag{4.46b}$$

In the following a strict monolayer adsorption caused by a lowering in surface tension is assumed. Describing the monolayer as a two-dimensional gas, the lowering of surface tension results in a two-dimensional film pressure [71], Π_1:

$$\Pi_1 = \gamma_0 - \gamma_1 \tag{4.47a}$$

where γ_0 and γ_1 are the surface tension of free and filled surfaces, respectively. In the case of an ideal two-dimensional gas, the following equation of state is obtained:

$$\Pi\sigma = RT \tag{4.48a}$$

where $\sigma = 1/n_s$ is the area per mol of sorbed molecules. Taking into account cohesive interactions between sorbed molecules, the equation of state may be written [72]:

$$\Pi\sigma = aRT \tag{4.48b}$$

where the parameter a gives a measure of the cohesive forces. Using this equation of state, the following relationship results:

$$-(\partial\gamma_1/\partial n_s)_T = aRT \tag{4.49}$$

Combining equations (4.46b) and (4.49) results in the total differential:

$$\mathrm{d}\ln(n_s) = (1/a)\mathrm{d}\ln(p_v/p_0) \tag{4.50a}$$

Integration of this equation leads to an adsorption isotherm equivalent to the Freundlich isotherm:

$$n_s = k(p_v/p_0)^{1/a} \tag{4.50b}$$

where k is the integration constant. With the boundary condition of a completely filled layer ($n_s = n_m$) at saturation ($p_v = p_0$), it follows:

$$k = n_m \tag{4.50c}$$

Normalizing the size of the adsorbing area by the unit weight of the solid, the amount of gas adsorbed per unit weight of the solid becomes v. The integration constant is then equal to the amount of gas in a completely filled monolayer per weight of the solid, $\mathrm{v_m}$. The sorption isotherm can then be written:

$$\mathrm{v} = \mathrm{v_m}(p_v/p_0)^{1/a} \tag{4.50d}$$

Rearranging, the following relationship results:

$$\ln \mathrm{v} = \ln \mathrm{v_m} + (1/a)\ln(p_v/p_0) \tag{4.50e}$$

where the constants $\mathrm{v_m}$ and a may be obtained from a $\ln \mathrm{v}$ vs. $\ln(p_v/p_0)$ plot of experimental data.

While $\mathrm{v_m}$ is directly proportional to the surface area, v is directly proportional to the filled surface area, and the difference $\mathrm{v_0} = \mathrm{v_m} - \mathrm{v}$ is directly proportional to the free surface area. Therefore, the following equation holds:

$$\mathrm{v} = \mathrm{v_0}(p_v/p_0)^{1/a}/[1 - (p_v/p_0)^{1/a}] \tag{4.51}$$

which describes the thermodynamic equilibrium between the free and filled surface at monolayer sorption.

The parameter a describes the change in surface tension caused by adsorption. Assuming the change in molar surface energy of sorbed molecules, $e_1 = (\gamma_0 - \gamma_1)/n_s^m$, to be independent of temperature within a suitable temperature interval, the following relationship holds:

$$aRT = (\gamma_0 - \gamma_1)/n_s^m = e_1 = \text{const.} \tag{4.52}$$

The parameters a_1 and a_2 are inversely related to the temperatures T_1 and T_2 of the corresponding sorption isotherms

$$a_1/a_2 = T_2/T_1 \tag{4.53}$$

It should be pointed out that adsorption will occur whenever the gain in Gibbs surface free energy (= surface tension) is negative, which corresponds to a positive value of a. As v_m is always positive, the amount of adsorbed gas v increases monotonically with gas pressure p_v. In a v vs. p_v/p_0 plot, the curve obtained is thus convex for $a > 1$, linear for $a = 1$, and concave for $a < 1$. When $a = 0$, v vanishes for all gas pressures $p_v < p_0$. Only at $p_v = p_0$ does v abruptly become equal to v_m. Such an adsorption without any gain in surface energy takes place in the case of condensation or sublimation of a gas.

4.5.2(*b*) *Modified multilayer model*

In general, multi- rather than monolayer sorption has to be considered. The number of layers which are allowed to be adsorbed is taken to be n. This number n may become infinite. In addition, a model describing multilayer sorption must contain the corresponding monolayer isotherm for $n = 1$. Allowing for a finite number of sorbed layers, n, and employing the Langmuir isotherm for each layer, Dent [73] recently developed a modified multilayer model for the sorption of vapours by polymers. By utilization of assumptions similar to those used in the BET model, another modified sorption model was developed by Burghoff & Pusch. [44] In this, in contrast to the BET model and its modifications by Dent, the Freundlich isotherm rather than the Langmuir one is applied to each of the layers. The film pressure of the first layer is taken to be Π_1, while all succeeding layers are assumed to be adsorbed with a different but constant film pressure Π_2 as a consequence of cohesive interactions between molecules in the higher layers, which are different from those between the molecules sorbed in the first layer:

$$\Pi_1 = \gamma_0 - \gamma_1 \tag{4.47a}$$

$$\Pi_i = \gamma_{i-1} - \gamma_i, \qquad i > 1 \tag{4.47b}$$

Adsorption at the ith layer is assumed to take place only on that part of the surface which is covered by $(i - 1)$ layers. Desorption from the ith layer is assumed to take place only at that part of the surface where the $(i + 1)$th–nth layers are uncovered. As illustrated in figure 4.30, the multilayer is broken down into several parts; a part where the surface is uncovered v_0, a part where the surface is covered by a monolayer v_1, a part where the surface is covered by a double layer v_2, and so on. The adsorbed molecules which contribute to thermodynamic equilibrium are shaded. The equilibrium between v_0 and v_1 is given by the Freundlich isotherm (equation (4.51)):

$$v_1 = v_0 x$$

$$x = (p_v/p_0)^{1/a}/[1 - (p_v/p_0)^{1/a}]$$

and

$$a = e_1/RT = \sigma\Pi_1/RT \tag{4.54a}$$

Similarly, the following relation holds for all succeeding layers:

$$v_i = v_{i-1} y, \qquad i > 1$$

$$y = (p_v/p_0)^{1/b}/[1-(p_v/p_0)^{1/b}]$$

and

$$b = e_2/RT = \sigma\Pi_2/RT \tag{4.54b}$$

From equations (4.54a) and (4.54b) it follows that for v_i $(1 \leqslant i \leqslant n)$:

$$v_i = v_0 x y^{i-1} \tag{4.55}$$

The total surface area v_m is given by:

$$v_m = \sum_{i=0}^{n} v_i \tag{4.56a}$$

while the amount of adsorbate v is related to v_i by:

$$v = \sum_{i=1}^{n} i v_i \tag{4.56b}$$

Then:

$$v/v_m = x \sum_{i=1}^{n} i y^{i-1} \Bigg/ \left(1 + x \sum_{i=1}^{n} y^{i-1}\right) \tag{4.57}$$

leads to the final result:

$$v/v_m = x[1-(n+1)y^n + ny^{n+1}]/(1-y)[1-y+x(1-y^n)] \tag{4.58}$$

where v_m is the surface area, n is the maximum number of layers, and the parameters x and y are related to the molar surface energies by equations (4.54a) and (4.54b). With a monolayer the sum in equation (4.58) up to $n = 1$ yields the Freundlich isotherm.

The four parameters v_m, n, a and b can be obtained from experimental data by the following procedure: at very low gas pressures $(p_v \ll p_0)$, only

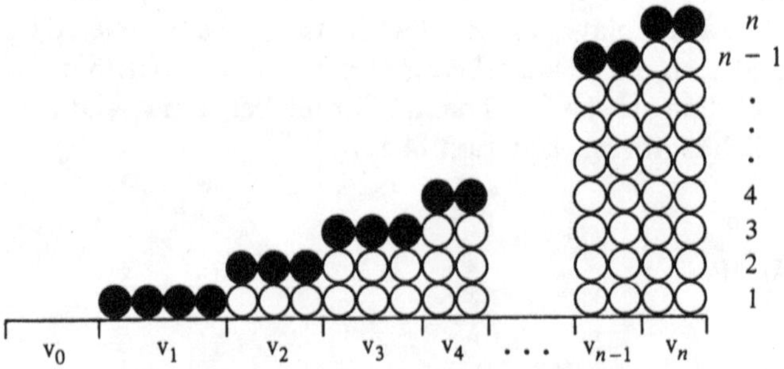

Fig. 4.30. Diagrammatic presentation of multilayer sorption. [44]

monolayer adsorption is assumed to take place. Thus, from a ln v vs. $\ln(p_v/p_0)$ plot of experimental data, the constants v_m and a can be determined. At saturation ($p_v = p_0$), equation (4.58) yields by use of the rule of l'Hospital:

$$\lim_{p_v \to p_0} v = n v_m = v^* \qquad (4.58a)$$

Use of the experimentally determined amount of gas (vapour) sorbed at $p_v = p_0$ and the known value of v_m results in the number of layers n. The parameter b may then be derived by a least-squares analysis of the experimental data employing equation (4.58). If there is any water sorbed by capillary phenomena, which may occur at relative humidities above 85%, both parameters n and b have to be determined by a least-squares analysis using only experimental data which do not include the effects of capillary phenomena.

Bearing in mind the results of the discussion on the partial thermodynamic potentials obtained with a Freundlich-type sorption process, one might reasonably describe the sorption of water in polymers by the modified multilayer model since the experimentally determined partial entropy of sorption is nearly independent of the water content for many polymers at least at low vapour pressures. [74] Using experimental data of water sorption on homogeneous CA membranes, the model parameters were calculated. The experimental results, the values of the model parameters obtained with these data, and the sorption isotherms calculated by means of equation (4.58) are presented in figures 4.31(*a*)–4.31(*c*). As is obvious from these figures, use of the modified multilayer model gives water sorption isotherms which agree well with experimental data for the whole range of relative humidity. It might be concluded that all the water within homogeneous CA membranes interacts to some extent with the polymer.

That there is a limited number of about two layers of sorbed water ($n \approx 1.8$ (T-900), 1.9 (K-700), 1.7 (E-320)) might be a consequence of the short-range forces responsible for the water sorption. The action of short-range forces is supported by the values of the surface energies of the first and second layers, as the surface energy of the second layer of sorbed water molecules is found to be much smaller than the surface energy of the first layer. The parameters a and b, related to the surface energy of the corresponding layer, satisfy the following equations:

$$a_1/a_2 = T_2/T_1 \quad \text{and} \quad b_1/b_2 = T_2/T_1$$

Therefore, the surface energies may reasonably be assumed to be independent of temperature within the temperature interval considered.

It should be noted that the sorption model described does not require the existence of a two-dimensional geometric surface within the membrane, but rather considers a distribution of the sorbed water within the entire bulk phase of the polymer. This picture of the sorption process is different from one

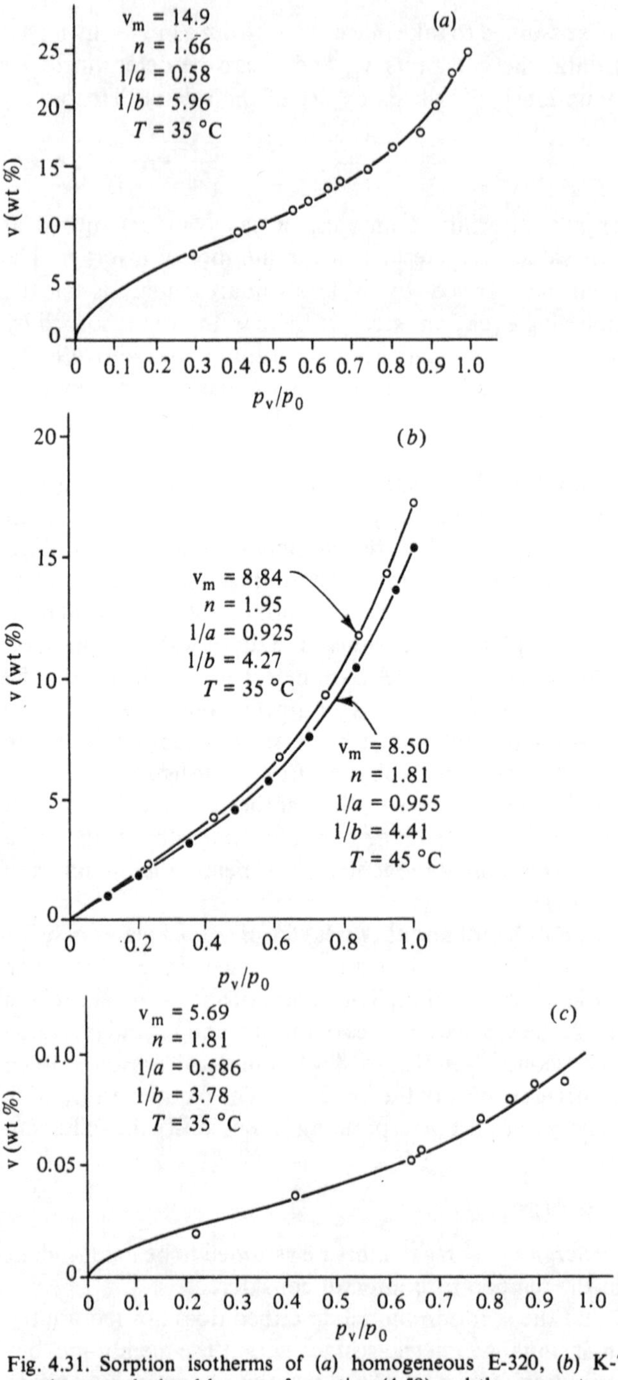

Fig. 4.31. Sorption isotherms of (*a*) homogeneous E-320, (*b*) K-700, and (*c*) T-900 CA membranes calculated by use of equation (4.58) and the parameter values v_m, n, $1/a$, and $1/b$ specified in the figures. The points represent experimental data. [44, 86]

assuming the existence of a real solution. The existence of a real solution requires certain degrees of freedom for both the solvent and solute molecules. In contrast, in the described sorption model only the water is considered to move in a plane, whereas the polymer is taken to be equivalent to a rigid matrix.

Water is dissolved in asymmetric CA membranes at relative humidities above 85% by capillary condensation. For this reason, only experimental data of water sorption obtained at relative humidities below 85% were used to determine the model parameters. The experimental data, as well as the resulting isotherms, are presented in figure 4.32. Excellent agreement again exists between calculated and experimental data at relative humidities below 85%.

Table 4.6 summarizes the values of the model parameters characterizing water sorption of various polymers. As capillary condensation can again appear with these polymers at relative humidities above 90%, only experimental data obtained at relative humidities below 90% have been used to determine the parameter values. Figure 4.33 shows the experimental data and the isotherms calculated by means of equation (4.58). The multilayer model characterizes water sorption of polymers and yields calculated sorption isotherms for all systems analysed in excellent agreement with the experimentally determined ones at relative humidities below 90%.

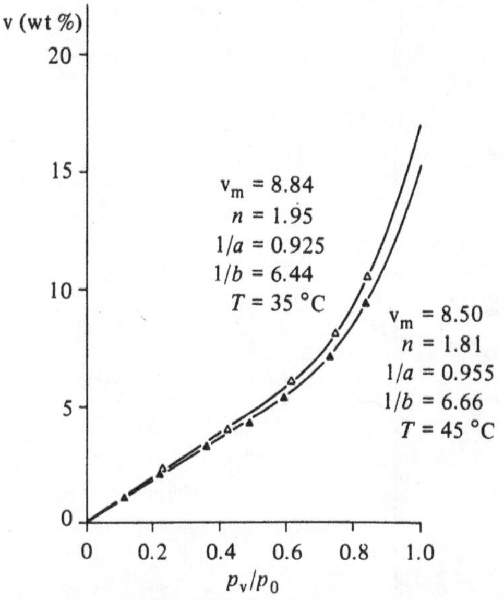

Fig. 4.32. Sorption isotherms of an asymmetric CA K-700 membrane calculated by use of equation (4.58) and the parameter values v_m, n, $1/a$ and $1/b$ specified in the figure. The points represent experimental data. [44]

Table 4.6. *Parameters of the modified multilayer model of water sorption for various polymers*

Polymer	T (°C)	v_m (wt %)	n	$1/a$	e_1 (cal/mol)	$1/b$	e_2 (cal/mol)
Cellophane	50	12.53	1.98	0.55	1168	3.76	171
Cellulose-2-acetate	50	10.47	1.77	0.84	765	3.48	185
Ethyl-cellulose	50	2.26	1.96	1.00	642	3.38	190
Fibrolan	30	14.88	2.17	0.55	1095	4.18	144
Nylon 6,6	30	4.00	1.84	0.69	873	3.37	179
E-320	25	15.5	1.61	0.640	926	5.62	105
	35	14.9	1.66	0.656	934	3.44	178
K-700	25	10.3	1.63	0.895	662	3.08	192
	35	8.84	1.95	0.925	662	4.27	143
	45	8.50	1.81	0.955	662	4.41	143
T-900	25	7.5	1.28	0.920	644	3.07	193
	35	5.69	1.81	0.936	654	4.02	152

As can be seen from table 4.6, the determination of the model parameters results in about two layers ($n \approx 2$) of sorbed water for all the polymers considered. Also, the fourth parameter, b, does not vary too much with polymer except for the E-320 membrane. The finite value of $n \approx 2$ for the number of sorbed water layers in polymers as well as the low surface energy of the second layer compared to the surface energy of the first layer are quite consistent with the concept of the short-range interactions.

The analysis of water sorption data, obtained with homogeneous and asymmetric CA membranes by means of a modified multilayer model, allows an interpretation of the different water sorption capacities of homogeneous and asymmetric CA membranes at intermediate regains (water vapour activities). As is obvious from the model parameters reported in figures 4.31 and 4.32, the difference in the water sorption isotherms of a homogeneous and an asymmetric K-700 membrane only finds expression in different b parameters at each temperature; b is larger for the homogeneous membrane. As the b parameter reflects the surface energy of the second layer, this model suggests that the surface energy of the second layer of water sorbed by asymmetric CA membranes is much less (about 34%) than that sorbed by homogeneous ones. This result agrees with the fact that water sorbed by asymmetric CA membranes is essentially sorbed by the fine and coarse porous matrix of these membranes since water sorbed by the dense active layer cannot be detected. This might indicate that the water of the second sorbed layer in

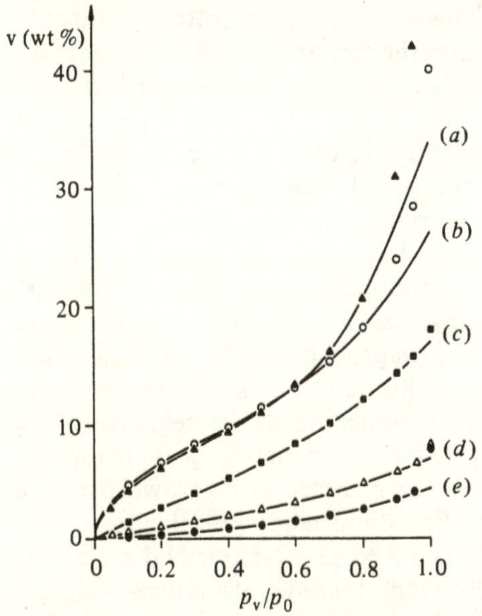

Fig. 4.33. Sorption isotherms of (*a*) Fibrolan, (*b*) cellophane, (*c*) cellulose-2-acetate, (*d*) nylon 6,6 and (*e*) ethyl-cellulose calculated by use of equation (4.58) and the parameter values v_m, n, $1/a$ and $1/b$ specified in table 4.6. The points represent experimental data taken from the literature. [44]

homogeneous CA membranes is in a stronger force field due to the greater density of this CA membrane and thus in a possibly more ordered and more compact state than in the certainly less dense matrix of asymmetric CA membranes. It is thus not cogent to assume capillary condensation to be responsible for the extra water sorbed by homogeneous CA membranes in comparison to asymmetric ones at intermediate humidities. However, the interpretation of the extra sorbed water in homogeneous CA membranes as capillary water would not conflict with the foregoing analysis since capillary water would certainly possess a higher density than water sorbed in the second layer of a more porous (open) matrix which does not possess equally narrow pores. Thus, the interpretation of the state of the extra sorbed water finally reduces to a question of semantics.

Furthermore, the analysis shows that there is a quite limited space available for sorbed water. Due to these space restrictions it is quite plausible that small water clusters will form within the CA membranes. Since, as already discussed, small water clusters exhibit thermodynamic and caloric properties different from those of bulk water, a consistent interpretation of the experimental findings regarding water sorption and partial molar heat of water in CA membranes is made possible on the basis outlined above.

The different state of water in CA membranes is also suggested by findings obtained from hyperfiltration (reverse osmosis) experiments using different membranes (e.g. CA and ion-exchange membranes), and different metal chloride, metal nitrate, and sodium halide brine solutions. The experimental findings are reproduced in figures 4.34–4.36 where $1/r$ is plotted as a function of $1/q$; r characterizes the salt rejection of the corresponding membrane and is defined by $r = (c_s' - c_s'')/c_s'$ where c_s' and c_s'' are the brine and product concentration of the salt. The slopes of the $1/r$ vs. $1/q$ plots are approximately equal to the normalized salt permeability, P_s/d, of the membrane for the corresponding salt provided the asymptotic salt rejection, r_∞, is nearly 1. [75] The asymptotic salt rejection characterizes the salt rejection of the membrane at infinite pressure difference which corresponds to an infinite volume flux q. As can be seen from figures 4.34–4.36, the normalized salt permeabilities of the CK-1 membrane vary with a change of the cation or anion of the corresponding salt just as does the Walden product $\Lambda\eta$ [76] (Λ = equivalent conductivity). However, the positions of Rb^+ and Cs^+ are exchanged in the presence of NO_3^- as the anion. On the other hand, the sequence of the normalized salt permeabilities of asymmetric CA membranes approximately follows the Hofmeister series only in the presence of NO_3^- whereas it is completely reversed in the presence of the sodium halides and alkali chlorides: CsCl < RbCl < KCl < NaCl < LiCl and NaCl < NaBr < NaI < $NaNO_3$. These experimental findings might also support the idea of a water structure in the skin of asymmetric CA membranes and in homogeneous ones which is different from that in bulk water.

All the information obtained from water sorption, calorimetric, and permeation measurements with synthetic membranes is consistent with the

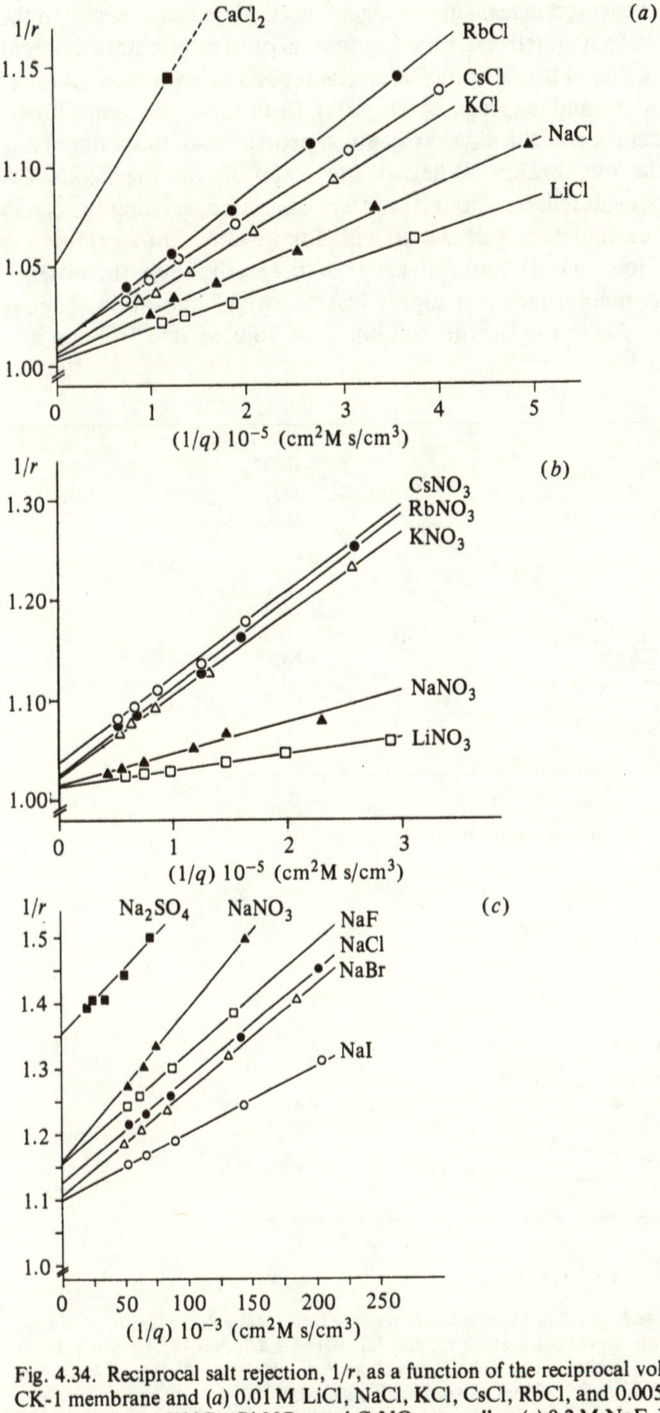

Fig. 4.34. Reciprocal salt rejection, $1/r$, as a function of the reciprocal volume flux, $1/q$, using a CK-1 membrane and (*a*) 0.01 M LiCl, NaCl, KCl, CsCl, RbCl, and 0.005 M $CaCl_2$; (*b*) 0.01 M $LiNO_3$, $NaNO_3$, KNO_3, $RbNO_3$, and $CsNO_3$; as well as (*c*) 0.2 M NaF, NaCl, $NaNO_3$, NaBr, NaI, and Na_2SO_4 brine solutions ($T = 298$ K; 20 bar $\leqslant P' \leqslant$ 100 bar). [80]

assumption that the water, at least in homogeneous CA membranes and in the skin (active layer) of asymmetric CA membranes, is present in a state different from that of bulk water. This state may be characterized as both 'bound' water and water existing in small aggregates (clusters). Both these concepts allow a consistent interpretation of the physicochemical properties of the water in fine porous and dense membranes. Whether the water in homogeneous CA membranes is considered as 'bound' water or water existing in small aggregates, it is certain that this water will, for instance, not exhibit the hydration capabilities of bulk water. Because of its specific state, the water in homogeneous CA membranes will supply less energy of hydration for ions than bulk water. Accordingly, the solubility of solutes (electrolytes and

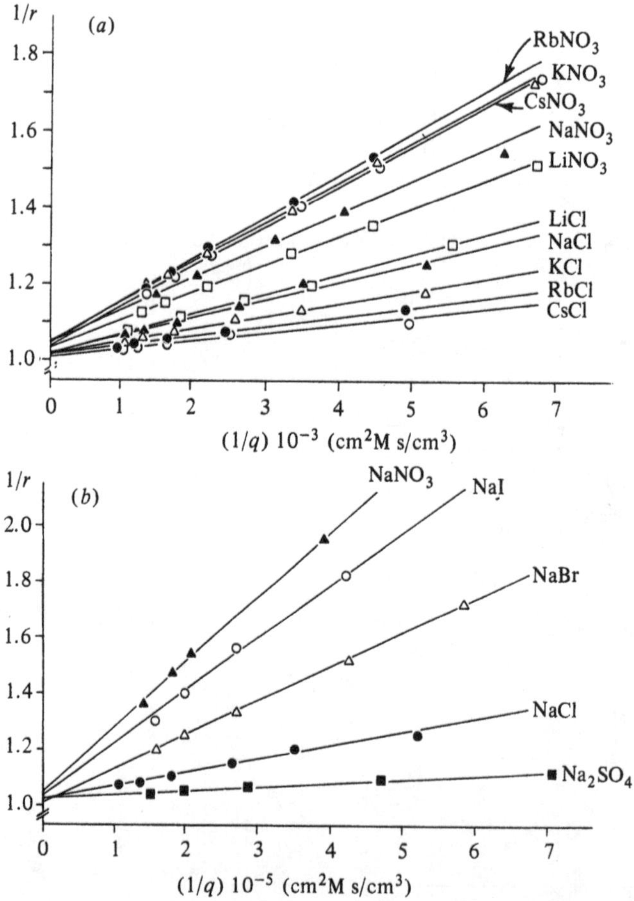

Fig. 4.35. Reciprocal salt rejection, $1/r$, as a function of the reciprocal volume flux, $1/q$, using an asymmetric CA membrane, annealed at 80 °C, and (a) 0.01 M LiCl, NaCl, KCl, CsCl, RbCl, $LiNO_3$, $NaNO_3$, KNO_3, $RbNO_3$, and $CsNO_3$ as well as (b) 0.2 M NaCl, $NaNO_3$, NaBr, NaI, and Na_2SO_4 brine solutions ($T = 298$ K; 20 bar $\leqslant P' \leqslant$ 100 bar). [80]

nonelectrolytes) in the membrane phase, which do not interact specifically with the membrane, such as phenol for instance, will be reduced in comparison to their solubilities in bulk water. The poor solute permeabilities of homogeneous CA membranes and of the active layer of asymmetric ones for most electrolytes and nonelectrolytes is thus mainly a consequence of the poor solubility of the corresponding solutes within the membrane phase.

4.6 Conclusions

The characterization of water and solute transport by means of membrane model-independent transport relationships yields the corresponding

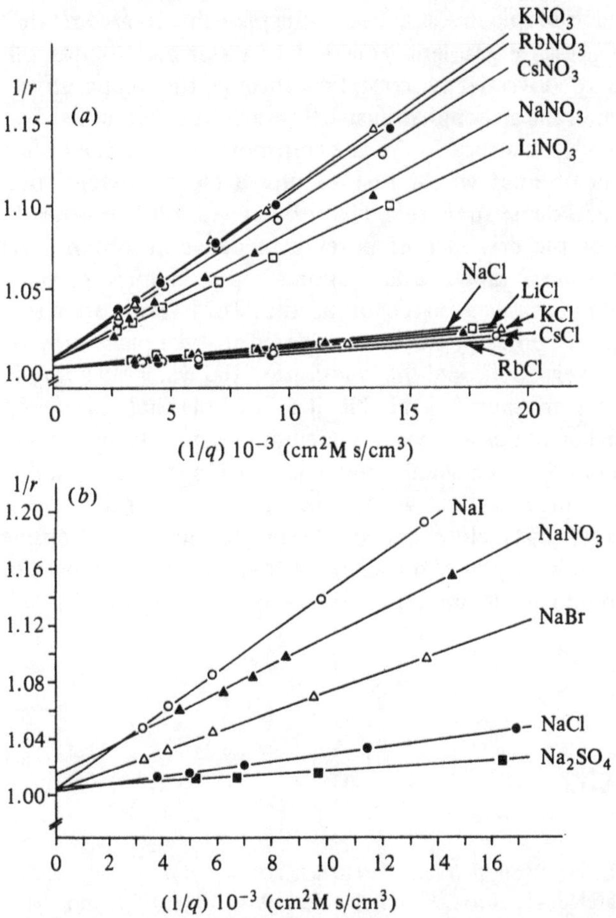

Fig. 4.36. Reciprocal salt rejection, $1/r$, as a function of the reciprocal volume flux, $1/q$, using an asymmetric CA membrane, annealed at 87.5 °C, and (*a*) 0.01 M LiCl, NaCl, KCl, CsCl, RbCl, $LiNO_3$, $NaNO_3$, KNO_3, $RbNO_3$, and $CsNO_3$ as well as (*b*) 0.2 M NaCl, $NaNO_3$, NaBr, NaI, and Na_2SO_4 brine solutions ($T = 298$ K; 20 bar $\leqslant P' \leqslant$ 100 bar). [80]

transport parameters without any information on the underlying physicochemical origin of transport. On the other hand, applying membrane model-dependent transport relationships yields the transport parameters inherent in these relationships again without enabling a conclusive clarification of the transport mechanisms although information on the type of flow can be obtained to some extent by the extraction of meaningful transport parameters such as water and solute diffusion coefficients and by a subsequent comparison of these parameters with corresponding transport coefficients in bulk solution. Consideration of the results of electron microscopic investigations of the corresponding membranes, in addition to the results of the transport measurements, makes the clarification of transport mechanisms in coarse porous membranes possible. If a membrane were proved to be coarse porous by electron microscopic investigations, it is plausible to assume that in the presence of a pressure gradient transport of water and solutes will essentially take place by convective (viscous) flow through the comparatively large pores of the membrane although diffusive flow will be superimposed on this. On the other hand, a clarification of the transport mechanisms in fine porous and dense membranes would first require a clear differentiation between fine porous and dense (pore free) membranes. Although it is nearly impossible to rule out the existence of pores in a dense membrane, the assumption of a dense membrane is a fairly good approximation for thick homogeneous CA membranes ($d > 20$ Å), for instance. [67] Transport across a membrane without any pores might be considered to take place only by diffusion. When, however, additional information on the water structure in fine porous and dense membranes is available, it may be plausible to suggest that diffusive transport of matter will occur in membranes possessing water in a specific state (e.g. 'bound' water, small water clusters). On the other hand, if the existence of water similar to bulk water (free water) in the membrane is proved experimentally, convective transport, in addition to diffusive transport, cannot be completely ruled out since the so-called microviscosity of solutions passing through small capillaries, as discussed by Wirtz [77], will also come into play.

References

1. S. J. Singer & G. L. Nicolson, *Science* **175** (1972), 720.
2. O. Kedem & A. Katchalsky, *Trans. Farad. Soc.* **59** (1963), 1918, 1931, and 1941.
3. R. Schlögl, *Stofftransport durch Membranen*, Dr.-Dietrich-Steinkopff-Verlag, Darmstadt, 1964.
4. A. J. Staverman, *Rec. Trav. Chim. Pays-Bas* **70** (1951), 344.
5. U. Merten, *Desalination by Reverse Osmosis*, The MIT Press, Cambridge, Mass., 1966.

6. H.-G. Burghoff, K. L. Lee & W. Pusch, *J. Appl. Polymer Sci.* **25** (1980), 323.
7. I. Prigogine, *Etude Thermodynamique des Phénomènes irréversibles*, Desoer, Paris–Liège, 1947, p. 89.
8. P. Bo & V. Stannett, *Desalination* **18** (1976), 113.
9. G. Schmid, *Chemie-Ing.-Technik* **37** (1965), 616; *Z. Elektrochem.* **54** (1950), 424; **55** (1951), 229; **56** (1952), 181; G. Schmid & H. Schwarz, *Z. Elektrochem.* **55** (1951), 295, 684; **56** (1952), 35.
10. E. Manegold, *Kapillarsysteme, Chemie und Technik*, Vol. 1 and 2, Verlagsgesellschaft, Heidelberg, 1955.
11. G. Gouy, *J. Phys.* **9** (1910), 457; *Ann. Phys.* **7** (1917), 129; D. L. Chapman, *Phil. Mag.* **25** (1913), 475.
12. O. Stern, *Z. Elektrochem.* **30** (1924), 508.
13. H. v. Helmholtz, *Wied. Ann. Phys.* **7** (1919), 337.
14. M. v. Smoluchowski, *Hb. der Elektrizität und des Magnetismus*, Vol. 2. Verlag Greatz, Berlin, 1921, p. 336.
15. R. Schlögl, *Z. phys. Chem., NF* **1** (1954), 305.
16. A. D. MacGillivray & D. Hare, *J. Theor. Biol.* **25** (1969), 113.
17. E. A. Guggenheim, *Thermodynamics*. North Holland Publishing Corp., Amsterdam, 1959.
18. D. Tsimboukis & J. H. Petropoulos, *J. Chem. Soc., Farad. Trans.* I **75** (1979), 705.
19. G. N. Lewis & M. Randall, *Thermodynamics and The Free Energy of Chemical Substances*. McGraw-Hill Book Co. Inc., New York, 1923.
20. J. Koryta, J. Dvorák & V. Bohácková, *Electrochemistry*. Methuen & Co. Ltd, London, 1970.
21. K. S. Spiegler, *J. Electrochem. Soc.* **100** (1953), 303C.
22. H. Falkenhagen, *Theorie der Elektrolyte*. S.-Hirtzel-Verlag, Stuttgart, 1971.
23. R. Paterson. In: *Biological and Artificial Membranes and Desalination of Water*, (ed. R. Passino). Elsevier, Amsterdam, 1976, pp. 517–65.
24. K. Spurný, *Z. Biol. Aerosol-Forsch.* **12** (1965), 369.
25. J. S. Mackie & P. Meares, *Proc. Roy. Soc.* **A 232** (1955), 498.
26. T. Foley, J. Klinowski & P. Meares, *Proc. Roy. Soc. London* **A 336** (1974), 327.
27. R. Paterson, R. G. Cameron & I. S. Burke. In: *Charged & Reactive Polymers*, Vol. 3: *Charged Gels and Membranes*, Part I (ed. E. Sélégny). D. Reidel Publishing Co., Dordrecht, Holland, 1976, pp. 157–82.
28. W. Pusch, *Desalination* **16** (1975), 65.
29. W. Pusch, *Ber. Bunsenges. physik. Chem.* **81** (1977), 269, 854.
30. W. Kuhn, *Z. Elektrochem.* **55** (1951), 207.
31. W. A. P. Luck, D. Schiöberg & U. Siemann, *J. Chem. Soc., Farad. Trans.* II **76** (1980), 136.
32. C. Toprak, J. N. Agar & M. Falk, *J. Chem. Soc., Farad. Trans.* I **14** (1979), 803.
33. G. Belfort, J. Scherfig & D. O. Seevers, *J. Colloid Interface Sci.* **47** (1974), 106; E. Almagor & G. Belfort, *J. Colloid Interface Sci.* **66** (1978), 146.
34. C. A. J. Hoeve. In: *Water in Polymers* (ed. St. P. Rowland). ACS Symposium Ser. 127, Am. Chem. Soc., Washington, DC, 1980, p. 135.
35. S. Deodhar & P. Luner. In: *Water in Polymers* (ed. St. P. Rowland). ACS Symposium Ser. 127, Am. Chem. Soc., Washington, DC, 1980, p. 273.
36. H.-G. Burghoff & W. Pusch, *J. Appl. Polymer Sci.* **20** (1976), 789.
37. J. L. Williams, H. B. Hopfenberg & V. Stannett, *J. Macromol. Sci. Phys.* **B 3** (1969), 711.

38. J. W. McBain, *The Sorption of Gases and Vapours by Solids*. Routledge & Kegan Paul, London, 1932.
39. J. W. Gibbs, *The Collected Works of J. Willard Gibbs*, Vol. I. Yale University Press, New Haven, Conn., 1957.
40. W. Pusch, unpublished results.
41. R. L. Riley, J. O. Gardner & U. Merten, *Science* **143** (1964), 801.
42. G. J. Gittens, P. A. Hitchcock & G. E. Wakley, *Desalination* **8** (1970), 369; *Desalination* **12** (1973), 315.
43. R. E. Kesting, *Ion Exchange Membranes* **1** (1974), 197.
44. H.-G. Burghoff & W. Pusch, *J. Appl. Polym. Sci.* **24** (1979), 1479.
45. Y. Taniguchi & S. Horigome, *J. Appl. Polym. Sci.* **19** (1975), 2743.
46. H. Yasuda, H. G. Olf, B. Christ, C. E. Lamaze & A. Peterlin. In: *Water Structure at the Water Polymer Interface* (ed. H. H. G. Jellinek). Plenum Press, New York, 1972, pp. 39–55.
47. L. Ter-Minassian-Sarag & G. Mademont. In: *Biophysics of Water* (eds. F. Franks & Sh. Mathias). John Wiley & Sons, Chichester, 1982, pp. 127.
48. H. Pauly, *Biophysik* **10** (1973), 7.
49. A. T. Dick. In: *Water and Aqueous Solutions: Structure, Thermodynamics, Transport Processes*, ch. 7 (ed. R. A. Horne). Wiley-Interscience, New York, 1972, pp. 265–93.
50. M. N. Plooster & S. N. Gittlin, *J. Phys. Chem.* **75** (1975), 3322.
51. H.-G. Burghoff & W. Pusch, *J. Appl. Polym. Sci.* **23** (1979), 473.
52. W. Lukas & W. Pusch, 'Characterization of the state of water within synthetic membranes by water sorption isotherms and heat capacity measurements', in: *Biophysics of Water* (eds. F. Franks & Sh. Mathias). John Wiley & Sons, Chichester, 1982, pp. 27–31.
53. A. R. Haly & J. W. Snaith, *Biopolymers* **6** (1968), 1355.
54. V. V. Tarasov & G. A. Yunitskii, *Zhur. Fiz. Khim.* **39** (1965), 2077.
55. H. K. Lonsdale, U. Merten & R. L. Riley, *J. Appl. Polym. Sci.* **9** (1965), 1341.
56. F. Franks, *Water: A Comprehensive Treatise*, Vol. 5. Plenum Press, New York, 1975.
57. W. Kuhn, *Helv. Chim. Acta.* **39** (1956), 1071.
58. R. Defay & I. Prigogine, *Surface Tension and Adsorption*. Longmans, London, 1966.
59. H.-G. Burghoff, W. Pusch & E. Staude, 'Characterization of cellulose acetate membranes for organic solutes', *Proc. 5th Intl. Symposium on Fresh Water from the Sea*, Vol. 4 (eds. A. Delyannis & E. Delyannis). Published by the editors, Athens, 1976, pp. 143–56.
60. J. Clifford. In: *Water, A Comprehensive Treatise*, Vol. 5, Ch. 2 (ed. F. Franks). Plenum Press, New York, 1975, pp. 75–132; G. Belfort & N. Sinai. In: *Water in Polymers* (ed. St. P. Rowland). ACS Symposium Ser. 127, Am. Chem. Soc., Washington, DC, 1980, p. 323.
61. M. C. Phillips. In: *Water, A Comprehensive Treatise*, Vol. 5, Ch. 3 (ed. F. Franks). Plenum Press, New York, 1975, pp. 133–72.
62. R. J. Speedy & C. A. Angell, *J. Chem. Phys.* **65** (1976), 851.
63. C. A. Angell, J. Shuppert & J. C. Tucker, *J. Phys. Chem.* **77** (1973), 3092.
64. T. D. Gierke, *Ionic Clustering in Nafion Perfluorosulfonic Acid Membranes and Its Relationship to Hydroxyl Rejection and Chlor-Alkali Current Efficiency*, paper

presented to 152nd National Meeting of The Electrochemical Society, Atlanta, Georgia, 10–14 Oct. 1977, Diamond Shamrock Corp., Cleveland, Ohio.

65. F. H. Stillinger. In: *Water in Polymers* (ed. St. P. Rowland). ACS Symposium Ser. 127, Am. Chem. Soc., Washington, DC, 1980, p. 11.

66. St. P. Rowland. In: *Applied Fiber Science* (ed. F. Happey). Academic Press, London, 1979, pp. 206–37.

67. N. Choji, W. Pusch, M. Satoh, T.-M. Tak & A. Tanioka, 'Structure investigations of homogeneous cellulose acetate membranes by gas permeation measurements', *Desalination* **53** (1985), 347–61.

68. B. E. Conway, *Ionic Hydration in Chemistry and Biophysics*. Elsevier Publishing Co., Amsterdam, 1981, Ch. 2, p. 15.

69. T. L. Hill, *Thermodynamics of Small Systems*, Part I and II. W. A. Benjamin Inc., Publishers, New York, 1963.

70. St. Brunauer, *The Adsorption of Gases and Vapors*, Vol. I. Princeton University Press, Princeton, New Jersey, 1945.

71. P. T. Flood, *The Solid–Gas Interface*, Vol. I. Marcel Dekker, New York, 1953, p. 11; and A. W. Adamson, *Physical Chemistry of Surfaces*. Wiley, New York, 1967.

72. R. K. Schofield & E. K. Rideal, *Proc. Roy. Soc. London, Ser. A* **109** (1925), 57.

73. R. W. Dent, *Text. Res. J.* **47** (1977), 147, 188.

74. R. Jeffries, *J. Text. Inst.* **51** (1960), T339.

75. W. Pusch, *Desalination* **59** (1986), 105–98.

76. R. L. Kay. In: *Water: A Comprehensive Treatise*, Vol. 3 (ed. F. Franks). Plenum Press, New York, 1973, pp. 173–209.

77. A. Spernol & K. Wirtz, *Z. Naturforsch.* **28** (1946), 522.

78. W. Pusch & A. Walch, 'Membrane structure and its correlation with membrane permeability', *J. Membrane Sci.* **10** (1982), 325–60.

79. W. Pusch & A. Walch, 'Synthetic membranes – Preparation, structure, and application', *Ang. Chem.* **94** (1982), 670; *IE* **21** (1982), 660.

80. M. Gunkel & W. Pusch, *Ber. Bunsenges. Physik. Chem.* **83** (1979), 1089.

81. M. Ataka & W. Pusch, unpublished results.

82. W. Lukas & W. Pusch. In: *Biophysics of Water* (eds. F. Franks & Sh. Mathias). John Wiley & Sons, Chichester, 1982, pp. 27–31.

83. W. Lukas, *Thermodynamische Charakterisierung des in homogenen Celluloseacetat-Membranen gebundenen Wassers*. Diplomarbeit, Johann-Wolfgang-Goethe-Universität, Frankfurt am Main, Sept. 1980.

84. W. Lukas, *Kalorimetrische Untersuchungen des Wassers in synthetischen Membranen*, unpublished results.

85. H.-G. Burghoff, K. L. Lee & W. Pusch, *J. Appl. Polym. Sci.* **25** (1980), 323.

86. W. Lukas & W. Pusch, unpublished results.

For EU product safety concerns, contact us at Calle de José Abascal, 56–1°, 28003 Madrid, Spain or eugpsr@cambridge.org.

www.ingramcontent.com/pod-product-compliance
Lightning Source LLC
LaVergne TN
LVHW101457200926
848483LV00001B/154
* 9 7 8 0 5 2 1 3 5 0 1 4 3 *